Advances in Fluid Power Systems

Advances in Fluid Power Systems

Editors

Ryszard Dindorf
Jakub Takosoglu
Piotr Wos

MDPI • Basel • Beijing • Wuhan • Barcelona • Belgrade • Manchester • Tokyo • Cluj • Tianjin

Editors
Ryszard Dindorf
Department of Mechatronics and Armament, Faculty of Mechatronics and Mechanical Engineering
Kielce University of Technology
Kielce
Poland

Jakub Takosoglu
Department of Mechatronics and Armament, Faculty of Mechatronics and Mechanical Engineering
Kielce University of Technology
Kielce
Poland

Piotr Wos
Department of Mechatronics and Armament, Faculty of Mechatronics and Mechanical Engineering
Kielce University of Technology
Kielce
Poland

Editorial Office
MDPI
St. Alban-Anlage 66
4052 Basel, Switzerland

This is a reprint of articles from the Special Issue published online in the open access journal *Energies* (ISSN 1996-1073) (available at: www.mdpi.com/journal/energies/special_issues/Fluid_Power_Systems).

For citation purposes, cite each article independently as indicated on the article page online and as indicated below:

LastName, A.A.; LastName, B.B.; LastName, C.C. Article Title. *Journal Name* **Year**, *Volume Number*, Page Range.

ISBN 978-3-0365-5354-2 (Hbk)
ISBN 978-3-0365-5353-5 (PDF)

Contents

About the Editors

Ryszard Dindorf

Ryszward Dindorf is currently a Professor at the Department of Mechatronics and Armament, Faculty of Mechatronics and Mechanical Engineering, Kielce University of Technology, Kielce, Poland. His research interests include fluid power control systems, mechatronics engineering, actuators, as well as drives, control, and sensors in the automation and robotization of production processes. He was a member of the Committee on Automatic Control and Robotics of the Polish Academy of Sciences.

Jakub Takosoglu

Jakub Takosoglu is currently an Associate Professor at the Department of Mechatronics and Armament, Faculty of Mechatronics and Mechanical Engineering, Kielce University of Technology, Kielce, Poland. His main research interests include electro-pneumatic servo drives, especially fuzzy-logic and real-time control systems.

Piotr Wos

Piotr Wos is currently an Associate Professor at the Department of Mechatronics and Armament, Faculty of Mechatronics and Mechanical Engineering, Kielce University of Technology, Kielce, Poland. His main research interests include electro-hydraulic servo drives and sensory systems, especially adaptive and real-time control systems.

Preface to "Advances in Fluid Power Systems"

Fluid power systems (hydraulic and pneumatic drives and control) involve the use of fluid properties to generate, control, and transmit power using pressurized fluid flow. The fluid power system sector is undergoing tremendous expansion at this time, invigorating strategic companies - power engineering, oil and gas industry, mining industry, shipbuilding, steel industry, and machine industry, and metal processing. There are many challenges for fluid power systems, such as increasing energy efficiency, improving reliability, building smart components and systems, reducing the size and weight of components, reducing environmental impact, improving the capabilities for recovery and storage of energy for reuse, digitalization, and wireless connectivity of smart components. Advances in fluid power systems are leading to the creation of new smart devices that replace tried-and-true solutions from the past.

This Special Issue focuses on the recent advances and smart of fluid power systems in a wide range of areas.

Ryszard Dindorf, Jakub Takosoglu, and Piotr Wos

Editors

energies

MDPI

Editorial

Advances in Fluid Power Systems

Ryszard Dindorf *, Jakub Takosoglu and Piotr Wos

Department of Mechatronics and Armament, Faculty of Mechatronics and Mechanical Engineering, Kielce University of Technology, 25-314 Kielce, Poland; qba@tu.kielce.pl (J.T.); wos@tu.kielce.pl (P.W.)
* Correspondence: dindorf@tu.kielce.pl

Citation: Dindorf, R.; Takosoglu, J.; Wos, P. Advances in Fluid Power Systems. *Energies* **2021**, *14*, 8589. https://doi.org/10.3390/en14248589

Received: 3 December 2021
Accepted: 11 December 2021
Published: 20 December 2021

1. Special Issue Information

The main purpose of this special edition of "Advances in Fluid Power Systems" was to present new scientific work in the field of fluid power systems for the hydraulic and pneumatic control of machines and devices that are used in various industries. The development work of authors from various research centres have been published. Fluid power systems (hydraulic and pneumatic drives and their control) involve the use of fluid properties to generate, control, and transmit power using pressurized fluid flow. Fluid power systems are simple and easy to use, capable of accurately controlling the position, speed, force, and torque in machines, and are economical and safe to operate. The fluid power system research sector is currently undergoing tremendous expansion, invigorating strategic companies, power engineering, the oil and gas industry, the mining industry, the steel industry, and metal processing in the machine industry. High-pressure hydraulic forging presses are used for open-die forging of heavy steel forgings (carbon, alloy, high-alloy, stainless, and other steels) for large-size hot forged products, which include turbine shafts (water, gas, steam), rotors for wind and gas power generators. These products find buyers mainly in the shipbuilding, machinery, energy, and metallurgy industries. There are several main challenges for fluid power systems, such as increasing energy efficiency, improving reliability, building smart components and systems, reducing the size and weight of components, reducing environmental impact, improving the capabilities for recovery and storage of energy for reuse, digitalization, and 5G wireless connectivity of smart components. Advances in fluid power systems are leading to the creation of new smart devices that can replace tried-and-true solutions from the past. This special issue focuses on the recent advances and smart solutions for fluid power systems in a wide range of topics, including:

- Fluid power to the IoT and Industry 4.0: smart fluid power technology, wireless 5G connectivity in fluid power, smart components, and sensors.
- Fluid power in the renewable energy sector: hydraulic drivetrains for wind power and for wave and marine current power, and hydraulic systems for solar power.
- Fluid power hybrid: hybrid transmissions, energy recovery and accumulation, energy efficiency of hybrid drives.
- Industrial and mobile fluid power: industrial fluid power solutions, mobile fluid power solutions, energy efficiency solutions for fluid power systems.
- Environmental aspects of fluid power: hydraulic water control technology, noise and vibration of fluid power components, safety, reliability, fault analysis, and diagnosis of fluid power systems.
- Fluid power and mechatronic systems: servo-drive control systems, fluid power drives in manipulators and robots, fluid power in autonomous solutions.

2. Published Papers

Filo et al. presented the results of a computational fluid dynamics (CFD) analysis of an innovative directional control valve consisting of four poppet seat valves and two electromagnets enclosed within a single body [1]. The valve has a unique design, allowing the use of any poppet valve configuration. Seat valves that are normally opened (NO) and normally closed (NC) can be used. The combination of four universal valve seats and two electromagnets gives a wide range of flow path configurations. This significantly increases the possibility of practical applications. However, due to the significant miniaturization of the valve body and the requirement to obtain necessary connections between flow paths, multiple geometrically complex channels had to be made inside the body. Therefore, the main purpose of the work was to shape the geometry of the flow channels in such a way as to minimize pressure losses. During the CFD analyses, the dissociation velocity in the flow channels and the pressure distribution on the walls were determined. The results were used to obtain the pressure loss as a function of the flow rate, which was then verified by means of laboratory experiments conducted on a test bench.

The aim of Basiak's study was to develop a mathematical model for non-contact face seals to analyse the influence of thermoelastic phenomena on their operation [2]. The model was used to solve thermal conductivity and thermoelasticity problems. The primary goal was to calculate the thermal deformation values of the sealing rings in a non-contacting face seal with a flexibly mounted rotor (FMR) for a turbomachine. The model assumes the conversion of mechanical energy into heat in the fluid film. The heat flux generated in the fluid film is transferred, first to the sealing rings, and then to the fluid surrounding them. The asymmetric distribution of the temperature within the sealing rings leads to the occurrence of thermal stresses and, consequently, to a change in the geometry of the rings. The model is solved analytically. The distributions of the temperature fields for the sealing rings in the cross sections are calculated using the Fourier-Bessel series as a superficial function of two variables (r, z). The thermo-elastic problems described by the Navier equations are solved by applying the Boussinesq harmonic functions and Goodier's thermo-elastic displacement potential function. The proposed method involves solving various theoretical and practical problems of thermoelasticity in FMR type non-contacting face seals. The solution for the mathematical model made use of analytical methods, and the most important results obtained are presented in graphical form, such as the temperature distributions and axial thermal distortions in cross-sections of the rings. The calculated thermal deformations of the sealing rings are used to determine the most important seal performance parameters, such as leakage rate and power loss. The article also presents a multi-criteria analysis of seal-ring materials and geometry, making it easier to choose the type of material used for sliding rings.

Szumska et al. presented driver behaviour as one of the most important factors affecting road safety [3]. Many traffic situations require that a driver recognize a possible danger. In numerous works, aggressive driving is understood as unsafe and as a hazard entailing the risk of potential crashes. However, traffic safety is not the only thing affected by the driving style of a vehicle operator. Driver behaviour also impacts the operating costs of a vehicle and the emission of environmental air pollutants. This is confirmed by numerous works devoted to the examination of the effect of driving style on fuel economy and air pollution. The objective of this study was to investigate the influence of aggressive driving on fuel consumption and air pollutants emission. The simulation was carried out based on real velocity profiles collected in real-world tests under urban and motorway driving conditions. The results of the simulations confirmed that an aggressive driving style causes a significant increase in both fuel consumption and emission of air pollutants. This was particularly apparent in urban test cycles, where aggressive driving style resulted in higher average fuel consumption and pollutant emissions, as much as 30% to 40% above the average for calm driving.

Grybos et al. demonstrated how to reduce the level of noise from the expanded air from pneumatic tools [4]. Instead of a muffler, an expanded collecting system was proposed, where the air expands through the pneumatic tube and expansion collector. A mathematical model was developed which illustrates the dynamics of the air flow, as well as the acoustic pressure at the end of the tube. The computational results were compared with experimental data to check the air dynamics and sound pressure. Furthermore, the study presents the methodology for measuring noise generated with a pneumatic screwdriver in a quiet back room and on a window fitting stand in a production hall. In addition, noise measurements were made for the pneumatic screwdriver and the pneumatic screwdriver on an industrial scale. These measurements demonstrated a reduction in the noise of pneumatic tools when the expanded collection system was used. When the expanded collecting system was applied to the screwdriver, the measured sound pressure level (SPL) decreased from 87 to 80 dB (A).

Sliwinski described a method for determining the theoretical and actual working volume of a hydraulic motor [5]. The new proposed method is based on the characteristics of the effective absorbency of the motor. Effective absorbency has been defined as the ratio of the flow rate of a motor to the rotational speed of the motor shaft. It has been shown that effective absorbency is a non-linear function of rotational speed and a non-linear function of pressure drop in motor working chambers. Furthermore, it has been proven that the actual working volume of a motor is a function of a third degree of pressure drop in the working chamber of the motor. The actual working volume should be taken to assess the mechanical and volumetric energy losses in the motor. Furthermore, the influence of the flowmeter location in the measurement system and the compressibility of the liquid on the result of the theoretical and actual working volume calculation was also considered and is described in this article. Differences in the assessment of the volumetric efficiency were also shown by assuming theoretical and actual working volumes.

Krzysztofik et al. investigated the sensitivity (resistance) of a quadcopter on-board gyroscope system, for the observation and tracking of a moving ground target, to changing parameters of its regulator under interference conditions [6]. It was shown that the gain in matrix elements is very sensitive, and even a slight deviation from optimal values can lead to reduced target tracking efficiency and the loss of control system stability. Furthermore, the authors studied the energy expenditure at various gyroscope system control parameter values, while homing a quadcopter onto a ground target. A Matlab/Simulink environment was used to perform simulations of the dynamics of the controlled gyroscope system. Selected test results are shown in graphic form.

Fabis-Domagala et al. proposed the use of matrix analysis and statistical methods to perform a simplified RCA (Root Cause Analysis) and to classify potential failures for a variable delivery vane pump [7]. The presented methodology is an extension of matrix FMEA and allows for prioritizing potential failures and their causes in relation to functions performed by pump components, the end effects, and the defined symptoms of failure of the vane pump.

Gapinski et al. presented the concept of controlling the designed optoelectronic tracking and scanning seeker [8]. The device described above is intended for the so-called passive guidance of short-range antiaircraft missiles to various types of air manoeuvring targets. In the presented control method, the modified linear quadratic regulator (LQR) and the estimation of input signals using the extended Kalman filter (EKF) were used. LQR regulation utilizes linearization of the mathematical model of the seeker mentioned above by means of the so-called Jacobians. To improve the stability of the seeker control, the selection of vectors of signals received by the optoelectronic system was used, which also utilized EKF. The results of the research are presented in a graphical form.

Pan et al. presented a new type of electronic oxygen regulator (EOR) for use in oxygen equipment for aviation, that uses a technology where electronic servo control of the gas pressure of controls the gas pressure during breathing [9]. In this paper, the control method of EOR was studied and a dynamic model of the aircraft oxygen system was established. A disturbance-observer-based controller (DOBC) was designed using the backstepping method to achieve the goal of stable and fast breath pressure control. A sensitivity function was proposed to describe the effect of inspiratory flow on breath pressure. Combined with the analysis of the frequency domain of the input sensitivity function, the parameters of the DOBC were analysed and designed. Simulation and experiment studies were carried out to examine the control performance of the DOBC in respiratory resistance and the positive pressurization process under the influence of noise and time delay in the discrete electronic control system, which could meet the physiological requirements of aviation. The research results not only verified the rationality of the application of the DOBC EOR in EOR breath control, but also proved the effectiveness of the control parameter design method according to frequency domain analysis, which provided an important design basis for the subsequent study of EOR.

Luo et al. presented the simulation platform of the molten salt reactor (MSR) plant developed to study the control characteristics under normal and accident conditions [10]. In the platform, a nonlinear dynamic model of the whole system was built consisting of a liquid fuel reactor with a graphite moderator, an intermediate heat exchanger, and a steam generator. A new control strategy based on a combined feed-forward and feedback scheme is presented, where a power control system and a steam temperature control system were designed to regulate load changes in the plant. Three different types of operation conditions were simulated with the control systems, including transients of normal load-follow operation, a reactivity insertion accident, and a loss-of-flow accident. The simulation results showed that the developed control system not only had a fast load-following capability during normal operation, but also had good control performance under accident conditions.

Sliwinski et al. presented new methods to determine the pressure drop in the internal channels of a hydraulic motor [11]. Mathematical models of pressure losses in internal channels are also described. Experimental tests of the satellite motor were carried out according to one of the proposed methods. The tests were carried out for two liquids, i.e., water and mineral oil. Experimental studies have shown that at a high flow rate in a motor supplied with water, the pressure losses are a dozen or so percent greater than in a motor supplied with oil. However, at low flow rates, the inverse is true; that is, the pressure losses in the motor supplied with water are about 10% lower than in the motor supplied with oil. A CFD calculation of the pressure drop in the internal channel of the motor was also performed. It was assumed that the holes in the commutation unit plate are placed face-to-face and that the liquid did not cause changes in the volume of the working chambers. In this way, it was proven that those simplified assumptions can have up to a 50% difference in relation to the experimental tests.

Fabis-Domagala et al. presented a FMEA analysis as a quality improvement tool that has been widely used for decades [12]. Its classical version prioritizes the risk of failure by risk priority number (RPN). The RPN is a product of severity (S), occurrence (O), and detection (D), where all factors have equal levels of significance. This assumption is one of the most criticized drawbacks, as it has given unreasonable results for real-world applications. RPN can produce equal values for combinations of risk factors with different risk implications. Another issue is the uncertainties and subjectivities of the information employed in the FMEA analysis that may arise from the lack of knowledge, experience, and linguistic terms used. Many alternatives to risk assessment methods have been proposed to overcome the weaknesses of classical FMEA risk management, in which we can distinguish modification methods by RPN numbers of employing new tools. In this study, we proposed a modification of the traditional RPN number. The main difference is that severity and occurrence are valued based on subfactors. The detection number remained unchanged. Additionally, the proposed method prioritizes risk in terms of implied risk to systems by implementing functional failures (effects of potential failures). A typical fluid power system was used to illustrate the application of this method. The method showed the correct failure classification, which meets the industrial experience and other results of research on failures of fluid power systems.

Liu et al. presented a small modular dual fluid reactor as a novel variant of the Generation IV molten salt reactor and the liquid metal fast reactor [13]. In the primary circuit, molten salt, or the liquid eutectic metal (U-Pu-Cr), is used as fuel, and liquid lead works as the coolant in the secondary circuit. To design the control system of such an advanced reactor, the uncertainties of the computer model, and the physicochemical properties of the materials, must both be considered. In this paper, a one-dimensional model of a core is established, based on the equivalent parameters achieved via the coupled three-dimensional model and accounting for delayed neutron precursor drifting, and a power control system is developed. The performance of the designed controllers is assessed, taking into account the model and property uncertainties. The results achieved show that the designed control system can maintain the stability of the system and regulate the power as expected. Among the uncertain parameters considered, the fuel temperature reactivity coefficients have the greatest influence on the performance of the control system. The most optimized configuration of the control system is delivered based on uncertainty propagation characteristics using the particle swarm optimization method.

3. Conclusions

Trends in the development of fluid power systems are related to electronics, information technology, artificial intelligence, miniaturization, energy efficiency, environmental protection, and greater efficiency. The fluid power system industry aims to increase energy efficiency, improve operational reliability, use smart components and systems, reduce size and weight, and recover, store, and reuse energy.

Author Contributions: Conceptualization, R.D., J.T. and P.W.; methodology, R.D.; data curation, J.T. and P.W., writing—original draft preparation, R.D.; writing—review and editing, R.D.; supervision, J.T. and P.W. All authors have read and agreed to the published version of the manuscript.

Funding: This research received no external funding.

Institutional Review Board Statement: Not applicable.

Informed Consent Statement: Not applicable.

Data Availability Statement: The data presented in this study are available on the website https://www.mdpi.com/journal/energies/special_issues/Fluid_Power_Systems, accessed on 11 December 2021.

Acknowledgments: The guest editors thank the MDPI Publishing House and the "Energies" editorial staff for inviting us to act as guest editors of the special edition of "Advances in Fluid Power Systems".

Conflicts of Interest: The authors declare no conflict of interest.

References

1. Filo, G.; Lisowski, E.; Rajda, J. Pressure Loss Reduction in an Innovative Directional Poppet Control Valve. *Energies* **2020**, *13*, 3149. [CrossRef]
2. Blasiak, S. Influence of Thermoelastic Phenomena on the Energy Conservation in Non-Contacting Face Seals. *Energies* **2020**, *13*, 5283. [CrossRef]
3. Szumska, E.; Jurecki, R. The Effect of Aggressive Driving on Vehicle Parameters. *Energies* **2020**, *13*, 6675. [CrossRef]
4. Gryboś, D.; Leszczyński, J.; Czopek, D.; Wiciak, J. Exhaust Noise Reduction by Application of Expanded Collecting System in Pneumatic Tools and Machines. *Energies* **2021**, *14*, 1592. [CrossRef]
5. Sliwinski, P. Determination of the Theoretical and Actual Working Volume of a Hydraulic Motor—Part II (The Method Based on the Characteristics of Effective Absorbency of the Motor). *Energies* **2021**, *14*, 1648. [CrossRef]
6. Krzysztofik, I.; Koruba, Z. Study on the Sensitivity of a Gyroscope System Homing a Quadcopter onto a Moving Ground Target under the Action of External Disturbance. *Energies* **2021**, *14*, 1696. [CrossRef]
7. Fabis-Domagala, J.; Domagala, M.; Momeni, H. A Matrix FMEA Analysis of Variable Delivery Vane Pumps. *Energies* **2021**, *14*, 1741. [CrossRef]
8. Gapiński, D.; Koruba, Z. Control of Optoelectronic Scanning and Tracking Seeker by Means the LQR Modified Method with the Input Signal Estimated Using of the Extended Kalman Filter. *Energies* **2021**, *14*, 3109. [CrossRef]
9. Pan, R.; Lin, G.; Shi, Z.; Zeng, Y.; Yang, X. The Application of Disturbance-Observer-Based Control in Breath Pressure Control of Aviation Electronic Oxygen Regulator. *Energies* **2021**, *14*, 5189. [CrossRef]
10. Luo, R.; Liu, C.; Macián-Juan, R. Investigation of Control Characteristics for a Molten Salt Reactor Plant under Normal and Accident Conditions. *Energies* **2021**, *14*, 5279. [CrossRef]
11. Sliwinski, P.; Patrosz, P. Methods of Determining Pressure Drop in Internal Channels of a Hydraulic Motor. *Energies* **2021**, *14*, 5669. [CrossRef]

12. Fabis-Domagala, J.; Domagala, M.; Momeni, H. A Concept of Risk Prioritization in FMEA Analysis for Fluid Power Systems. *Energies* **2021**, *14*, 6482. [CrossRef]
13. Liu, C.; Luo, R.; Macián-Juan, R. A New Uncertainty-Based Control Scheme of the Small Modular Dual Fluid Reactor and Its Optimization. *Energies* **2021**, *14*, 6708. [CrossRef]

energies

MDPI

Article

Influence of Thermoelastic Phenomena on the Energy Conservation in Non-Contacting Face Seals

Slawomir Blasiak

Department of Manufacturing Engineering and Metrology, Faculty of Mechatronics and Mechanical Engineering, Kielce University of Technology, Aleja Tysiaclecia Panstwa Polskiego 7, 25-314 Kielce, Poland; sblasiak@tu.kielce.pl; Tel.: +48-41-34-24-756

Received: 17 September 2020; Accepted: 8 October 2020; Published: 12 October 2020

Abstract: The purpose of this study was to develop a mathematical model for non-contacting face seals to analyze how their performance is affected by thermoelastic phenomena. The model was used to solve thermal conductivity and thermoelasticity problems. The primary goal was to calculate the values of thermal deformations of the sealing rings in a non-contacting face seal with a flexibly mounted rotor (FMR) for a turbomachine. The model assumes the conversion of mechanical energy into heat in the fluid film. The heat flux generated in the fluid film is transferred first to the sealing rings and then to the fluid surrounding them. Asymmetric distribution of temperature within the sealing rings leads to the occurrence of thermal stresses and, consequently, a change in the geometry of the rings. The model is solved analytically. The distributions of temperature fields for the sealing rings in the cross-sections are calculated using the Fourier-Bessel series as a superficial function of two variables (r,z). The thermoelasticity problems described by the Navier equations are solved by applying the Boussinesq harmonic functions and Goodier's thermoelastic displacement potential function. The proposed method involves solving various theoretical and practical problems of thermoelasticity in FMR-type non-contacting face seals. The solution of the mathematical model was made use of analytical methods, and the most important obtained results are presented in graphical form, such as the temperature distributions and axial thermal distortions in cross-sections of the rings. The calculated thermal deformations of the sealing rings are used to determine the most important seal performance parameters such as the leakage rate and power loss. The article also presents a multi-criteria analysis of seal rings materials and geometry, which makes it easier to choose the type of materials used for the sliding rings.

Keywords: mechanical seal; non-contacting face seal; heat transfer; thermal analysis

1. Introduction

This article provides an analytical solution to an axisymmetric thermoelastic problem for sealing rings in non-contacting face seals. The basic requirement concerning the performance of non-contacting seals is to maintain the height of the radial clearance within the limits determined at the design stage. This is difficult to achieve because of different disturbances affecting the seal performance. The most important are disturbances to the equilibrium of forces acting on the system of rings, which may be caused, for example, by thermal deformations of these elements.

There is plenty of research into the behavior of non-contacting seals. This article reviews only studies focusing on heat transfer and thermal deformations.

Some of the first research papers on the subject provided mathematical descriptions of the heat transfer phenomena in non-contacting seals in the form of one-dimensional models; they analyzed only the distributions of pressure and temperature within the sealing rings and the fluid film (e.g., [1]). Then, more complex thermohydrodynamic and thermos-elasto-hydro-dynamic models were proposed. Refs. [2–6], for instance, discuss numerical solutions to advanced two- and three-dimensional mathematical models.

Studies on thermoelastic problems for non-contacting seals include Ref. [7], which provides numerical calculations of thermal deformations of the sealing rings to analyze their effect on the seal performance. Ref. [8] describes two types of macroscopic thermoelastic deformations: those taking place under quasi-steady-state conditions and those typical of unsteady-state conditions. The latter, referred to as thermoelastic instability, are sudden uncontrolled deformations of the surface of the sealing rings. The experiments confirmed the occurrence of both types of thermal deformation.

The geometry of the clearance between the sealing rings changes when the elements are deformed; this disturbs the equilibrium of forces and causes changes in the leakage rate and power loss. In a publication (Ref. [9]) was presented the uncoupled thermoelastic brittle fracture problem is discussed in terms of the types of stress fields produced by surface heating or cooling and the generic characteristics of the thermally generated stress intensity factors. Also presented an Examples of experimental measurements and numerical calculations with demonstrating these general characteristics.

This article proposes a two-dimensional model to describe the heat transfer and thermal deformations in non-contacting face seals. The equations with partial derivatives are solved using the technique of separation of variables. For both rings, the equation of energy and equations of conductivity are written in a cylindrical coordinate system, and the analytical solutions are based on the Bessel functions of the first and second kind. The thermoelasticity problems described by the Navier equations are solved using the Boussinesq harmonic functions as well as Goodier's thermoelastic displacement potential function.

Because of their specific design and varied operating conditions, non-contacting seals are vital elements of sealing systems in a variety of machines. The proposed solution of the complex mathematical model is used to analyze the effect of thermal deformations on the basic seal performance parameters, i.e., leakage rate and power loss.

The main aim of the study was to determine the influence of the selection of materials on the working rings on heat transfer, the accompanying thermoelastic deformations, power loss and leakage rate.

2. Mathematical Model

Figure 1 shows a schematic diagram of an FMR-type non-contacting face seal. The seal consists of two rings: a rotor (2), flexibly mounted to the shaft (6) of the turbomachine, and a stator (1), fixed to the housing.

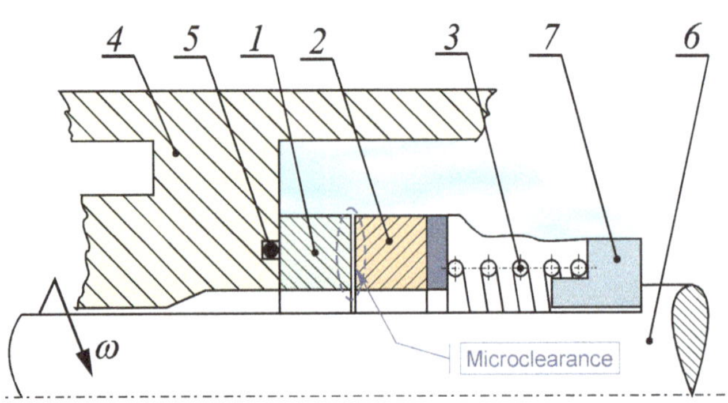

Figure 1. Non-contacting face seal: 1—stator, 2—rotor, 3—spring, 4—housing, 5—O-ring, 6—shaft, 7—steady pin.

The mathematical model describing the physical phenomena occurring in the sealing rings–fluid film system of a non-contacting face seal was developed using some simplifications, like in [10–13]. The formulated model was then solved analytically.

2.1. Function of the Radial Clearance Height

A key parameter of non-contacting face seals is the nominal height of the radial clearance h_o, which is dependent on the equilibrium of forces acting on the sealing rings. The geometry of the fluid film separating the rings can be written as a difference between the functions of the surface topographies of the rotor and stator. The general form of the function is:

$$h = h(r) = h_o + h^r(r) - h^s(r) + h(r)^r_{def} - h(r)^s_{def} \quad (1)$$

Although there are many theoretical studies on the subject providing models of heat transfer in the rotor-stator system, they do not take into consideration changes in the geometry of the radial clearance resulting from the thermal deformations of the rings. In Equation (1), the relationships $h(r)^r_{def}$ and $h(r)^s_{def}$ represent deformations of the seal faces caused by nonuniform distributions of temperature within the sealing rings.

2.2. Reynolds Equation

The distribution of the fluid pressure in the clearance was determined by solving the following one-dimensional Reynolds equation for the laminar flow:

$$\frac{d}{dr}\left(\frac{\rho h^3}{\mu}\frac{dp}{dr}\right) = 0 \quad (2)$$

Solving Equation (2) requires specifying the boundary conditions for the inner and outer radii of the radial clearance. Variables p, ρ and μ are functions dependent on the changes in the fluid temperature in the clearance, whereas the h function is the height of the radial clearance described by Equation (1).

2.3. Equation of Energy

According to the classical hypothesis on thin viscous fluid films, with the fluid being incompressible and in the steady state, the equation of energy for the process fluid takes the following form [2]:

$$\rho c_v \left\{ v_r \frac{\partial T^f}{\partial r} + \frac{v_\theta}{r}\frac{\partial T^f}{\partial \theta} + v_z \frac{\partial T^f}{\partial z} \right\} = \mu \left\{ \left(\frac{\partial v_r}{\partial z}\right)^2 + \left(\frac{\partial v_\theta}{\partial z}\right)^2 \right\} + \lambda \frac{\partial^2 T^f}{\partial z^2} \quad (3)$$

After simplifications made for the model of fluid flow in a given type of seal, the equation of energy takes the form [11,14]:

$$\mu\left(\frac{\partial v_\phi}{\partial z}\right)^2 + \lambda^f \frac{\partial^2 T^f}{\partial z^2} = 0 \quad (4)$$

Once the equation of energy (4) is solved, it is possible to determine the distribution of temperature within the fluid film. The face seal model requires making a simplification concerning the distribution of fluid velocity in the fluid film v_ϕ and an assumption that the velocity changes linearly according to the following relationship [10]:

$$\frac{\partial v_\phi}{\partial z} = \frac{\omega r}{h} \quad (5)$$

Equation (5) describes changes in the fluid velocity along the radial clearance height.

2.4. Fluid Dynamic Viscosity

The fluid dynamic viscosity is largely dependent on the distribution of temperature within the fluid separating the rings; this relationship is provided in [10]. For water, the relationship between fluid viscosity and temperature can be expressed by:

$$\mu = \mu_o \exp(-b(T_m - T_o)) \tag{6}$$

The average fluid temperature is defined as:

$$T_m = \frac{1}{h(r)} \int_0^{h(r)} T^f dz \tag{7}$$

Equation (6) describes the distribution of fluid dynamic viscosity $\mu(r)$ in the radial direction.

2.5. Distribution of Temperature

With axisymmetric heat transfer conditions and constant coefficients of conductivity, the distributions of temperature within the sealing rings can be written in the general form of the equation of conductivity for the steady-state model:

$$\frac{1}{r}\frac{\partial\theta}{\partial r} + \frac{\partial^2\theta}{\partial r^2} + \frac{\partial^2\theta}{\partial z^2} = 0 \tag{8}$$

Suppose that the temperature changes θ has no effect on the material coefficients. These can thus be treated as constant. The superscript notation is used, e.g., θ^s and θ^r, for the stator and rotor, respectively.

2.6. Thermoelasticity

The term thermoelasticity refers to a wide range of phenomena. Thermoelasticity represents a generalization of the classical theory of elasticity and the theory of thermal conductivity. In this section, the basic equations describing the thermoelasticity of a homogeneous isotropic body are formulated.

The changes in temperature of the continua, i.e., the ring materials, attributable to the uneven heating of the ring faces, lead to the occurrence of stresses σ_{ij} and strains ε_{ij}; the former can be expressed in the general form using Hooke's law [15]:

$$\sigma_{ij} = 2\vartheta\varepsilon_{ij} + (\lambda e - \beta\theta)\delta_{ij}(i, j = 1, 2, 3), \tag{9}$$

where the Kronecker symbol is defined as: $\delta_{ij} = \begin{cases} 1 & for\, i = j \\ 0 & for\, i \neq j \end{cases}$.

From the equations of equilibrium in the form:

$$\sigma_{ji,j} + F_i = 0 \quad (i, j = 1, 2, 3), \tag{10}$$

as well as Equation (9) and the defined strains, we can formulate equations of displacement in the general form:

$$\vartheta\, \nabla^2 u_i + (\lambda + \vartheta) u_{k,ki} - \beta\, \theta_{,i} + F_i = 0, \tag{11}$$

where: $\beta = \frac{\tau E}{(1-2v)}$, $G = \vartheta$, with the values of the coefficients E, v are Young's modulus and Poisson's coefficient respectively.

Introducing a cylindrical coordinate system (r, ϕ, z) for the axisymmetric problem, we can write the equations of equilibrium of forces Equation (10) in the r and z directions as [15]:

$$\begin{aligned} \tfrac{\partial \sigma_{rr}}{\partial r} + \tfrac{\partial \sigma_{zr}}{\partial z} + \tfrac{\sigma_{rr} - \sigma_{\phi\phi}}{r} + F_r = 0 \\ \tfrac{\partial \sigma_{rz}}{\partial r} + \tfrac{\partial \sigma_{zz}}{\partial z} + \tfrac{\sigma_{rz}}{r} + F_z = 0 \end{aligned} \tag{12}$$

Substituting Relationships (9) and the strain components into Equation (12) yields a solution to the Navier equations for axisymmetric problems of thermoelasticity, which, when external forces are omitted, can be expressed by means of the thermoelastic displacement potential function Φ and the Boussinesq harmonic functions φ and ψ. The functions must satisfy the following equations:

$$\nabla^2 \Phi = K\theta \tag{13}$$

$$\nabla^2 \varphi = 0, \nabla^2 \psi = 0 \tag{14}$$

with the coefficient K being: $K = \frac{1+v}{1-v}\tau$.

The Michell function M, which is dependent on the Boussinesq harmonic functions:

$$M = -\int (\varphi + z\,\psi)\, dz \tag{15}$$

can be used to write the displacements in the cylindrical coordinate system as [15]:

$$\begin{aligned} u_r = \tfrac{\partial \Phi}{\partial r} - \tfrac{\partial^2 M}{\partial r \partial z} = \tfrac{\partial \Phi}{\partial r} + \tfrac{\partial \varphi}{\partial r} + z\tfrac{\partial \psi}{\partial r} \\ u_z = \tfrac{\partial \Phi}{\partial z} + 2(1-v)\nabla^2 M - \tfrac{\partial^2 M}{\partial z^2} = \tfrac{\partial \Phi}{\partial z} + \tfrac{\partial \varphi}{\partial z} + z\tfrac{\partial \psi}{\partial z} - (3-4v)\psi \end{aligned} \tag{16}$$

The Michell solution is a biharmonic function satisfying the equation:

$$\nabla^2 \nabla^2 M = -2\nabla^2 \psi = 0 \tag{17}$$

Thus, the stress components are represented by the thermoelastic displacement potential function Φ and the Michell solution M:

$$\begin{aligned} \sigma_{rr} = 2G\left[\tfrac{\partial^2 \Phi}{\partial r^2} - K\theta + \tfrac{\partial}{\partial z}\left(v\nabla^2 M - \tfrac{\partial^2 M}{\partial r^2}\right)\right] \\ \sigma_{\phi\phi} = 2G\left[\tfrac{1}{r}\tfrac{\partial \Phi}{\partial r} - K\theta + \tfrac{\partial}{\partial z}\left(v\nabla^2 M - \tfrac{1}{r}\tfrac{\partial M}{\partial r}\right)\right] \\ \sigma_{zz} = 2G\left[\tfrac{\partial^2 \Phi}{\partial z^2} - K\theta + \tfrac{\partial}{\partial z}\left((2-v)\nabla^2 M - \tfrac{\partial^2 M}{\partial z^2}\right)\right] \\ \sigma_{rz} = 2G\left[\tfrac{\partial^2 \Phi}{\partial r \partial z} + \tfrac{\partial}{\partial r}\left((1-v)\nabla^2 M - \tfrac{\partial^2 M}{\partial z^2}\right)\right] \end{aligned} \tag{18}$$

$$\sigma_{zz} = \sigma_{rz} = 0 \tag{19}$$

Like in [15], the fields of stresses and displacements are determined using the boundary conditions for Equation (19), which are such that the surfaces of the rotor $(z = 0)$ and the stator $(z = -L)$ are free from stresses.

3. Boundary Conditions

The system of coupled equations describing the distributions of temperature within the sealing rings and the radial clearance separating them is solved by imposing the boundary conditions.

The equations of conductivity and equations of energy are solved using suitable boundary conditions, like in [2,11,16]. The boundary conditions for the system considered are shown in Figure 2.

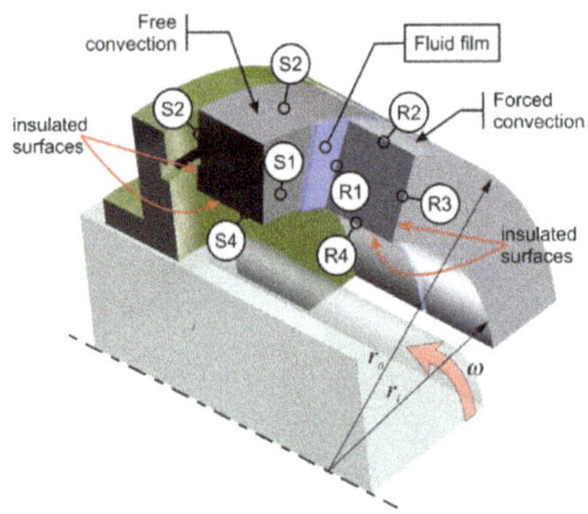

Figure 2. Heat transfer conditions for the FMR-type non-contacting face seal, where S_1, S_2, S_3, S_4—surfaces of the stator; and R_1, R_2, R_3, R_4—surfaces of the rotor.

It is assumed that the surfaces S_3 and S_4 of the stator mounted to the housing and the surfaces R_3 and R_4 of the rotating ring are completely separated so there is no transfer of heat to the fluid surrounding the rings. This condition for both the stator and rotor can be written in the general form as:

$$\frac{\partial \theta}{\partial n} = 0. \tag{20}$$

Another assumption is that the heat transfer across the surfaces S_1 and R_1 (faces) of the stator and rotor, respectively, which are in direct contact with the fluid filling the radial clearance, occurs by conduction. This implies that the value of the heat flux at the surface of the ring is equal to that of the fluid in the clearance:

$$\lambda^s \frac{\partial \theta^s}{\partial z} = \lambda^f \frac{\partial \theta^f}{\partial z} = q^s{}_v(r) \text{ with } \theta^s = \theta^f \text{ for } S_1 \tag{21}$$

and

$$q^r{}_v(r) = \lambda^f \frac{\partial \theta^f}{\partial z} = -\lambda^r \frac{\partial \theta^r}{\partial z} \text{ with } \theta^f = \theta^r \text{ for } R_1 \tag{22}$$

It is also assumed that the heat transfer across the outer surfaces of the rings S_2 and R_2 (Figure 2), being in contact with the process fluid, takes place by convection. For the stator and rotor, the condition can be written in the general form:

$$-\lambda \left.\frac{\partial \theta}{\partial r}\right|_{r=r_o} = \alpha\,\theta|_{r=r_o} \tag{23}$$

In the analytical calculations, an additional assumption is made that the heat transfer coefficients for the stator the rotor have different values. This is due to the fact the transfer of heat to the process fluid is higher from the rotating ring than from the stationary ring, with the latter occurring by free convection.

It is assumed that the heat transfer coefficient for the stationary ring has a constant value; the value is given in Table 1.

Table 1. Geometrical and performance parameters.

Geometry		Performance Parameters	
Inner radius r_i	0.040 (m)	Angular velocity ω	200–1500 (rad/s)
Outer radius r_o	0.045 (m)	Nominal seal clearance h_o	1×10^{-6} (m)
Ring thickness L^s and L^r	0.005 (m)	Fluid temperature T_o	20 (°C)
Thermal conductivity λ^f	0.65 (W/m K)	Heat transfer coefficient α^s	18,000 (W/m^2 K)

For the rotor, the heat transfer coefficient is calculated using relationships [11]:

$$\alpha^r = 0.133\,\mathrm{Re}_D{}^{2/3}\,\mathrm{Pr}^{1/3}\frac{\lambda^f}{D} \tag{24}$$

where D is the outer diameter of the seal and Re_D is the Reynolds number based on this diameter; the parameters λ^f and Pr are the thermal conductivity and the Prandtl number of the fluid, respectively.

The Reynolds and Prandtl numbers are described as [17]:

$$\mathrm{Re}_D = \frac{\omega\,D^2\rho}{\mu} \tag{25}$$

and

$$\mathrm{Pr} = \frac{C_p\,\mu}{\lambda^f} \tag{26}$$

The next section describes the analytical solution of the mathematical model of the heat transfer in non-contacting face seals.

3.1. Analytical Solution of the Heat Transfer Model

The formulated mathematical model is solved analytically using the technique of separation of variables, like in [18,19]. The first step of the problem-solving process consists of determining the distributions of temperature within the sealing rings by defining the general form of the functions satisfying the differential Equation (8) for both the stator and the rotor. Subsequently, a general solution is found to satisfy the specific boundary conditions defined by relationships (20)–(23). The equations describing the distributions of temperature are as follows:

for the stator:

$$T^s = T_0 + \sum_{n=1}^{\infty} B_n^s \cosh(s_n^s z^s)\left(J_0(s_n^s r) - \frac{J_1(s_n^s r_i)}{Y_1(s_n^s r_i)}\,Y_0(s_n^s r)\right) \tag{27}$$

for the fluid film separating the rings:

$$\begin{aligned} T^f = {} & T_0 + \tfrac{1}{2}\cdot\tfrac{\mu}{\lambda^f}\cdot\tfrac{\omega^2 \cdot r^2}{h^2}\cdot\left(h^2 - \left(z^f\right)^2\right) + \\ & + \sum_{n=1}^{\infty} B_n^r \cosh(s_n^r z^r)\left(J_0(s_n^r r) - \tfrac{J_1(s_n^r r_i)}{Y_1(s_n^r r_i)}\,Y_0(s_n^r r)\right) \end{aligned} \tag{28}$$

and for the rotor:

$$T^r = T_0 + \sum_{n=1}^{\infty} B_n^r \cosh(s_n^r z^r)\left(J_0(s_n^r r) - \frac{J_1(s_n^r r_i)}{Y_1(s_n^r r_i)}\,Y_0(s_n^r r)\right) \tag{29}$$

The above relationships describing the distributions of temperature within the stator–rotor system and the fluid film will be used to determine the stress fields.

3.2. Analytical Solution of the Thermoelasticity Model

The model describing thermoelastic phenomena is relatively complex and its analytical solution requires performing complex calculations. A general method of the model solution will be used for both rings. The first stage involves determining the thermoelastic displacement potential function using Equation (27) or Equation (29), with the superscripts for the stator and rotor being omitted:

$$\theta(r,z) = \sum_{n=1}^{\infty} B_n \cosh(s_n z)\left(J_0(s_n r) - \frac{J_1(s_n r_i)}{Y_1(s_n r_i)}\,Y_0(s_n r)\right) \tag{30}$$

The thermoelastic displacement potential function is written as:

$$\Phi = \frac{K}{2}\sum_{n=1}^{\infty}\frac{B_n}{s_n}(z-L)\sinh(s_n z)\left(J_0(s_n r) - \frac{J_1(s_n r_i)}{Y_1(s_n r_i)}Y_0(s_n r)\right) \quad (31)$$

As assumed, the above equation needs to satisfy Relationship (13).

The second step requires finding the Michell function. It is predicted to have the following form:

$$M = \frac{K}{2}\sum_{n=1}^{\infty}\frac{B_n}{{s_n}^3}\left(J_0(s_n r) - \frac{J_1(s_n r_i)}{Y_1(s_n r_i)}Y_0(s_n r)\right)\left(\begin{array}{c}(1-2v)\sinh(s_n z)+ \\ +(z-L)s_n\cosh(s_n z)\end{array}\right) \quad (32)$$

Substituting these relationships into Equation (18) yields the components of the stress tensor:

$$\sigma_{rr} = 4KG\sum_{n=1}^{\infty}\left(\frac{B_n\cosh(s_n z)\left(\begin{array}{c}-Y_1(s_n r_i)J_1(s_n r) + Y_1(s_n r_i)J_1(s_n r)\,v \\ +J_1(s_n r_i)Y_1(s_n r) - J_1(s_n r_i)Y_1(s_n r)\,v\end{array}\right)}{Y_1(s_n r_i)\,s_n\,r}\right) \quad (33)$$

$$\sigma_{\phi\phi} = 4GK\sum_{n=1}^{\infty}\left(\frac{B_n\cosh(s_n z)}{Y_1(s_n r_i)\,s_n\,r}\left(\begin{array}{l}-s_n r\,Y_1(s_n r_i)J_0(s_n r) + s_n r\,J_1(s_n r_i)Y_0(s_n r)+ \\ +v\,s_n r\,J_0(s_n r)Y_1(s_n r_i)+ \\ -v\,s_n r\,J_1(s_n r_i)Y_0(s_n r)+ \\ +Y_1(s_n r_i)J_1(s_n r) - v\,Y_1(s_n r_i)J_1(s_n r)+ \\ -J_1(s_n r_i)Y_1(s_n r) + v\,J_1(s_n r_i)Y_1(s_n r)\end{array}\right)\right) \quad (34)$$

The relationships describing displacements are given as:

$$u_r = 2K\sum_{n=1}^{\infty}\frac{B_n(-1+v)(-Y_1(s_n r_i)J_1(s_n r) + J_1(s_n r_i)Y_1(s_n r))\cosh(s_n z)}{Y_1(s_n r_i)s_n} \quad (35)$$

$$u_z = 2K\sum_{n=1}^{\infty}\frac{B_n(-1+v)(-J_0(s_n r)Y_1(s_n r_i) + J_1(s_n r_i)Y_0(s_n r))\sinh(s_n z)}{Y_1(s_n r_i)s_n} \quad (36)$$

The above relationships were used to graphically represent fields of stresses and displacements occurring in the sealing rings due to uneven distributions of temperature.

4. Results and Discussion

A numerical analysis was conducted to verify the relevant hypotheses and assess the influence of the selected parameters on the behavior of non-contacting face seals. In the reference case, water is used as the fluid. Water as a working medium is commonly used in various types of utility installations. The sealing rings are assumed to be in alignment and their faces create a radial clearance with a constant height h_o. Another assumption is that the process fluid is not in contact with the outer surfaces of the rings, except for the cylindrical surfaces, where the heat transfer to the fluid occurs by convection (Figure 2). The geometrical and performance parameters of the seal under consideration are defined in Table 1.

One of the first problems mechanical designers need to deal with is selecting the right seal, i.e., a seal that meets the criteria specified for the sealing fluid to be used in the turbomachine, including its temperature and pressure. They must also predict the dry running condition, which takes place during the machine startup and shutdown. Another important problem design engineers have to consider is the chemical durability of the sealing rings and secondary seals. The materials they select for these elements need to have appropriate physicochemical properties. The most common materials used for mechanical seals are characterized in Table 2.

Table 2. Properties of materials used for the sealing rings.

Material	Young's Modulus E (GPa)	Poisson's Coefficient ν	Thermal Conductivity λ (W m^{-1} K^{-1})	Linear Thermal Expansion Coefficient τ (10^{-6} °C)
SiC (Silicon Carbide)	400	0.17	150	4.3
Resin-impregnated carbon	24	0.12	10	4.9
Alumina (Ceramic)	350	0.23	30	7.5
Tungsten carbide (WC)	630	0.24	85	5.4

The materials used for the sealing rings need to be suitable for the seal operating conditions.

In the case of non-contacting face seals, the sealing rings are usually made of dissimilar materials, i.e., ones differing in physicochemical properties. This, however, leads to the occurrence of a sequence of undesirable phenomena, starting with uneven heat transfer from the radial clearance to the surrounding fluid, followed by asymmetric thermoelastic deformations of the rings, then a change in the geometry of the radial clearance and, finally, an increase in the leakage rate.

Non-contacting face seals, also called mechanical face seals, are used in many mechanical sealing systems. The correct performance of non-contacting seals is conditioned by the properly selected parameters of the sealing rings. One of the most significant parameters that have a direct effect on the distribution of pressure within the fluid film and the heat flux generated in the clearance is the angular velocity of the rotor.

The key feature of mechanical seals is the ability to operate in 'no lubrication' conditions. In practice, this does not last long, but the materials used for the sealing rings must have sufficient hardness and thermal resistance to withstand an increase in temperature caused by excessive friction forces. Two types of material configurations are used for the sealing rings: 'hard-hard' and 'hard–soft'.

Seals with both rings made of hard materials, e.g., silicon carbide-silicon carbide or tungsten carbide–tungsten carbide, exhibit high resistance to many active chemical compounds, but they are characterized by little resistance to dry running. Seals with a hard-soft ring configuration, on the other hand, e.g., carbon–tungsten carbide, are more resistant to dry running but have limited chemical resistance to fluids other than water. As the selection of ring materials to match the seal operating conditions is an open problem, a numerical analysis is necessary. This study was conducted for two types of rotor materials, i.e., silicon carbide and tungsten carbide, and three types of stator materials, i.e., silicon carbide, tungsten carbide, and alumina (ceramic). The numerical analysis was thus performed for six pairs of sealing rings.

The diagrams in Figure 3 illustrate temperature distributions in the cross-sections of the sealing rings and the fluid film separating them. The results show that in all the cases considered, the maximum temperature of the fluid film is registered along the inner radius r_i. Depending on the kind of used materials and theirs combinations, the temperature range is 61.4–71.6 °C for the inner radius $r_i = 0.04$ (m) of the fluid film.

According to Figure 3, the temperature of the water is close to the saturation temperature of the water under the condition of atmospheric pressure. For this reason, the temperature of the water is limited and should be below the boiling temperature. The evaporation of the working medium can lead to loss of the lubricating film stability and consequently to working with dry friction conditions between the sliding rings.

The lowest temperature of the fluid film is reported for the Alumina-SiC pair (Figure 3a), while the highest for the RiC-WC and WC-WC pairs (Figure 3e,f). Another observation is a large difference in temperature along the outer radius r_o. It should be noted that the analyzed model assumes forced convection for the rotor, which implies increased heat transfer to the fluid surrounding this ring. From Figure 3b,e, it is evident that, for the RiC-SiC and RiC-WC pairs, the temperature along the outer radius is up to 5 °C lower for the stator than for the rotor. When the stator and rotor are made of the same material, i.e., for the SiC-SiC and WC-WC pairs (Figure 3c,f), the situation is the opposite; the temperature of the stator along the outer radius is 15 °C higher than that of the rotor. Another important observation is that the use of silicon carbide in a combination of material types lowers the temperature of the working fluid, in all cases. This is due to the increased thermal conductivity of silicon carbide.

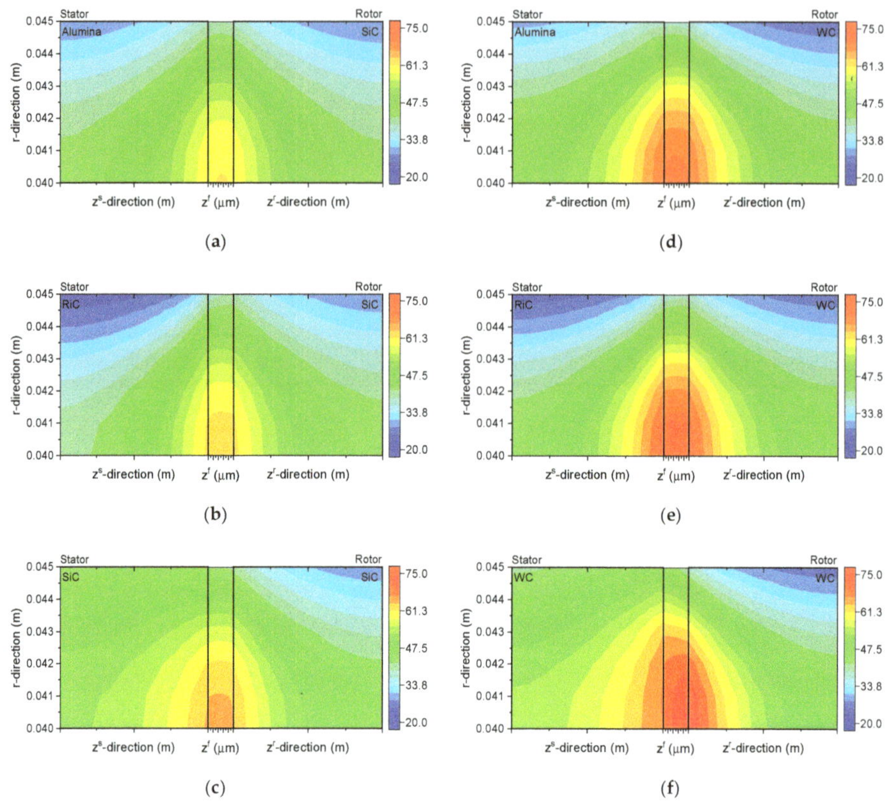

Figure 3. Temperature distributions in the cross-sections of the sealing rings and the fluid film at $\omega = 1500\,(\mathrm{rad/s})$, (**a**) Alumina-SiC; (**b**) RiC-SiC; (**c**) SiC-SiC; (**d**) Alumina-WC; (**e**) RiC-WC; (**f**) WC-WC pairs of matrices.

The temperature differences in the ring cross-sections are reduced to a minimum to ensure minimum thermal stresses.

Figure 4 shows the fields of angular displacements in the sealing rings. Since the displacements were very small, of the order of $10^{-6}(\mathrm{m})$, the deformations of the rings caused by thermal stresses were magnified 2000-fold. Thanks to that, it is possible to observe warps on the sealing rings. The numerical data indicate that the highest displacements of the order of $1.2 \times 10^{-6}(\mathrm{m})$ occurrence along the inner radii of the rings in a seal where both the stator and the rotor are made of tungsten carbide (Figure 4f).

A change in the geometry of the sealing rings causes a change in the height of the radial clearance, taken into consideration in Relationships (1), and a change in the distribution of pressure, calculated using the Reynolds equation.

Figure 5 provides a graphical representation of a change in the function of the radial clearance height, taking account of the thermal deformations of the sealing rings resulting from the asymmetric distribution of temperature in these elements. For each case considered, there is an increase in the minimum height of the radial clearance; the highest value of $1.72 \times 10^{-6}(\mathrm{m})$ is reported for the

Alumina–WC pair, whereas the lowest of $1.43 \times 10^{-6}(\mathrm{m})$ is observed for the SiC-SiC pair. A change in the geometry of the radial clearance has a direct effect on the distribution of pressure in the fluid film (Figure 5b), which causes an increase in the opening force.

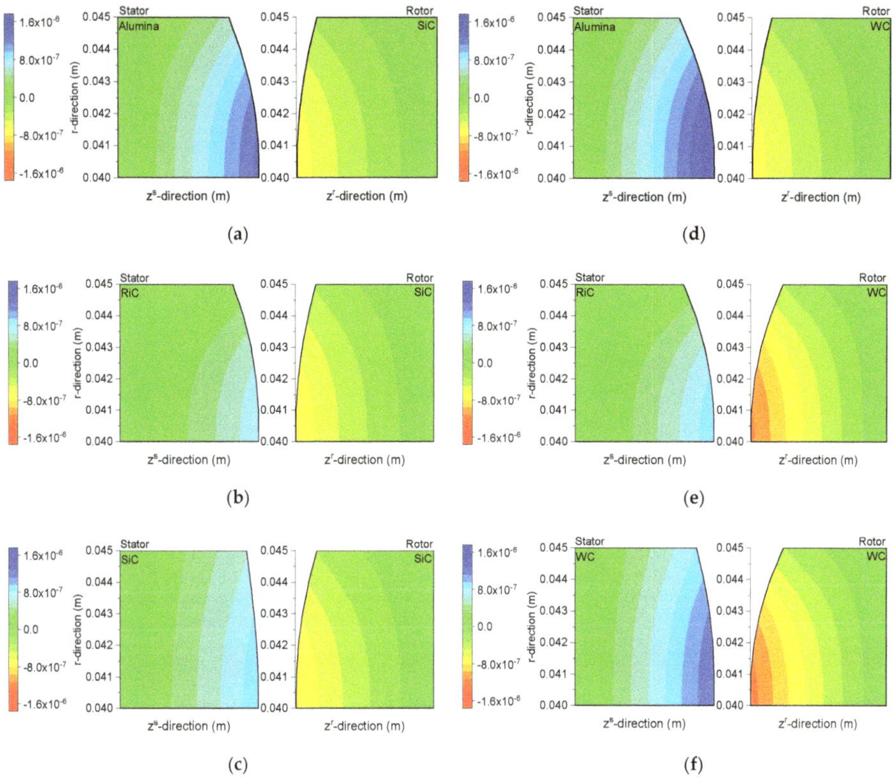

Figure 4. Distributions of axial displacements in the cross-sections of the sealing rings at $\omega = 1500$ (rad/s), (**a**) Alumina-SiC; (**b**) RiC-SiC; (**c**) SiC-SiC; (**d**) Alumina-WC; (**e**) RiC-WC; (**f**) WC-WC pairs of matrices.

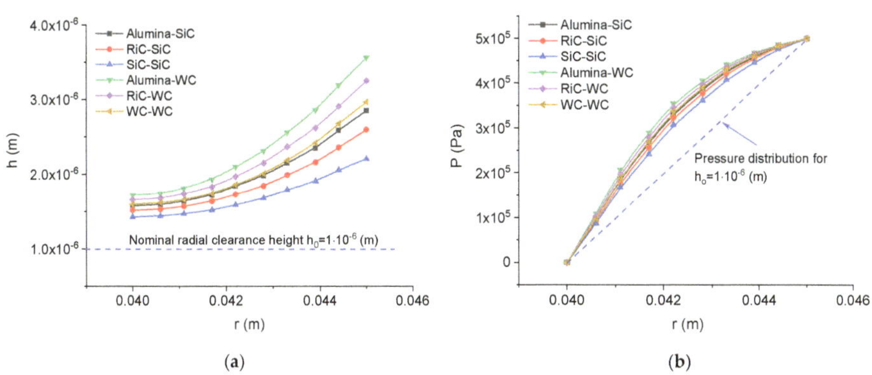

Figure 5. Clearance height and the fluid pressure distribution versus the ring radius at $\omega = 1500$ (rad/s), (**a**) Clearance height; (**b**) fluid pressure distribution, versus the ring radius at $\omega = 1500$ (rad/s)

Figure 6 shows changes in the minimum height of the radial clearance and the opening force against the angular velocity of the machine shaft. An increase in this parameter causes an increase in the heat flux generated in the fluid film, according to Relationship (4). As a consequence, there is a rise in the temperature of the sealing rings-fluid film system, which is responsible for thermal deformations of the sealing rings, then a change in the geometry of the radial clearance and, finally, an increase in the fluid film pressure. All this leads to a greater radial force generated in the fluid film and a greater minimum distance between the sealing rings, as shown in Figure 6.

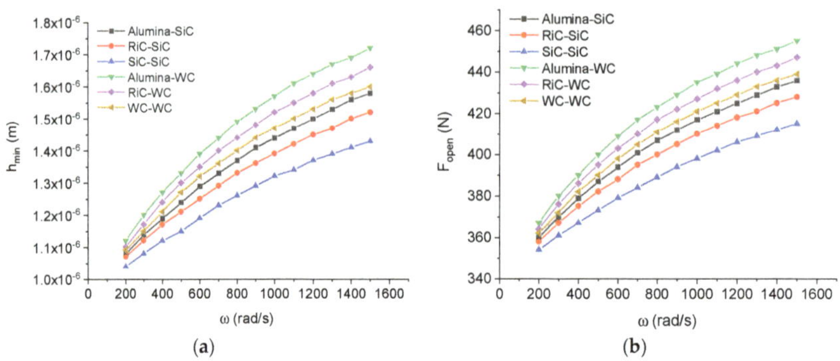

Figure 6. Minimum radial clearance height and opening force, against the angular velocity of the shaft, (**a**) Minimum radial clearance height; (**b**) opening force, against the angular velocity of the shaft.

Power loss results mainly from the viscous shearing of the fluid film between the mating surfaces; it can be calculated from the following relationship [20]:

$$P_{loss} = \int_{0}^{2\pi}\int_{r_i}^{r_o} \frac{\mu\omega^2 r^2}{h} r dr d\theta \tag{37}$$

The key parameter of the performance of non-contacting face seals is the fluid leakage rate, which is largely attributable to a change in the geometry of the radial clearance and it is described with the following formula:

$$Q_v = \left.\left(r\int_{0}^{2\pi} -\frac{h^3}{12\mu}\frac{\partial p}{\partial r} d\theta \right)\right|_{r=r_i} \tag{38}$$

As shown in Figure 7, the geometry of the clearance changes with changing temperature. Figure 7 shows the curves of the volumetric flow rate and power loss against the angular velocity of the shaft.

From the curves in Figure 7, it is apparent that, when the angular velocity is low, i.e., $\omega = 300$ (rad/s), the leakage rate and the power loss are very small for all the material pairs considered. An increase in the angular velocity causes an almost linear increase in the power loss in the stator-rotor system and an increase in the leakage rate. At $\omega = 1500$ (rad/s), the maximum leakage rate is: $Q_v = 2.15 \times 10^{-8}$ (rad/s), for the SiC-SiC pair and $Q_v = 5.2 \times 10^{-8}$ (rad/s), for the Alumina-WC pair. At that angular velocity, i.e., $\omega = 1500$ (rad/s), there may be a double difference in the leakage rate between some pairs of materials.

In the case of power loss, the situation is the opposite. An increase in the height of the radial clearance leads to a decrease in power loss, as illustrated in Figures 6a and 7b.

The results suggest that omitting thermal deformations is too great a simplification of the model of non-contacting seals.

It should be noted that the values and character of thermal deformations are in agreement with the numerical results presented, for example, in [7].

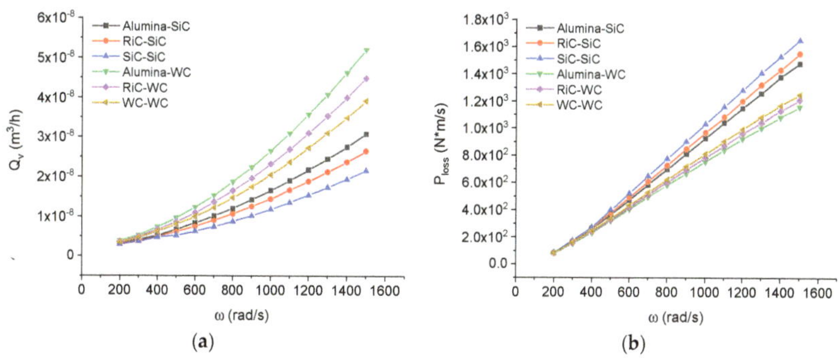

Figure 7. Volumetric flow rate and power loss, versus the angular velocity of the shaft, (**a**) Volumetric flow rate; (**b**) power loss, versus the angular velocity of the shaft.

5. Multi-Criteria Analysis

The multi-criteria analysis is a method used to facilitate the decision-making process when many variants are analyzed to select the best. In this case, the most appropriate pair of materials for the sealing rings is chosen. It is important to note that the criteria used in this analysis were selected and given weights subjectively by the author. The purpose of this multi-criteria analysis was to choose the best material pair variant to achieve the optimum seal performance, ring geometry and system temperature. Table 3 summarizes the criteria selected for the analysis.

Table 3. Criteria for Analysis.

Criteria	Unit	Weight ω	Materials					
			Alumina-SiC	RiC-SiC	SiC-SiC	Alumina-WC	RiC-WC	WC-WC
Performance								
Volumetric flow rate	m^3/h	0.25(−)	3.08×10^{-8}	2.64×10^{-8}	2.15×10^{-8}	5.20×10^{-8}	4.48×10^{-8}	3.91×10^{-8}
Power loss	Nm/s	0.15(−)	1480	1550	1650	1160	1210	1250
Opening force	N	0.2(+)	436	428	415	455	447	439
Geometry								
Minimum radial clearance	m	0.2(−)	1.58×10^{-6}	1.52×10^{-6}	1.43×10^{-6}	1.72×10^{-6}	1.66×10^{-6}	1.6×10^{-6}
Temperature								
Difference in fluid temperature	°C	0.2(−)	22.2	23.9	22.5	32.3	33.7	28.5

The criteria used in the multi-criteria analysis can be expressed by means of measurable or unmeasurable parameters. All the criteria are given dimensionless numerical values to assess and compare the selected variants. The process of replacing dimensional values with dimensionless values is called normalization. Normalization may involve maximizing the variables (stimulants (+)) or minimizing them (destimulants (−)). Signs (+) and (−) are given in Table 3 next to the values of the weights.

The variants are rated on the basis of the synthetic estimates S_i, which are calculated using the summation index, taking into consideration the weights of the particular criteria determined according to the formula:

$$S_i = \sum_{j=1}^{m} x_{ij}^* \omega_j \tag{39}$$

The final estimate of the analyzed variants is calculated by reducing the sum of S_i to a unity according to the following formula:

$$S_i^* = S_i / \sum_{i=1}^{n} S_i \tag{40}$$

The most favorable variant is that with the highest S_i^*.

Table 4 explains the method for criteria normalization with the notations being as follows: x_{ij}^*—normalized value of the i-th variant satisfying the j-th criterion; and x_{ij}—value of the *i*-th variant satisfying the *j*-th criterion.

Table 4. Method for the criteria normalization.

Normalization Method	Maximization (Stimulants)	Minimization (Destimulants)
Van Delft and Nijkamp approach	$x_{ij}^* = \frac{x_{ij}}{\sqrt{\sum_{i=1}^{n} x_{ij}^2}}$	$x_{ij}^* = 1 - \frac{x_{ij}}{\sqrt{\sum_{i=1}^{n} x_{ij}^2}}$

The calculation results obtained using the Van Delft and Nijkamp normalization method indicate that, for the predetermined heat transfer conditions and seal performance parameters, the volumetric flow rate is the lowest when both rings made of silicon carbide (Figure 8).

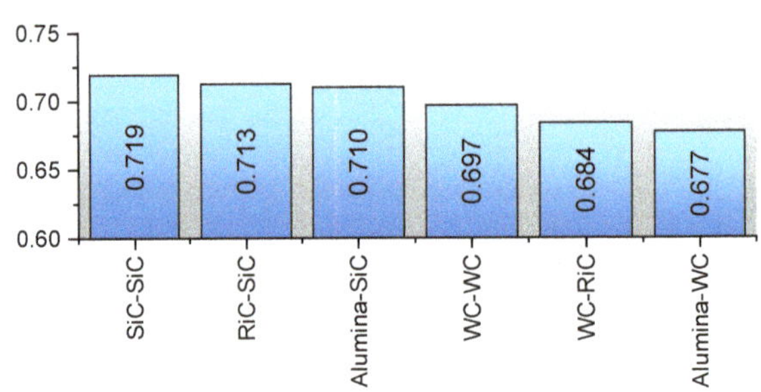

Figure 8. Results of the multi-criteria analysis.

The multi-criteria analysis is suitable for facilitating the decision-making process in situations when there are many variants to consider and/or the results do not give a clear best choice.

6. Conclusions

Sealing systems, which are crucial elements of many industrial machines, need to be designed in such a way that they conform to stringent safety standards if hazardous substances are used as process fluids.

Non-contacting face seals have numerous applications mainly because they meet divergent demands concerning both efficient performance and environmental safety. The major requirement to be met by seals is to maintain leak-tightness, irrespective of changes in the internal factors. Meeting this condition; however, is extremely difficult and one of the reasons for this may be changes in the geometry of the radial clearance caused by thermal deformations. Such deformations are likely to result in a disturbance to the equilibrium of forces, i.e., an increase in the opening force, which contributes to a higher leakage rate.

There is a direct relationship between the radial clearance geometry, the equilibrium of forces acting on the sealing rings, the leakage rate and the power loss. A change in any of these parameters may result in the disturbances to the performance of the whole sealing system, and consequently a failure of the machine in which the seal is installed. One of the main influencing the operation of mating slide rings in non-contact face seals is power loss (Figure 7b). Choosing the appropriate materials for the mating rings can provide a balance between reducing power loss and leakage. The decrease of power losses has a significant impact on the reduction of production costs etc.

The main aim of the research was realized. Based on simulation tests, the influence of material selection on working rings on power loss and leakage rate was confirmed.

The effectiveness of a seal depends on some factors. First of all, it is essential to select the right type of seal for a given application. Then, at the design stage, it is necessary to specify the desired temperature of the fluid film and the rings. It is also vital to predict the thermal deformations of the rings in order to prevent excessive wear of their surfaces as well as uncontrolled fluid leakage or power loss.

Funding: This research received no external funding.

Conflicts of Interest: The authors declare no conflict of interest.

Nomenclature

b	scale factor (1/°C); for water $b = 0.0175$,
C_p	specific heat,
e	dilatation $e = \varepsilon_{kk}$,
E	Young's modulus
G	elastic constant, $G = \vartheta$,
h	clearance height along the radial coordinate r,
h_o	nominal clearance height,
L^s,L^r	thickness of the stator and rotor, respectively,
n	normal vector to the surface,
p	fluid pressure,
p_i,p_o	fluid pressure at the inner and outer radii, respectively,
P_r	Prandtl number,
R_{eD}	Reynolds number,
r_i,r_o	inner and outer radii, respectively,
s_n	constant $(\frac{1}{\mathrm{m}})$,
T	absolute temperature at a specific point of the body,
T^f	fluid temperature along the r and z coordinates,
T_m	average temperature of the fluid film,
T_o	temperature of the process fluid (generally assumed to be constant),
u_r,u_z	ring displacements along the r and z coordinates, respectively.
α^s	free convection heat transfer coefficient,
λ, ϑ	Lame constants, where $\lambda = \frac{\nu E}{(1+\nu)(1-2\nu)}$ and $\vartheta = \frac{E}{2(1+\nu)}$,
λ^f	heat conduction coefficient for fluid,
λ^s, λ^r	heat conduction coefficient for the stator and rotor, respectively,
μ_o	fluid dynamic viscosity at T_o
v_ϕ	distribution of fluid velocity in the clearance,
$\theta = T - T_o$	change in the fluid temperature,
θ^s, θ^r	changes in the stator and rotor temperatures along the r and z coordinates, respectively
ρ	fluid density,
$\sigma_{rr}, \sigma_{\phi\phi}$	stresses in the r and ϕ directions, respectively,
ω	angular velocity.

References

1. Pascovici, M.D.; Etsion, I. A thermohydrodynamic analysis of a mechanical face seal. *J. Tribol.* **1992**, *114*, 639–645. [CrossRef]
2. Tournerie, B.; Danos, J.C.; Frêne, J. Three-Dimensional Modeling of THD Lubrication in Face Seals. *J. Tribol.* **2001**, *123*, 196. [CrossRef]
3. Brunetière, N.; Tournerie, B.; Frene, J. TEHD Lubrication of Mechanical Face Seals in Stable Tracking Mode: Part 2—Parametric Study. *J. Tribol.* **2003**, *125*, 617. [CrossRef]
4. Brunetière, N.; Tournerie, B.; Frene, J. TEHD Lubrication of Mechanical Face Seals in Stable Tracking Mode: Part 1—Numerical Model and Experiments. *J. Tribol.* **2003**, *125*, 608. [CrossRef]
5. Blasiak, S.; Kundera, C. A Numerical Analysis of the Grooved Surface Effects on the Thermal Behavior of a Non-Contacting Face Seal. *Procedia Eng.* **2012**, *39*, 315–326. [CrossRef]
6. Blasiak, S.; Kundera, C.; Bochnia, J. A Numerical Analysis of the Temperature Distributions in Face Sealing Rings. *Procedia Eng.* **2012**, *39*, 366–378. [CrossRef]
7. Li, C.-H. Thermal Deformation in a Mechanical Face Seal. *Am. Soc. Lubr. Eng. Trans.* **1976**, *19*, 146–152. [CrossRef]
8. Banerjee, B.N. The influence of thermoelastic deformations on the operation of face seals. *Wear* **1980**, *59*, 89–110. [CrossRef]
9. Emery, A.F. Thermal Stress Fracture in Elastic-Brittle Materials. In *Thermal Stresses in Severe Environments*; Hasselman, D.P.H., Heller, R.A., Eds.; Springer: Boston, MA, USA, 1980; pp. 95–121. ISBN 978-1-4613-3158-2.
10. Etsion, I.; Pascovici, M.D. A Thermohydrodynamic Analysis of a Misaligned Mechanical Face Seal. *Tribol. Trans.* **1993**, *36*, 589–596. [CrossRef]
11. Gu, B.; Zhou, J.; Chen, Y.; Sun, J. Frictional heat transfer regularity of the fluid film in mechanical seals. *Sci. China Technol. Sci.* **2008**, *51*, 611–623. [CrossRef]
12. Blasiak, S. Time-fractional heat transfer equations in modeling of the non-contacting face seals. *Int. J. Heat Mass Transf.* **2016**, *100*, 79–88. [CrossRef]
13. Blasiak, S. Heat transfer model for the wavy-tilt-dam mechanical seals using Green's Function Method. In *Engineering Mechanics*; Fuis, V., Ed.; Brno University of Technology: Brno, Czech Republic, 2017; pp. 158–161. ISBN 978-80-214-5497-2.
14. Zhou, J.F.; Gu, B.; Chen, Y. An Improved Design of Spiral Groove Mechanical Seal. *Chin. J. Chem. Eng.* **2007**, *15*, 499–506. [CrossRef]
15. Noda, N.; Hetnarski, R.B.; Tanigawa, Y. *Thermal Stresses*, 2nd ed.; Taylor & Francis Group: Abingdon, UK, 2003; ISBN 9781560329718.
16. Strąk, K.; Piasecka, M.; Maciejewska, B. Spatial orientation as a factor in flow boiling heat transfer of cooling liquids in enhanced surface minichannels. *Int. J. Heat Mass Transf.* **2018**, *117*, 375–387. [CrossRef]
17. Du, Q.; Gao, K.; Zhang, D.; Xie, Y. Effects of grooved ring rotation and working fluid on the performance of dry gas seal. *Int. J. Heat Mass Transf.* **2018**, *126*, 1323–1332. [CrossRef]
18. Kukla, S.; Siedlecka, U. Fractional heat conduction in a sphere under mathematical and physical Robin conditions. *J. Theor. Appl. Mech.* **2018**, *56*, 339. [CrossRef]
19. Maciejewska, B.; Piasecka, M. An application of the non-continuous Trefftz method to the determination of heat transfer coefficient for flow boiling in a minichannel. *Heat Mass Transf.* **2017**, *53*, 1211–1224. [CrossRef]
20. Ruan, B. Numerical Modeling of Dynamic Sealing Behaviors of Spiral Groove Gas Face Seals. *J. Tribol.* **2002**, *124*, 186. [CrossRef]

energies

MDPI

Article

A Concept of Risk Prioritization in FMEA Analysis for Fluid Power Systems

Joanna Fabis-Domagala [1], Mariusz Domagala [1,*] and Hassan Momeni [2]

[1] Faculty of Mechanical Engineering, Cracow University of Technology, Al. Jana Pawla II 37, 31-864 Cracow, Poland; joanna.fabis-domagala@pk.edu.pl

[2] Department of Mechanical and Marine Engineering, Western Norway University of Applied Sciences, N5020 Bergen, Norway; Hassan.Momeni@hvl.no

* Correspondence: domagala@mech.pk.edu.pl

Abstract: FMEA analysis is a tool of quality improvement that has been widely used for decades. Its classical version prioritizes risk of failure by risk priority number (RPN). The RPN is a product of severity (S), occurrence (O), and detection (D), where all of the factors have equal levels of significance. This assumption is one of the most commonly criticized drawbacks, as it has given unreasonable results for real-world applications. The RPN can produce equal values for combinations of risk factors with different risk implications. Another issue is that of the uncertainties and subjectivities of information employed in FMEA analysis that may arise from lack of knowledge, experience, and employed linguistic terms. Many alternatives of risk assessment methods have been proposed to overcome the weaknesses of classical FMEA risk management in which we can distinguish methods of modification of RPN numbers of employing new tools. In this study, we propose a modification of the traditional RPN number. The main difference is that severity and occurrence are valued based on subfactors. The detection number remained unchanged. Additionally, the proposed method prioritizes risk in terms of implied risk to the systems by implementing functional failures (effects of potential failures). A typical fluid power system was used to illustrate the application of this method. The method showed the correct failure classification, which meets the industrial experience and other research results of failures of fluid power systems.

Keywords: FMEA; risk analysis; fluid power systems

Citation: Fabis-Domagala, J.; Domagala, M.; Momeni, H. A Concept of Risk Prioritization in FMEA Analysis for Fluid Power Systems. *Energies* **2021**, *14*, 6482. https://doi.org/10.3390/en14206482

Academic Editors: Ryszard Dindorf, Jakub Takosoglu and Piotr Wos

Received: 19 August 2021
Accepted: 2 October 2021
Published: 10 October 2021

1. Introduction

Fluid power systems find wide industrial applications as drive or control systems. Due to their advantage, they can perform various functions, including those that may directly impact human safety or as a crucial component in applications that require high reliability. They have been used in aerospace applications as a vital component of aircraft, rockets, and spaceships to actuate critical flight components: actuating flaps, brakes, and landing gears, opening/closing doors, etc. The marine industry uses hydraulics for controlling ships and deck appliances such as winches, cranes, or hatch covers. They found applications in metal making machinery, in production lines, as presses, or other machine tools. They are broadly used in mobile machinery such as cranes, excavators, earth-moving equipment, and automobiles. The mining industry utilized hydraulics in drilling equipment for oil and gas extraction. The energy industry employs hydraulic systems in control systems of wind and water turbines and other processes. Depending on the applications, they can have different levels of complexity and use purely mechanical–hydraulic or electromechanical–hydraulic systems. High power density and almost unconstrained flexibility are the main advantages of these systems, making them peerless to mechanical or electrical power systems. Fluid power is not a new technology, but, complemented with the electronic control system [1], it offers new possibilities and can be one of the leaders among drive systems. Fluid power systems and their components must be highly reliable, their potential failures must be recognized early and investigated, and corrective plans must be prepared. Research on their failures and reliability is conducted using qualitative and quantitative tools and methods individually or together [2–4]. One of the widely used methods is the FMEA (failure modes and effects analysis). The origin of the FMEA analysis is dated to the 1950s when it was formalized in the military standard [5]. It found practical application during the NASA Apollo mission and in the automotive industry in the 1980s and finally became a part of international standards: ISO 9000 and SAE 1793 [6]. The FMEA method was an inspiration for other methods and tools such as RCM (reliability center management), concept FMEA and FMEDA [7].

One of the steps of FMEA analysis is a risk evaluation, which, in its conventional form, is calculated using three risk factors: severity (S), occurrence (O), and detection (D). Severity is defined as the ranking of the end effect of failure mode to the system. Occurrence is defined as the likelihood of failure occurrence. Detection is the possibility of failure detection. These risks factors are quantified by experts with integer numbers from 1 to 10 each. The risk of individual failure mode is evaluated using risk priority number (RPN), which is a product of the aforementioned risk factors. A higher RPN value defines a higher risk of related failure modes. However, this approach is widely questioned due to several limitations. The equally weighted risk factors may produce the same RPN number for various combinations of S, O, and D, which may have different risk implications. The utilization of only three factors and related failure modes is also criticized as ineffective. Another problem of conventional FMEA is data uncertainty and subjectivity. All information in FMEA analysis is delivered by a group of experts who have to assign linguistic terms to exact numbers and rely on their own knowledge and experience. All shortcomings of traditional FMEA analysis are widely summarized in previous work [8]. To overcome the FMEA drawbacks, alternatives methods are implemented. The systematic literature review of employed methods was presented by Liu et al. [9]. The main trends which are observed to overcoming drawbacks of traditional FMEA are:

1. Modification of classical RPN by implementing customer perspective [10], performing risk evaluation in terms of risk factors and their implication to risk scenario [11], other factors [9].
2. Extension of S, O, D with new factors e.g.:
   - expected cost [12];
   - corrective actions [13];
   - maintainability criticality index [14];
   - division of the main risk factors into subfactors [15].
3. Implementing risk factors weights: [16,17].

Risk assessment is performed by different methods and with the utilization of various tools. The most common methods that concurrently resolves problems with information uncertainty are:

1. Fuzzy set theory: [18–20];
2. Reasoning theory: [21,22];
3. Linguistic theory: [23,24];
4. Grey theory: [25,26].

The FMEA is a hierarchical multicriteria decision-making (MCDM) process that can utilize decision-making tools. Literature review indicates that the following methods mainly find application in risk analysis:

1. Distance-based methods:
   - distance operator;
   - technique for order of preference by similarity to ideal solution (TOPSIS) [27];
   - multiattributive border approximation area comparison (MACBAC) [28];
   - shortest distance algorithms [29].
2. Piecewise comparison method (analytical hierarchy process: AHP and others).

3. Aggregation operator-based methods.
4. Relation analysis methods.

The drawbacks of FMEA analysis were not only recognized by scholars but also by industry, which led to employing ACP (action priority number) in FMEA analysis in the latest automotive standard [30].

Although many attempts have been made and new methods implemented to defeat the weaknesses of classical RPN, a risk assessment in FMEA is still a challenge. All information used in FMEA is delivered by experts who have to rely on their knowledge and/or experience what along with linguistic evaluation methods that can lead to a high level of uncertainty and subjectivity. It also implies that the results of FMEA analysis are unique for specific problems and can not be extended on similar cases. Methods proposed by scholars are too complex or computationally intensive to be employed in the practice. To overcome the aforementioned shortcomings, we proposed a method that combines known methods as a division of severity and occurrence into subfactors and risk prioritization based on associated risk to the system. The severity in the proposed method is calculated based on component importance, failure effect, and factor, which define a relationship to the other failures. The occurrence number was replaced by a failure predictor, which uses a base failure rate, and modification factors, which take into account the influence of size, load, working conditions, and operating time. The detection number is estimated in the traditional way. Risk is evaluated for classified functional failures that directly correspond with system risk. The proposed method is analogous to the conventional FMEA, easy to use, and can reduce uncertainties of severity and occurrence caused by expert subjectivities.

2. Methodology

2.1. Assumptions

The primary purpose of this study was a qualitative analysis of failure modes, failures, and their end effects for fluid power systems. Failures in fluid power components are complex, and primary failure may only trigger the final failure. In this method, only primary forms of failures were assumed. The analysis was carried out only for primary mechanical failures. The electrical components were omitted.

Investigated fluid power system utilizes typical components for mobile fluid power systems without any diagnostic systems. The primary form of failure detection is a visual inspection. Additionally, access to system components is relatively easy and can be compared to a hydraulic system for mobile machinery (e.g., excavator).

We assume that investigated system already exists and only a few essential data are available.

2.2. Method

In the presented method, we define criticality C_R in a similar way to traditional risk priority number (RPN) as a product of severity S, a failure predictor P, and detection D:

$$C_R = S \cdot P \cdot D \tag{1}$$

where $S = c_i \cdot c_e \cdot f_i$, $P = \lambda_b \cdot m_f \cdot t_f$. All three are valued with numbers from 1 to 20.

We can calculate severity S as a product of component importance c_i, failure which can appear in the component f_i and modification factor c_e:

$$S = c_i \cdot c_e \cdot f_i \tag{2}$$

Component importance c_i defines how an individual component is important to the system for a specified criterion. It can be safety or ability to perform the specific function. We assumed that components are valued in the way similar to that which was presented in our previous study [31]:

- Main components. They are essential for performing the intended function.

- Major components. They ensure the proper operation of the system. Their possible failures may cause the system to malfunction, but its main task is still maintained.
- Additional components. Their failure has little effect on the main task of the system.

The component importance c_i reaches values 1–4, where 1 is the lowest importance.

Failure f_i describes the final effect that failure brings in the component. Potential failures were classified and valued in the following way:

- Catastrophic: major damages with component destruction–4.
- Critical: component malfunction with severe damages–3.
- Marginal: component malfunction with minor damages–2.
- Minor: less than minor damages–1.

Modification factor c_e is an influencing factor that tells which component may influence others in the system in case of failure. It reaches value 1–1.25, where value 1 means that potential failures in the component do not influence others. Failure predictor P is expressed as:

$$P = \lambda_b \cdot m_f \cdot t_f \tag{3}$$

where λ_b is a base failure rate, m_f is failure modification factor, t_f is a time factor.

$$m_f = s \cdot p \cdot t_e \cdot w_e \tag{4}$$

We utilize the failure rate value that is commonly used in reliability and maintainability to rank components in the system. We assumed that the failure rate for individual components is the estimator of its possible failure. The real value of base failure rates for the system components [32] were assigned to individual components and renormalized to scale 1–5. Value 1 indicates a component that unlikely fails, while 5 indicates a component with highly expected failure. The value of failure rate for fluid power components according to handbook [32] depends on the following:

- Size (s).
- Operating pressure (p).
- Leakage value (t_e).
- Temperature.
- Fluid contamination.

Factors "Temperature" and "Fluid Contamination" were joined to one factor, "Environment" (w_e), which considers working conditions as an equivalent of both. Besides primary and internal (in-operation) oil contamination, the ingressed (or external) contamination is also recognized as one of the main sources as states by industry [33–35].

To include the above factors, we implemented an equivalent scale defined in the following way:

- Very high.
- High.
- Moderate.
- Minor.

Due to components' different structures and performed functions, we can not use one uniform evaluations scale. However, the scale mentioned above was adapted to the individual features of the components. Factor "Size" for hydraulic pumps depends on their volumetric displacement:

- Very high, volumetric displacement > 125 dm^3.
- High, volumetric displacement 80–125 dm^3.
- Moderate, volumetric displacement 21–79 dm^3.
- Minor, volumetric displacement 1–20 dm^3.

Factor "Size" for hydraulic actuators depends on their piston diameter:

- Very high, piston diameter > 200 mm.
- High, piston diameter 160–200 mm.

- Moderate, piston diameter 41–159 mm.
- Minor, piston diameter < 40 mm.

Factor "Size" for valves depends on their nominal size:

- Very high, nominal size > 16 mm.
- High, nominal size 12–16 mm.
- Moderate, nominal size 6–10 mm.
- Minor, nominal size < 6 mm.

Factor "Size" for accumulators depends on their nominal volume:

- Very high, nominal volume > 50 dm^3.
- High, nominal volume 24–49 dm^3.
- Moderate, nominal volume 6–23 dm^3.
- Minor, nominal volume < 6 dm^3.

Factor "Size" for pipes and hoses depends on their nominal or outside diameter:

- Very high, pipe outside diameter > 30 mm, hose nominal diameter > 38 mm.
- High, pipe outside diameter 22–30 mm, hose nominal diameter 20– 32 mm.
- Moderate, pipe outside diameter 12–20 mm, hose nominal diameter 10–16 mm.
- Minor, pipe outside diameter <12 mm, hose nominal diameter <10 mm.

Factor "Operating pressure" was defined in the following way:

- Very high, operating pressure > 50 MPa.
- High, operating pressure 21–49 MPa.
- Moderate, operating pressure 6–20 MPa.
- Minor, operating pressure < 6 MPa.

Factors "Leakage" is directly connected with manufacturing aspects (tolerances, surface quality, manufacturing methods, materials) and is described as:

- Very high, very high tolerances and surface finishing for providing internal tightness: piston pump, slide-type valves.
- High, high tolerances and surface finishing: gear and vane pumps, poppet-type valves, hydraulic actuators.
- Moderate, higher than typical manufacturing requirements: pipes, hoses, accumulators.
- Minor, typical manufacturing requirements: tanks, filters.

The factor "Environment" is defined in the following way:

- Very high, extreme dusty or chemically aggressive environmental conditions with large temperature variations.
- High, dusty environment, temperature variations.
- Moderate, moderate environmental conditions.
- Minor, small influence of environmental conditions.

In general, we can calculate the failure rate for the system based on the failure rate of system components [36]. For a serial system, the total reliability is a sum of the failure rate for individual components:

$$\lambda_b = \sum_{i=1}^{n} \lambda_i \tag{5}$$

For a parallel system, it is a product:

$$\lambda_b = \prod_{i=1}^{n} \lambda_i \tag{6}$$

For the majority of fluid power components, base failure rate λ_b is available [32]. The failure rate for the accumulator can be calculated as a sum of individual components (as a serial system):

$$\lambda_b AC = \lambda_{bSSE} + \lambda_{bSP} + \lambda_{bPC} + \lambda_{bV} + \lambda_{bCW} \tag{7}$$

where λ_{bSSE} is a failure rate for static sealing, λ_{bSP} is a failure rate for spring, λ_{bPC} is a failure rate for piston–cylinder interface, λ_{bV} is a failure rate for valve, and λ_{bCW} is a failure rate for cylinder wall.

For a diaphragm-type accumulator, the failure rate can be simplified to the following formula:

$$\lambda_b AC = \lambda_{bSSE} + \lambda_{bV} + \lambda_{bCW} \tag{8}$$

For hydraulic actuator, the flow rate can be calculated from the following formula [32]:

$$\lambda_{bHC} = \lambda_{bPC} + 2\lambda_{bSD} \tag{9}$$

where

$$\lambda_{bPC} = \frac{10 \times 10^6}{N} \tag{10}$$

and λ_{bSD} is a failure rate for a dynamic seals. N is a number of wear cycles.

The number of cycles was estimated assuming that the equipment pressure should withstand the number of infinite fatigue strength, which is 2×10^6 according to [37]. Research shows that fatigue failures may occur much sooner [38,39]. Another reported failure is leaking due to the sealing failure [40]. The values of base failure rates (in failures/million cycles) for other hydraulic components is presented below [32]:

1. Valves:
    - spool type: $\lambda_{bVS} = 3.75$;
    - poppet type: $\lambda_{bVP} = 3.9$.

    The values mentioned above are addressed to direct-operated valves. For pilot-operated, we can calculate failure rate as a sum of the pilot valve and main valve. In the simplest case (pilot and main valve are the same types), we can assume that the pilot-operated valve failure rate is twice more than a direct-operated value.
2. Seals:
    - static: $\lambda_{bSS} = 2.4$;
    - dynamic: $\lambda_{bSD} = 22.8$.
3. Pumps:
    - piston: $\lambda_{bPP} = 1.05$;
    - gear: $\lambda_{bPG} = 0.75$;
    - vane: $\lambda_{bPV} = 0.4$.
4. Pipe: $\lambda_{bP} = 0.57$.
5. Hose: $\lambda_{bH} = 1.95$.
6. Accumulator (diaphragm type): $\lambda_{bAC} = 3.81$.
7. Hydraulic oil: as a common practice, hydraulic oil has to be replaced much sooner than any failure in components may occur. Therefore, the value of the failure rate was set as maximal from other components in the system.

All of the above-mentioned factors were evaluated using the below scale:

- Very high: 9–10.
- High: 6–8.
- Moderate: 3–5.
- Minor: 1–2.

and then were renormalized to new scales according to Formula [41]:

$$x_{norm} = a + \frac{x - min(x)(b - a)}{max(x) - min(x)} \tag{11}$$

The "Size" (s) factor value is 1–1.25, the "Operating pressure" (p) factor value is 1–1.25, the "Leakage" factor (t_e) value is 1–1.12, and the "Environmental" factor (w_e) is 1.14. The range of factors range is not equal because operating pressure and size plays more important role than leakage and environmental conditions. The time factor (t_f) allows for distinguishing components with different operating time regimes. The range is 1–2. To make all data easier to identify and recognize, we prepared a practical chart presented in Figure 1.

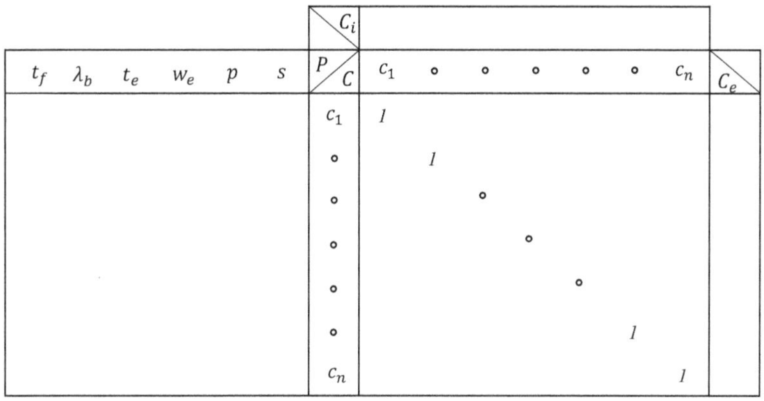

Figure 1. Data chart.

The chart above includes the pairwise comparison matrix, which is used to evaluate component modification factor c_e. The weights factors were calculated in the following way:

$$w_{ij} = \sum_{k,l=1}^{n} c_{kl}, k = 1...n \quad (12)$$

were next renormalized to range 1–1.25.

We evaluated the detection D in a similar way to classic FMEA analysis with values 1–20:

- detection of failure is almost certain: 1–2
- detection is very high: 3–5
- detection is high: 6–8
- detection is moderate high: 9–11
- detection is moderate: 12–13
- detection is very low: 14–15
- detection is remote: 16–17
- very remote: 18–19
- detection is nearly not possible: 20.

3. Case Study

An example of typical hydraulic system, presented in Figure 2, that includes typical fluid power components was used in this study. It is a hydraulic system that utilizes an accumulator as an auxiliary power source.

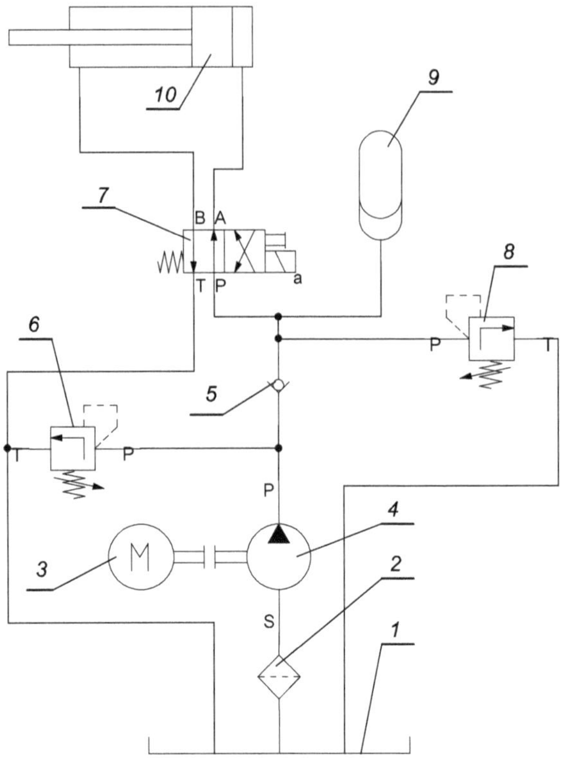

Figure 2. Hydraulic circuit: 1—reservoir; 2—filter; 3—motor; 4—pump; 5—check valve; 6 and 8—relief valve; 7—four way, two position directional control valve; 9—accumulator; 10—hydraulic cylinder.

The presented system's main task is to convert pressure energy into linear displacement of the hydraulic cylinder and actuate the component of a mechanical system (e.g., boom or arm of an excavator). The motor (3) drives a pump (4) and sucks hydraulic oil from the reservoir (1). Before passing the pump, the fluid is cleaned in the filter (2). The pump displaces the liquid to the hydraulic cylinder (10) through the check valve (5) and the directional control valve (7). The check valve (5) secures the system against the back flow. The hydraulic cylinder is a double-action actuator in which fluid acts on both sides of the piston. One side is connected with the supply line (with the pump) while the other is the drain (reservoir). Both ports can be alternatively connected to the supply line by switching a spool position of the control valve (7). In that way, the direction of piston rod movement is determined. The relief valve (6) plays a role of a safety valve in the system and secures the pump against excessive pressure. The other relief valve (8) sets the working pressure. The accumulator (9) is connected parallel to the supply line and is an auxiliary power source. During each operating cycle, the accumulator is charged and release fluid on power demand. Fluid is delivered to all components via rigid pipes or hoses, or both. The directional control valve (7) and relief valve (8) are components of the control system. Reservoir (1), filter (2), and pump (4) are components that generate pressure energy. Pipes or/and hoses transmit energy to the hydraulic cylinder (10), where fluid energy is converted into mechanical energy.

The following components were used in further analysis:

1. tank (reservoir) c_1 (item 1);
2. filter c_2 (item 2);
3. pump c_3 (item 4);

4. check valve c_4 (item 5);
5. relief valve (direct-operated) c_5 (item 6);
6. relief valve (pilot-operated) c_6 (item 8);
7. directional control valve (spool type) c_7 (item 7);
8. pipe c_8;
9. hose c_9;
10. accumulator c_{10} (item 9);
11. actuator (hydraulic cylinder) c_{11} (item 10);
12. oil (as a component of energy transfer) c_{12}.

Failures that can occur for the above-mentioned components are as follows [42]:

1. fracture f_1;
2. deformation f_2;
3. loosening f_3;
4. extreme contamination f_4;
5. properties f_5;
6. wear f_6;
7. corrosion f_7;
8. ageing/hardening f_8;
9. aeration f_9;
10. contamination f_{10};
11. properties changes f_{11};
12. cavitation f_{12}.

At the next step, we defined the system functional failures ff_i. The functional failure is understood as a system state (failure) that is categorized in the following way:

1. System is not able to perform intended function ff_1.
2. System is partially able to perform intended function. Major failure occurs ff_2.
3. System is able to perform intended function. Minor failure occurs ff_3.

Symptoms, failure modes, and failures matched with components corresponding to the above-mentioned functional failures were used to create relation tables presented below on Tables 1–3.

Table 1. Functional failure (ff_1): System is unable to perform intended function.

Symptom	Failure Mode	Failure	Components
Actuator: no motion	Loss of oil fm_1	Fracture f_1	Tank c_1 Pipe c_8 Hose c_9 Pump c_3 Valves c_4, c_5, c_6, c_7 Actuator c_{11} Accumulator c_{10}
	No output at pump fm_2	Fracture f_1	Pump c_3 Relief valve c_5
		Deformation f_2	Pump c_3
		Loosening f_3	Pump c_3
		Extreme level of contamination f_4	Oil c_{12}

Table 1. *Cont.*

Symptom	Failure Mode	Failure	Components
		Properties f_5	Oil c_{12}
	Fluid not delivered to actuator fm_3	Fracture f_1	Pipe c_8 Hose c_9 Accumulator c_{10} Valves c_4, c_5, c_6, c_7
		Deformation f_2	Valve c_7
	Blocked actuator fm_4	Fracture f_1 Deformation f_2	Actuator c_{11} Actuator c_{11}

Table 2. Functional failures (ff_2): System is partially able to perform the intended function.

Symptom	Failure mode	Failure	Components
Actuator: insufficient speed or force	Component malfunction fm_5	Wear f_6	Valves c_4, c_5, c_6, c_7 Pump c_3 Actuator c_{11}
		Corrosion f_7	Valves c_4, c_5, c_6, c_7 Pump c_3 Actuator c_{11}
		Deformation f_2	Actuator c_{11} Valve c_7
		Ageing/Hardening f_8	Accumulator c_{10}
		Aeration f_9 Contamination f_{10} Properties changes f_{11}	Oil c_{12}

Table 3. Functional failures (ff_3). System is able to perform intended function with minor malfunction.

Symptom	Failure Mode	Failure	Components
Noisy operation	Component malfunction fm_5	Wear f_6	Valves c_4, c_5, c_6, c_7 Pump c_3 Actuator c_{11}
		Deformation f_2	Pump c_3
		Corrosion f_7	Pump c_3
		Aeration f_9	Oil c_{12}
		Cavitation f_{12}	Pump c_3 Valves c_4, c_5, c_6, c_7
Overheating	Component malfunction fm_5	Wear f_6	Pump c_3
		Cavitation f_{12}	Pump c_3 Valves c_4, c_5, c_6, c_7
		Ageing/Hardening f_8	Accumulator c_{10}
Leaks	Component malfunction fm_5	Wear f_6	Component seals
		Ageing/Hardening f_8	Component seals

4. Results

Data from Table A1 along with failure values were allowed for calculating the criticality number C_r. The ranking lists were created for each functional failure and are presented on Table 4–6. Oil contamination or aeration appears six times among first five ranks for all functional failures. Pump and actuator failures were also recognized as components, which can lead to presented system failures.

Table 4. Functional failures ff_1. Criticality.

Rank	C_R	Failure	Component
1	1368	extreme oil cont. f_4	oil c_{12}
2	1310.1	deformation f_2	pump c_3
3	928.2	deformation f_2	actuator c_{11}
4	912.0	properties f_5	oil c_{12}
5	409.4	loosening f_3	pump c_3
6	309.4	fracture f_1	actuator c_{11}
7	83.7	fracture f_1	relief valve c_6
8	82.8	fracture f_1	directional control valve c_7
9	81.9	fracture f_1	pump c_3
10	81.0	fracture f_1	check valve c_4
15	71.6	fracture f_1	hose c_9
11	59.1	fracture f_1	pipe c_8
16	43.6	fracture f_1	accumulator c_{10}
14	33.8	fracture f_1	tank c_1
13	29.7	fracture f_1	relief valve c_5

Table 5. Functional failures ff_2. Criticality.

Rank	C_R	Failure	Component
1	2907.0	contamination f_{10}	oil c_{12}
2	2784.6	corrosion f_7	actuator c_{11}
3	2736.0	aeration f_9	oil c_{12}
4	2052.0	properties changes f_{11}	oil c_{12}
5	2784.6	wear f_6	actuator c_{11}
6	982.2	deformation f_2	actuator c_{11}
7	859.8	corrosion f_7	pump c_3
8	753.6	corrosion f_7	relief valve c_6
9	745.3	corrosion f_7	directional control valve c_7
10	728.7	corrosion f_7	check valve c_4
11	654.3	ageing/hardening f_8	accumulator c_{10}
12	628.0	wear f_6	relief valve c_6
13	621.1	wear f_6	flow control valve c_7
14	607.2	wear f_6	check valve c_4
15	267.3	corrosion f_7	relief valve c_5
16	248.5	deformation f_2	flow control valve c_7
17	297.0	wear f_6	relief valve c_5

Table 6. Functional failures ff_3. Criticality.

Rank	C_R	Failure	Component
1	2784.6	wear f_6	actuator c_{11}
2	2736.0	aeration f_9	oil c_{12}
3	1146.4	deformation f_2	pump c_3
4	1044.0	cavitation f_{12}	pump c_3
5	942.0	cavitation f_{12}	relief valve c_6
6	931.7	cavitation f_{12}	directional control valve c_7
7	910.8	cavitation f_{12}	check valve c_4
8	859.8	wear f_6	pump c_3
9	859.8	corrosion f_7	pump c_3
10	828.7	aeging/hardening f_8	accumulator c_{10}
11	628.0	wear f_6	relief valve c_6
12	621.1	wear f_6	directional control valve c_7

Table 6. *Cont.*

Rank	C_R	Failure	Component
13	607.2	wear f_6	check valve c_4
14	334.2	cavitation f_{12}	relief valve c_5
15	222.7	wear f_6	relief valve c_5

5. Discussion

Criticallity number for pair contamination–oil reached the highest value among all failure–component pairs. Almost all failures and related components for functional failure (ff_1) that are critical to the system are valued lower than for others functional failures (ff_2, ff_3). There are two reasons responsible for this situation. The first one is the detection number. The "fracture" failure for almost all components can be easily detected—for some, even without any instruments. The second reason is the components prioritization method that is realized by performed function, failure rate, and working regime. The tank, which is recognized as a highly reliable component, can be a source of critical failure for the system. Another aspect is that the components which play auxiliary functions and being in use occasionally (e.g., valve c_6) when failing make the system unable to operate. Failures that cause fluid losses are critical to the whole system and even if they occur in the component graded as secondary or minor. Treating such components with the highest importance can lead to overestimating other failures for these components. The nature of almost all failures for critical system failure ff_1 is random and unpredictable caused by sudden and extreme overload or hidden material/manufacturing and design flaws.

The first two criticality numbers for ff_3, which is less risky for the system than others, reached higher values than fourth rank in ff_2. Other pairs of failures and the corresponding components are higher for ff_3 than ff_2. This is caused by the intensity and exposure which determine the final effect, which can be marginal (at initial stages of wear, cavitation, and corrosion) or major after long exposure.

Presented results show that risk prioritization assigned to the system risk allows avoiding underestimating or overestimating potential failures for related components.

In all three functional failures, oil contamination plays the main role and should be recognized as a main problem of fluid power systems. It agrees with practical experience and also with the results of research presented in study [2] where failure analysis was conducted for a hydraulic system of a heavy-duty machine. The failure analysis in this work was conducted with more sophisticated methods and tools than the presented method: fault tree analysis, Dempster–Shafer theory, and rough set theory to fill were implemented to eliminate the incompleteness and the uncertainty delivered by experts.

The comparison of the presented method and conventional RPN approach was conducted for the severity factor, which in the proposed method is calculated according to Equation (2). To the comparison, we used failure: fracture that can occur in all of the system' components. The end effect of this failure for all of the components leads to critical system failure, which is unable to perform the intended function. As such, it should be ranked with maximal value according to the traditional RPN approach. The comparison of RPN value is presented in the Table 7.

Table 7. Severity number.

Component	Classical RPN (max 10)	Proposed Method (max 20)
Pump	10	19.9
Tank	10	16.9
Check valve	10	16.8
Directional control valve	10	16.7
Actuator	10	16.4
Filter	10	16.2
Hose	10	16.1
Pipe	10	16.0
Relief valve (c_6)	10	12.4
Relief valve (c_5)	10	12.2
Accumulator	10	8.1

The occurrence number may mainly depend on a subjective opinion of an expert (or experts), their knowledge and/or experience, and/or availability of relevant data. It means that the occurrence number may be scattered inside a wide range. The proposed method adopts quantitative data, including the essential information about components, which produce the occurrence number with less sensitivity to expert knowledge and/or experience.

The detection number in both methods are estimated in the same way.

Limitations of the method.

Although the structural design of fluid power components has remained unchanged for decades, the trend to implementing electronic control and diagnostic system can be observed recently. It brings new possibilities for system diagnostic and its management and makes the components are no longer purely mechanical/fluid but are more sophisticated. Additionally, new manufacturing methods and materials are implemented; thus, base failure rates from this study may not fit the latest components design. Furthermore, the values of base failure rates do not recognize differences in component structures or material and cannot evaluate the same component with various solutions separately. Another limitation of the presented method is its applicability only to components for which base failure rate is available. Those for which is unavailable would have to be estimated what can increase the level of uncertainty.

The proposed method can calculate the same criticality numbers for a few different components and corresponding failures. It can be overcome by detailed failures definition, which is sometimes problematic. Fluid power components encounter mechanical and fluid flow failures, which are too complex for easy identification.

6. Conclusions

FMEA analysis and its modification play an essential role in increasing reliability and safety despite the drawbacks, which, in classical FMEA analysis, is undoubtedly risk evaluation and uncertainties. In this study, a proposal of risk assessment for fluid power systems has been presented; its main aim was implementing a prioritization method of failures based on quantitative data. A classical risk priority number has been extended with modification factors for severity, while occurrence was replaced by a failure predictor, which uses failure rate value and corrective factors. The detection remained unchanged with classical risk prioritization. The severity number in the proposed method is calculated as a product of component importance, their influence on other components, and failure effects. It allows for prioritizing components that can be nondistinguishable in the classical RPN method. The proposed method's main application is a design stage or situation where details of the system components are unavailable. Therefore, the occurrence number is replaced by a failure predictor, which defines the likelihood of failure based on failure rate value and modification factors. These were determined based on specifications of typical components systems for mobile fluid power systems. If relevant data about components are available, the failure rate can be more precisely described. The proposed method was employed in a typical fluid power system which consists of common components and can be extended to any fluid power system. Modification factors presented in this study are universal and applicable for other systems of fluid power. Failure modes of individual components were classified into system functional failure (effects of component failure modes) to avoid underestimating failures whose consequences are catastrophic to the system. Obtained results allowed identify the most common failure for a considered hydraulic circuit, which agreed with research conducted by more sophisticated tools and methods and proved the usefulness presented method.

Author Contributions: Conceptualization, J.F.-D. and M.D.; methodology, J.F.-D.; software, H.M.; validation, M.D., J.F.-D. and H.M.; formal analysis, J.F.-D.; investigation, J.F.-D.; resources, J.F.-D.; data curation, M.D.; writing—original draft preparation, M.D.; writing—review and editing, H.M.; visualization, H.M.; supervision, M.D.; project administration, J.F.-D.; funding acquisition, H.M. All authors have read and agreed to the published version of the manuscript.

Funding: This research received no external funding.

Institutional Review Board Statement: Not applicable.

Conflicts of Interest: The authors declare no conflict of interest.

Abbreviations

The following abbreviations are used in this manuscript:

MDPI	hlMultidisciplinary Digital Publishing Institute
DOAJ	Directory of open access journals
MIL-STD	US Defense Standard
FMEA	Failure Modes and Effects Analysis
FMEDA	Failure Modes, Effects, and Diagnostic Analysis
FMCA	Failure Modes, Effects, and Criticality Analysis
NASA	National Aeronautics and Space Administration
RPN	Risk Priority Number
ISO	International Organization for Standardization
MCDM	Multicriteria Decision-Making

Appendix A

Table A1. Data chart.

						c_i	4.0	4.0	4.0	4.0	3.0	3.0	4.0	4.0	4.0	2.0	4.0	4.0	
t_f	λ_b	t_e	w_e	p	s		c_1	c_2	c_3	c_4	c_5	c_6	c_7	c_8	c_9	c_{10}	c_{11}	c_{12}	c_e
2	1.00	1.00	1.00	1.00	1.00	c_1	1.0	0.5	0.0	0.0	0.0	0.0	0.0	0.0	0.0	0.0	0.0	0.7	1.06
2	1.36	1.02	1.00	1.00	1.00	c_2	0.0	1.0	0.0	0.0	0.0	0.0	0.0	0.0	0.0	0.0	0.0	0.3	1.01
2	1.15	1.12	1.02	1.25	1.25	c_3	0.0	0.0	1.0	0.7	0.7	0.7	0.7	0.3	0.1	0.5	0.7	0.9	1.25
2	1.56	1.08	1.02	1.25	1.13	c_4	0.0	0.0	0.0	1.0	0.0	0.3	0.3	0.0	0.0	0.0	0.0	0.5	1.05
1	1.56	1.09	1.02	1.25	1.13	c_5	0.0	0.0	0.0	0.0	1.0	0.0	0.0	0.0	0.0	0.0	0.0	0.3	1.01
2	2.12	1.11	1.02	1.25	1.13	c_6	0.0	0.0	0.0	0.0	0.0	1.0	0.0	0.0	0.0	0.0	0.3	0.5	1.04
2	1.54	1.12	1.02	1.25	1.13	c_7	0.0	0.0	0.0	0.0	0.0	0.0	1.0	0.0	0.0	0.0	0.5	0.5	1.05
2	1.08	1.03	1.05	1.25	1.25	c_8	0.0	0.0	0.0	0.0	0.0	0.0	0.0	1.0	0.0	0.0	0.0	0.0	1.00
2	1.28	1.03	1.06	1.25	1.25	c_9	0.0	0.0	0.0	0.0	0.0	0.0	0.0	0.0	1.0	0.0	0.0	0.1	1.00
2	1.55	1.03	1.06	1.25	1.25	c_{10}	0.0	0.0	0.0	0.0	0.0	0.0	0.0	0.0	0.0	1.0	0.3	0.0	1.01
2	5.00	1.12	1.14	1.25	1.25	c_{11}	0.0	0.0	0.0	0.0	0.0	0.0	0.0	0.0	0.0	0.0	1.0	0.5	1.02
2	5.00	1.00	1.11	1.00	1.00	c_{12}	0.0	0.3	0.9	0.7	0.1	0.7	0.7	0.5	0.5	0.3	0.7	1.0	1.25

References

1. Kushwaha, P.; Dasgupta, K.; Ghoshal, S.K. A comparative analysis of the pump controlled, valve controlled and prime mover controlled hydromotor drive to attain constant speed for varying load. *ISA Trans.* **2021**, *14*, in press.
2. Li, S.; Yang, Z.; Tian, H.; Chen, C.; Zhu, Y.; Lu, S. Failure Analysis for Hydraulic System of Heavy-Duty Machine Tool with Incomplete Failure Data. *Appl. Sci.* **2021**, *14*, 1249. [CrossRef]
3. Lee, Y.; Lee, G.; Yang, J.; Park, J.; Baek, D. Failure analysis of a hydraulic power system in the wind turbine. *Eng. Fail. Anal.* **2020**, *107*, 104218. [CrossRef]
4. Guo, Y.; Xiong, G.; Zeng, L.; Li, Q. Modeling and Predictive Analysis of Small Internal Leakage of Hydraulic Cylinder Based on Neural Network. *Energies* **2021**, *14*, 2456. [CrossRef]
5. MIL-P 1629. *USA Military Standard, Procedure for Performing a Failure Mode, Effects and Criticality Analysis (MIL-P 1629, USA)*; Military Specifications and Standards: Washington, DC, USA, 1949.
6. J1739 200901. *Potential Failure Mode and Effects Analysis in Design (Design FMEA), Potential Failure Mode and Effects Analysis in Manufacturing and Assembly Processes (Process FMEA)*;SAE International: Warrendale, PA, USA, 2009.
7. Carlson, C.S. *Effective FMEAs: Achieving Safe, Reliable, and Economical Products and Processes Using Failure Mode and Effects Analysis*; John Wiley & Sons, Inc.: Hoboken, NJ, USA, 2012.
8. Liu, H.C.; Liu, L.; Liu, N. Risk evaluation approaches in failure mode and effects analysis: A literature review, *Expert Syst. Appl.* **2013**, *40*, 828–838. [CrossRef]
9. Liu, H.C.; Chen, X.; Dua, C.; Wang, Y. Failure mode and effect analysis using multi-criteria decision making methods: A systematic literature review. *Comput. Ind. Eng.* **2019**, *135*, 881–897. [CrossRef]
10. Koomsap, P.; Charoenchokdilok, T. Improving risk assessment for customer-oriented FMEA. *Total Qual. Manag. Bus. Excell.* **2018**, *29*, 1563–1579. [CrossRef]
11. Anes, V.; Henriques, E.; Freitas, M.; Reis, L. A new risk prioritization model for failure mode and effects analysis. *Qual. Reliab. Eng. Int.* **2018**, *34*, 516–528. [CrossRef]
12. Lo, H.W.; Liou, J.J.H. A novel multiple-criteria decision-making-based FMEA model for risk assessment. *Appl. Soft Comput.* **2016**, *73*, 684–696. [CrossRef]
13. Chen, J.K. Prioritization of Corrective Actions from Utility Viewpoint in FMEA Application. *Qual. Reliab. Eng. Int.* **2018**, *33*, 883–894. [CrossRef]
14. Pancholi, N.; Bhatt, M. FMECA-based maintenance planning through COPRAS-G and PSI. *J. Qual. Maint. Eng.* **2018**, *24*, 224–243. [CrossRef]
15. Wang, Z.; Gao, J.; Wang, R.; Chen, K.; Gao, Z.; Zheng, W. Failure Mode and Effects Analysis by Using the House of Reliability-Based Rough VIKOR Approach. *IEEE Trans. Reliab.* **2018**, *67*, 230–248. [CrossRef]
16. Tang, Y.; Zhou, D.; Chan, F.T.S. AMWRPN: Ambiguity Measure Weighted Risk Priority Number Model for Failure Mode and Effects Analysis. *IEEE Access* **2018**, *6*, 27103–27110. [CrossRef]
17. Zheng, H.; Tang, Y. Deng Entropy Weighted Risk Priority Number Model for Failure Mode and Effects Analysis. *Entropy* **2020**, *22*, 280. [CrossRef]
18. Park, J.; Park, C.; Ahn, S. Assessment of structural risks using the fuzzy weighted Euclidean FMEA and block diagram analysis. *Int. J. Adv. Manuf. Technol.* **2018**, *99*, 2071–2080. [CrossRef]
19. Godina, R.; Silva, B.G.R.; Espadinha-Cruz, P. A DMAIC Integrated Fuzzy FMEA Model: A Case Study in the Automotive Industry. *Appl. Sci.* **2021**, *11*, 3726. [CrossRef]
20. Filo, G.; Fabis-Domagala, J.; Domagala, M.; Lisowski, E.; Momeni, H. The idea of fuzzy logic usage in a sheet-based FMEA analysis of mechanical systems. *MATEC Web Conf.* **2018**, *183*, 03009. [CrossRef]
21. Yener, Y.; Can, G.F. A FMEA based novel intuitionistic fuzzy approach proposal: Intuitionistic fuzzy advance MCDM and mathematical modeling integration. *Expert Syst. Appl.* **2021**, *183*, 115413. [CrossRef]
22. Qin, J.; Xi, Y.; Pedrycz, W. Failure mode and effects analysis (FMEA) for risk assessment based on interval type-2 fuzzy evidential reasoning method. *Appl. Soft Comput.* **2020**, *89*, 106134. [CrossRef]
23. Chang, C.; Kontovas, C.; Yu, Q.; Yang, Z. Risk assessment of the operations of maritime autonomous surface ships. *Reliab. Eng. Syst. Saf.* **2021**, *207*, 107324. [CrossRef]
24. Shi, S.; Fei, H.; Xu, X. Application of a FMEA method combining interval 2-tuple linguistic variables and grey relational analysis in preoperative medical service process. *IFAC-PapersOnLine* **2019**, *52*, 1242–12473. [CrossRef]
25. Ouyang, L.; Zhu, Y.; Zheng, W.; Yan, L. An information fusion FMEA method to assess the risk of healthcare waste. *J. Manag. Sci. Eng.* **2021**, *6*, 111–124. [CrossRef]
26. Li, Z.; Chen, L. A novel evidential FMEA method by integrating fuzzy belief structure and grey relational projection method. *Eng. Appl. Artif. Intell.* **2019**, *77*, 136–147. [CrossRef]
27. Bashan, V.; Demirel, H.; Gul, M. An FMEA-based TOPSIS approach under single valued neutrosophic sets for maritime risk evaluation: The case of ship navigation safety. *Soft Comput.* **2020**, *24*, 18749–18764. [CrossRef]
28. Biswas, S.; Bandyopadhyay, G.; Guha, B.; Bhattacharjee, M. An ensemble approach for portfolio selection in a mutli-criteria descision making framework. *Decis. Making Appl. Manag. Eng.* **2019**, *2*, 138–158.
29. Ebrahimi, H.; Milos, T. Optimization of dangerous goods transport in urban zone. *Decis. Making Appl. Manag. Eng.* **2018**, *1*, 131–152. [CrossRef]

30. *AIAG & VDA FMEA Handbook*; AIAG & VDA: Southfield, MI, USA, 2019.
31. Fabis-Domagala, J.; Domagala, M.; Momeni, H. A Matrix FMEA Analysis of Variable Delivery Vane Pumps. *Energies* **2021**, *14*, 1741. [CrossRef]
32. *Handbook of Reliability Prediction for Mechanical Equipment*; Naval Surface Warfare Center: West Bethesda, MD, USA, 2011.
33. *Hydraulic Filtration Technical Reference*; Donaldson Filtration Solution, Bloomington, MN, USA, 2012.
34. *The Systematic Approach to Contamination Control*; Eaton Corporation: Andover, MA, USA, 2002.
35. *Hydraulic Filtration Pocket Guide*; Cummins Filtration: Nashville, TN, USA, 2020.
36. Aggarwal, K.K. *Reliability Engineering*; Springer Science+Business Media: Dordrecht, The Netherlands, 1993.
37. ISO EN 13445-1. *Unfired Pressure Vessels. Part 3: Design*; PKN: Warszawa, Poland, 2017.
38. Gianni, N.; Tito, M. Failure of a heavy-duty hydraulic cylinder and its fatigue re-design. *Eng. Fail. Anal.* **2011**, *18*, 1030–1036.
39. Roquet, P.; Gamez-Montero, P.J.; Castilla, R.; Raush, G.; Codina, E. A Simplified Methodology to Evaluate the Design Specifications of Hydraulic Components. *Appl. Sci.* **2018**, *8*, 1612. [CrossRef]
40. Mahankar, P.S.; Dhoble, A.S. Review of hydraulic seal failures due to effect of medium to high temperature. *Eng. Fail. Anal.* **2021**, *127*, 105552. [CrossRef]
41. Han, J. *Data Mining*, 3rd ed.; Morgan Kaufmann: Burlington, MA, USA, 2012.
42. Watton, J. *Modelling, Monitoring and Diagnostic Techniques for Fluid Power Systems*; Springer: London, UK, 2007.

energies

MDPI

Article

A Matrix FMEA Analysis of Variable Delivery Vane Pumps

Joanna Fabis-Domagala [1], Mariusz Domagala [1,*] and Hassan Momeni [2]

1 Institute of Applied Informatics, Cracow University of Technology, Al. Jana Pawla II 37, 31-864 Cracow, Poland; joanna.fabis-domagala@pk.edu.pl

2 Department of Mechanical and Marine Engineering, Western Norway University of Applied Sciences, N5020 Bergen, Norway; Hassan.Momeni@hvl.no

* Correspondence: domagala@mech.pk.edu.pl

Abstract: Hydraulic systems are widely used in the aeronautic, machinery, and energy industries. The functions that these systems perform require high reliability, which can be achieved by examining the causes of possible defects and failures and by taking appropriate preventative measures. One of the most popular methods used to achieve this goal is FMEA (Failure Modes and Effects Analysis), the foundations of which were developed and implemented in the early 1950s. It was systematized in the following years and practically implemented. It has also been standardized and implemented as one of the methods of the International Organization for Standardization (ISO) 9000 series standards on quality assurance and management. Apart from wide application, FMEA has a number of weaknesses, which undoubtedly include risk analysis based on the RPN (Risk Priority Number), which is evaluated as a product of severity, occurrence, and detection. In recent years, the risk analysis has been very often replaced by fuzzy logic. This study proposes the use of matrix analysis and statistical methods for performing simplified RCA (Root Cause Analysis) and for classification potential failures for a variable delivery vane pump. The presented methodology is an extension of matrix FMEA and allows for prioritizing potential failures and their causes in relation to functions performed by pump components, the end effects, and the defined symptoms of failure of the vane pump.

Keywords: FMEA; failure analysis; matrix FMEA; vane pump

Citation: Fabis-Domagala, J.; Domagala, M.; Momeni, H. A Matrix FMEA Analysis of Variable Delivery Vane Pumps. *Energies* **2021**, *14*, 1741. https://doi.org/10.3390/en14061741

Academic Editor: Ryszard Dindorf

Received: 25 February 2021
Accepted: 17 March 2021
Published: 21 March 2021

1. Introduction

Due to a high power transmission density, relatively simple movement transformation, and smooth control, hydraulic systems find wide industrial application. Such systems equipped with automation components enable high positioning accuracy and programming movement trajectory. Hydraulic drive systems use accumulated pressure energy of working fluids generated in the hydraulic pump, which is transferred to the receivers by rigid or flexible pipes. This feature creates flexibility, which is unreachable for other systems, and almost unrestricted design capabilities. In addition, compared to purely mechanical systems, any dynamic load can be relatively simply dissipated. The abovementioned advantages make hydraulic systems widely used in drive and control systems in the aeronautic (control systems in airplanes and helicopters), mining, energy (control systems), and automotive industries, which require high reliability and safety. In order to meet these requirements, knowledge about possible failures of such systems as well as their causes is needed. This might be obtained using various methods and tools. Y. Lee et al. [1] carried out an analysis of failures of hydraulic systems used in wind turbines to control pitch blades. The failures of a hydraulic system in wind turbines were also investigated by Y. Lee et al. [2] using Fault Tree Analysis (FTA) and fuzzy logic. Research on the influence of vibrations on hydraulic pipeline failures was conducted by P. Gao et al. [3], while G. Wang et al. [4] studied the influence of failures on shifting quality and driving safety for a continuously variable transmission (CVT) tractor. Research was conducted not only for the entire system but also for individual components among which pumps as an energy source are the most critical one. Z. Ma et al. [5] led research on hydraulic pump faults by using a model-based diagnosis system, while T. Li et al. [6] used multiscale information for predicting the remaining useful time of a piston pump. D. Hast et al. [7] also worked on fault specification for the detection of a piston pump fault. C. Lu et al. [8] developed a fault diagnosis method for hydraulic pumps used in aeronautic application based on the evidence theory. Relationships between failure modes and their causes and effects can be developed on the basis of the history of failures and/or using methods of failure analysis among which FMEA (Failure Modes and Effects Analysis) is one of the most successful. There are numerous variations of this method or methods such as fault tree analysis, event tree analysis, or other quality improvement methods and tools [9] that might be used for this purpose. The FMEA fundamental principles found many applications such as RCM (Reliability-Centered Maintenance), hazard analysis, concept FMEA, and FMEDA (Failure Modes, Effects, and Diagnostic Analysis) [10]. RCM is an evolution of FMEA that is used for planning preventive maintenance activity. Increasing demands triggered research on failure predictions and estimation of useful life. It can lead to predicting maintenance plans and, in consequences, achieves the required operation conditions. Fault prognosis methods are not formally classified yet. The work of [11] defined health monitoring systems in a hierarchical structure that uses expert knowledge, physical model, and data driven knowledge. Physical model approaches require an analytical model of the system, while data-driven knowledge approaches define detection and diagnosis as a classification task. The most common tool used in data-driven approaches are neutral network, control charts, and principal component analysis [12]. Another method is presented in Reference [13], in which fault diagnosis and related fault prognosis allow us to estimate the remaining useful life for data available in a normal operation. Research on failures and its causes has been carried out for many years, has found practical application, and was firstly implemented by the US military in 1949: MIL-P 1629 "Procedure for Performing a Failure Modes, Effects and Criticality Analysis" [14]. The next step in the practical implementation of FMEA was done by NASA during the Apollo mission program to estimate the influence of a potential failure on mission success and crew safety. In the next decade, the FMEA was adopted in the automotive industry by one of the largest players: Ford Motor Company. The FMEA was fully formulated and standardized in the 1980s and 1990s in the military standard MIL-STD 1629A, included in the International Organization for Standardization (ISO) 9000 series standards and implemented by SAE J1739 [15]. FMEA supplemented with criticality analysis is known as FMECA (Failure Modes Effect and Criticality Analysis). The FMEDA, for which the foundations were defined in the 1980s, also found practical application, initially in electrical and electronic systems [16]. It supplements the FMEA with two pieces of information: failure rates and the distribution of failure modes, and the ability of detection of failures via an online diagnostic system. It is used as a tool for supporting the IEC61508 standard [17]. The applicability of this method was later extended to electromechanical and mechanical systems [18].

The classic FMEA method has many weaknesses, the most important of which is the use of natural language [19] and erroneous risk estimation using the Risk Priority Number (RPN). Liu HC et al. in [20,21] analyzed the limitations in estimating the risk analysis with the use of RPN. Currently, in the literature, various approaches are used in the evaluation of risk analysis. One of the alternatives is fuzzy logic, which was used by Liu HC et al. [22], S. Bahrebar et al. [23], S. Liu et al. [24], Y. Lv et al. [25], and G. Filo et al. [26]. The weakness of the classical RPN model was also noticed by the industry and recently was replaced by action-priority [27]. In the case of a vane pump, an additional factor that makes evaluation of the probability of a failure detection relatively small is the diagnostic possibilities. During pump operation, they are limited to the measurement of several physical quantities, such as pressure, volumetric flow rate at the pump outlet, and the level of vibration and noise [28]. Additionally, the same symptoms may indicate various types of failures [29]. FMEA uses various tools, one of which is the matrix analysis, which was implemented by G. L. Barbour [30] and M. E. Stock et al. [31] and used in the developed function-failure design method. S. G. Arunajadai et al. [32] used the matrix analysis and clustering analysis for failure modes identification. This study proposes a method that is an extension of the method created by G. L. Barbour [30] and combines the aspects of classical FMEA, risk analysis, and RCA [33] by utilization matrix analysis and statistical tools for the failure analysis of a hydraulic vane pump with variable delivery. The presented method can be used in the conceptual, design, or early production stage or in case when expert knowledge base is not available or incomplete. The proposed approach eliminates some of the basic disadvantages of the traditional FMEA and takes into account the relationship between failures and their causes and end effects. It classifies the potential failures and their causes in relation to the importance of individual pump components. This is achieved by matrix transformations. It also allows for classifying potential failures that may occur during pump operation based on the symptoms of pump malfunction by using the Jaccard index [34]. In addition, a practical sheet was developed, which was used to define the necessary data for the analysis.

2. Research Methodology

Vane pumps are displacement-types pump in which the displacement chamber is created during shaft rotation between blades, stator, and the covers. An exemplary pump is shown in Figure 1. It is a variable delivery pump by Ponar Wadowice, Poland.

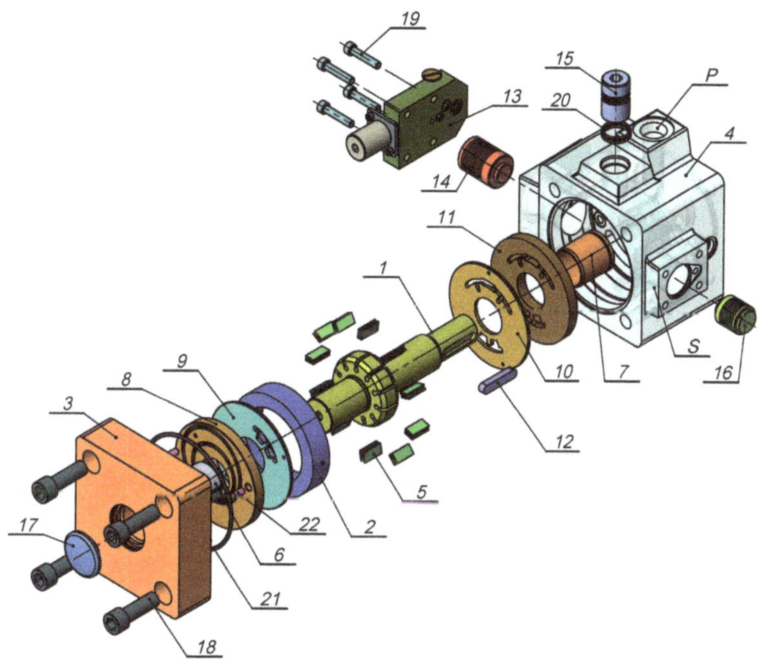

Figure 1. Vane pump: 1—shaft with rotor; 2—stator ring; 3—cover; 4—housing; 5—vanes; 6 and 7—plain bearings; 8–11—port plates; 12—groove key; 13—pressure controller; 14 and 16—control pistons; 15—adjustment screw; 17—plug; 18 and 19—bolts; 20 and 21—sealing rings; 22—pins; P—pressure port; S—supply port (at the bottom of housing).

Two vanes are placed in a single rotor socket, which is made as one part with the shaft. Hydraulic oil is supplied to the displacement chambers through a suction channel in the housing through the side covers. This pump is a double-vane design with relief channels that allow for balancing the pressure on both sides of the vanes and, in consequence, for reducing the frictional forces (vane/stator ring wear) and increasing the pump life time. Pump delivery is set by a regulator that sets the position of the stator via piston. The adjusting screw sets the maximal stator eccentricity, which defines the maximal pump delivery. The shaft is mounted on plain bearings in the cover and body, which are bolted together. The drive is transmitted through the key connection on the shaft. The proposed methodology is similar to classic FMEA [35] and consists of the following stages:

1. Main assumptions.
2. Preparation stage.
3. Specification of potential failure modes, and their causes and end effects for individual pump components.
4. Creating sheet.
5. Analysis.
6. Prioritization of failures, and their causes and end effects.

Assumptions.

The main purpose of this study was a qualitative analysis of failure modes, failures and their causes, prioritization of failures, and their causes and end effects. The analysis was carried out only for mechanical systems of the vane pump and without a delivery control system (valve). It was assumed that the analysis would use typical components that occur in vane pumps, not only those presented on Figure 1. In addition, it was assumed that the knowledge base of failures occurring in this type of pump, which may lead to pump malfunction or damage, and the probability of their occurrence is not available. Therefore, potential failures were not prioritized, but their occurrence was only indicated by a binary number system. The same method was used for indicating the cause of failures, the importance of pump components, and end effects. The end effects deal with individual components, not the pump as a whole system. It was assumed that the general form (primary form) of failure would be analyzed (each form of failure may have a variety of different mechanisms).

Preparatory stage.

In the preparatory stage, the pump was decomposed into individual components and their validity was determined based on the functional analysis of the pump. In contrast to the works of [31,33], the functions of the components were determined depending on their impact on the main task of the pump: pumping hydraulic oil to receivers. The following components of vane pumps were used: housing (c_1), cover (c_2), shaft and rotor as one part (c_3), stator (c_4), vane (c_5), port plate (c_6), bearing (c_7), piston (c_8), sealing ring (c_9), pin (c_{10}), spring (c_{11}), and fastener (c_{12}). The abovementioned components were categorized for the following functions:

1. Main function (components that perform the intended function: pumping fluid to receivers) (u_1). The following components were assigned to this function: housing, cover, shaft, stator, vane, and fastener.
2. Auxiliary functions. Components performing these functions ensure proper operation of the pump. Their possible failures may cause the pump malfunction, but the main task of the pump is still maintained (u_2). These components are the port plate, bearing, piston, and sealing ring.
3. Additional components. Their failure has little effect on the main task of the pump (u_3). These elements are the sealing ring, pin, and spring.

In the next step, potential failures that may occur for pump components were identified and classified. We assumed that the knowledge base about vane pump failures is not available; therefore, we used the analogy of failures that occur in typical mechanical systems. It was assumed that the vane pump experiences failures typical of machine shafts, plane and ball bearings, screw connections, sealing components, springs, and pumps (in general). The individual failure might be the result of a chain of several appearing failures, and consequently, determining the cause of the failure is a difficult task. The work of [36] shows examples of failures for which determining the root cause is not straightforward and requires detailed studies. Failure classification is a complex task, and various classification forms are used [37,38]. Donald J. Wulpi [39] defines the following primary failures: distortion or undesired deformation: fracture, corrosion, and wear. Wear mechanisms and modes were presented in [40] as mechanical, chemical, and thermal. Potential failures that can occur in vane pumps were classified according to the following mechanisms: fracture and separation, deformation, wear, erosion, corrosion, displacement, and material properties. The following types of failures were selected on the basis of the literature analysis concerning the wear of typical machine shafts [35]: fatigue, corrosion (stress corrosion cracking and corrosion fatigue), abrasive wear (erosion, polishing scratching, and gouging), adhesive wear (scoring, galling, seizing, scuffing, pitting, and fretting), corrosive wear, fretting corrosion, cavitation corrosion, and debris wear. The typical failures of screw connections [36] are fracture (under static load), fatigue, loosening (due to vibration), and corrosion. The typical failures of seals [41] are pattern failure, extrusion, hardening, and damage during installation. The following failures were selected for bearings [36]: deformation (brinelling), fatigue, corrosion (stress corrosion cracking and corrosion fatigue), abrassive wear (erosion, polishing scratching, and gouging), adhesive wear (scoring, galling, seizing, scuffing, pitting, and fretting), corrosive wear, fretting corrosion, cavitation corrosion, and debris wear.

In this study, potential failures were classified based on the wear mechanism in the following way:

1. Fracture and separation: brittle fracture (f_{11}), fatigue fracture (f_{12}), pitting (f_{13}).
2. Deformation: yielding (f_{21}), extrusion-shrinkage (f_{22}).
3. Wear: abrasive wear (f_{31}), adhesive wear (f_{32}).
4. Erosion: particles erosion (f_{41}), cavitation erosion (f_{42}).
5. Corrosion: general corrosion (f_{51}), chemical attack (f_{52}).
6. Displacement: loosening (f_{61}), seizing (f_{62}).
7. Material properties: aging (f_{71}), hardening (f_{72}).

In the next step, the primary root causes of failures were defined. The available literature [40] defines the following root causes: physical, human, and latent. For the vane pump, the primary root causes were set as follows:

1. Design/Specification (o_1).
2. Material/Manufacturing (o_2).
3. Assembly/Installation (o_3).
4. Maintenance/Fluid (o_4).
5. Operation: Overload (o_5).

At this stage, the end effects that can occur as a results of individual failure were also identified:

1. Catastrophic: major damages with component destruction (l_1).
2. Critical: component malfunction with severe damages (l_2).
3. Marginal: component malfunction with minor damages (l_3).
4. Minor: less than minor damages (l_4).

Creating sheet.

To perform the analysis, the sheet presented in Figure 2 was created. It consists of the relationships between the components, failures, causes, and end effects. These relation are binary numbers, where (1) indicates that the relation is true and (0) indicates that it is false.

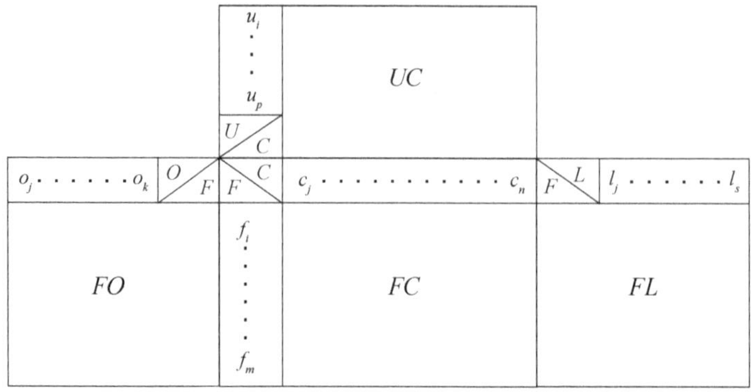

Figure 2. Sheet, C—component, U—component function, O—primary root causes, L—end effect, FC—relation failure–component, UC—relation function–component, FO—relation failure–cause, FL—relation failure–end effect).

Analysis.

The data from Figure 2 (the sheet) can be presented as sets of components, failures, causes of failures, end effects, and functions of components. We can write these sets as follows:

Set of components C, where

$$c_j \in C, j \in \{1 \ldots n\} \tag{1}$$

Set of failures F, where

$$f_i \in F, i \in \{1 \ldots m\} \tag{2}$$

Set of causes of failures O, where

$$o_j \in O, j \in \{1 \ldots k\} \tag{3}$$

Set of end effects L, where

$$l_j \in L, j \in \{1 \ldots s\} \tag{4}$$

Set of end effects, U where

$$u_j \in U, j \in \{1 \ldots p\} \tag{5}$$

We define the relationships between the listed sets. The relationship between the sets of failures F and components C can be defined as follows:

$$R_{fc} \subseteq F \times C = \begin{cases} 1, & \text{if failure occurs} \\ 0, & \text{no failure.} \end{cases} \tag{6}$$

The relationship between the sets of failures F and their causes O can be defined as follows:

$$R_{fo} \subseteq F \times O = \begin{cases} 1, & \text{if cause can make failure} \\ 0, & \text{otherwise} \end{cases} \tag{7}$$

The relationship between the sets of failures F and end effects L can be defined as follows:

$$R_{fl} \subseteq F \times L = \begin{cases} 1, & \text{if cause can make failure} \\ 0, & \text{otherwise} \end{cases} \tag{8}$$

The relationship between the sets of components C and their functions U can be defined as follows:

$$R_{cu} \subseteq C \times U = \begin{cases} 1, & \text{if component perform function} \\ 0, & \text{otherwise} \end{cases} \tag{9}$$

The obtained matrices can be used to prioritize potential failures in the vane pump for the functions performed by the elements by multiplying the matrix *UC* and the transposed matrix FC^T:

$$UF = UC \cdot FC^T \tag{10}$$

The relationship between the causes of potential failures for individual components is obtained by multiplying the transposed matrix FO^T and the matrix *FC*:

$$OC = FO^T \cdot FC \tag{11}$$

The relationship between the end effects inducted by failures and pump components is obtained by multiplying the transposed matrix FL^T and the matrix *FC*:

$$LC = FL^T \cdot FC \tag{12}$$

The relationship between the failure causes and the end effects is obtained by multiplying the transposed matrix FL^T and the matrix *FO*:

$$LO = FL^T \cdot FO \tag{13}$$

The relationship between components function and the end effects inducted by failures is obtained by multiplying the matrix *FU* and the matrix *FL*:

$$UL = FU \cdot FL \tag{14}$$

The relationship between components function and the causes of failures is obtained as follows:

$$UO = (UC \cdot FC^T) \cdot FL \cdot (FL^T \cdot FO) \tag{15}$$

The values of elements of the matrices *OC*, *LC*, *UF*, *LO*, *UL*, and *UO* can be written in the following general form:

$$c_{ij} = \sum_{m=1}^{n} a_{im} b_{mj} \tag{16}$$

where c_{ij} is the element of matrices *OC*, *LC*, *UF*, *LO*, *UL*, and *UO*, while a_{im} and b_{mj} are elements of the multiplied matrices *FC*, *UC*, *FO*, and *FL*.

The created matrices can also be used to determine the impact of potential failures on the pump failure modes. For realization of this purpose, we can assume a kind of test vector that we can define as follows:

$$T = [t_1, \ldots, t_n] \tag{17}$$

$$t_i \subseteq T = \begin{cases} 1, & \text{if failure occures} \\ 0, & \text{otherwise} \end{cases} \tag{18}$$

where t is the generalized failure for the i-th pump component, which belongs to set C.

Next, the Jaccard similarity index [34] is used for evaluating the influence of potential failure on given failure modes by measuring the similarity of test vector T with a row of matrix FC:

$$J(T, F_i) = \frac{|T \cap F_i|}{|T \cup F_i|} \tag{19}$$

where F_i is the i-th row of matrix FC.

3. Results

The sets of components, potential failures, causes, and end effects are expressed in the form presented below.

Set of components:

$$C = \begin{bmatrix} c_1 & c_2 & c_3 & c_4 & c_5 & c_6 & c_7 & c_8 & c_9 & c_{10} & c_{11} & c_{12} \end{bmatrix} \tag{20}$$

Set of potential failures:

$$F = \begin{bmatrix} f_{11} & f_{12} & f_{13} & f_{21} & f_{22} & f_{31} & f_{32} & f_{41} & f_{42} & f_{51} & f_{52} & f_{61} & f_{62} & f_{71} & f_{72} \end{bmatrix} \tag{21}$$

Set of failure causes:

$$O = \begin{bmatrix} o_1 & o_2 & o_3 & o_4 & o_5 \end{bmatrix} \tag{22}$$

Set of components function:

$$U = \begin{bmatrix} u_1 & u_2 & u_3 \end{bmatrix} \tag{23}$$

Set of end effects inducted by failures:

$$L = \begin{bmatrix} l_1 & l_2 & l_3 & l_4 \end{bmatrix} \tag{24}$$

Equation (6) gives matrix FC:

$$FC = \begin{bmatrix} 1 & 1 & 1 & 1 & 1 & 0 & 0 & 0 & 0 & 1 & 1 & 1 \\ 1 & 1 & 1 & 1 & 1 & 1 & 1 & 0 & 0 & 0 & 1 & 1 \\ 0 & 0 & 1 & 1 & 1 & 0 & 1 & 0 & 0 & 0 & 0 & 0 \\ 1 & 1 & 1 & 1 & 1 & 0 & 0 & 0 & 0 & 1 & 0 & 1 \\ 0 & 0 & 0 & 0 & 0 & 0 & 0 & 0 & 1 & 0 & 0 & 0 \\ 1 & 1 & 1 & 1 & 1 & 1 & 1 & 1 & 1 & 1 & 0 & 0 \\ 1 & 1 & 1 & 1 & 1 & 1 & 1 & 1 & 0 & 0 & 0 & 0 \\ 1 & 1 & 1 & 1 & 1 & 1 & 1 & 0 & 0 & 0 & 0 & 0 \\ 1 & 1 & 1 & 1 & 1 & 1 & 1 & 0 & 0 & 0 & 0 & 0 \\ 0 & 0 & 0 & 0 & 0 & 0 & 0 & 1 & 0 & 0 & 1 & 1 \\ 1 & 1 & 1 & 1 & 1 & 1 & 1 & 1 & 0 & 1 & 1 & 0 \\ 0 & 0 & 0 & 0 & 0 & 0 & 0 & 0 & 0 & 0 & 0 & 1 \\ 0 & 0 & 1 & 0 & 1 & 1 & 1 & 1 & 0 & 0 & 0 & 0 \\ 0 & 0 & 0 & 0 & 0 & 0 & 0 & 0 & 1 & 0 & 0 & 0 \\ 0 & 0 & 0 & 0 & 0 & 0 & 0 & 0 & 1 & 0 & 0 & 0 \end{bmatrix} \tag{25}$$

Equation (7) gives matrix FO, while Equation (8) gives matrix FL:

$$FO = \begin{bmatrix} 1 & 1 & 1 & 0 & 1 \\ 1 & 1 & 0 & 0 & 0 \\ 1 & 1 & 0 & 1 & 0 \\ 1 & 1 & 1 & 0 & 1 \\ 1 & 1 & 1 & 1 & 0 \\ 0 & 0 & 0 & 1 & 0 \\ 1 & 1 & 0 & 1 & 0 \\ 0 & 0 & 0 & 1 & 0 \\ 1 & 0 & 0 & 1 & 0 \\ 1 & 1 & 0 & 0 & 0 \\ 1 & 1 & 0 & 1 & 0 \\ 1 & 1 & 1 & 0 & 0 \\ 1 & 1 & 0 & 1 & 0 \\ 1 & 1 & 0 & 1 & 0 \\ 0 & 1 & 0 & 1 & 1 \end{bmatrix} FL = \begin{bmatrix} 1 & 1 & 1 & 1 \\ 1 & 1 & 1 & 1 \\ 0 & 0 & 1 & 1 \\ 1 & 1 & 1 & 1 \\ 1 & 1 & 1 & 1 \\ 0 & 0 & 1 & 1 \\ 0 & 0 & 1 & 1 \\ 0 & 0 & 1 & 1 \\ 0 & 0 & 1 & 1 \\ 0 & 0 & 1 & 1 \\ 0 & 0 & 1 & 1 \\ 0 & 0 & 1 & 1 \\ 0 & 0 & 1 & 1 \\ 0 & 0 & 1 & 1 \\ 0 & 0 & 1 & 1 \end{bmatrix} \quad (26)$$

Equation (9) gives matrix *UC*:

$$UC = \begin{bmatrix} 1 & 1 & 1 & 1 & 1 & 0 & 1 & 0 & 0 & 0 & 0 & 1 \\ 0 & 0 & 0 & 0 & 0 & 1 & 1 & 1 & 0 & 0 & 0 & 0 \\ 0 & 0 & 0 & 0 & 0 & 0 & 0 & 0 & 1 & 1 & 1 & 0 \end{bmatrix} \quad (27)$$

Equations (10)–(15) gives matrices *OC*, *LC*, *UF*, *LO*, *UL*, and *UO*, respectively. The matrix *UF* describes the relationships between failures and pump components:

$$UF = \begin{bmatrix} 6 & 6 & 3 & 6 & 0 & 5 & 5 & 5 & 5 & 1 & 5 & 1 & 2 & 0 & 0 \\ 0 & 2 & 1 & 0 & 1 & 4 & 3 & 2 & 2 & 1 & 3 & 0 & 3 & 1 & 1 \\ 2 & 1 & 0 & 1 & 1 & 2 & 0 & 0 & 0 & 1 & 2 & 0 & 0 & 1 & 1 \end{bmatrix} \quad (28)$$

The matrix *OC* describes the relationships between failure causes and pump components:

$$OC = \begin{bmatrix} 6 & 6 & 8 & 7 & 8 & 5 & 6 & 4 & 2 & 3 & 4 & 5 \\ 5 & 5 & 7 & 6 & 7 & 4 & 5 & 4 & 3 & 3 & 4 & 5 \\ 2 & 2 & 2 & 2 & 2 & 0 & 0 & 0 & 1 & 2 & 1 & 3 \\ 5 & 5 & 7 & 6 & 7 & 6 & 7 & 4 & 4 & 2 & 1 & 0 \\ 2 & 2 & 2 & 2 & 2 & 0 & 0 & 0 & 1 & 2 & 1 & 2 \end{bmatrix} \quad (29)$$

The matrix *LC* describes the relationships between pump components and end effects:

$$LC = \begin{bmatrix} 3 & 3 & 3 & 3 & 3 & 1 & 1 & 0 & 1 & 2 & 2 & 3 \\ 3 & 3 & 3 & 3 & 3 & 1 & 1 & 0 & 1 & 2 & 2 & 3 \\ 8 & 8 & 10 & 9 & 10 & 7 & 8 & 5 & 4 & 4 & 4 & 5 \\ 8 & 8 & 10 & 9 & 10 & 7 & 8 & 5 & 4 & 4 & 4 & 5 \end{bmatrix} \quad (30)$$

The matrices *LO*, *UL*, and *UO* describe the relationships between failure causes and end effects, component functions and end effects, and component functions and causes of failures, respectively.

$$LO = \begin{bmatrix} 4 & 4 & 3 & 1 & 2 \\ 4 & 4 & 3 & 1 & 2 \\ 12 & 12 & 4 & 10 & 3 \\ 12 & 12 & 4 & 10 & 3 \end{bmatrix} UL = \begin{bmatrix} 19 & 19 & 58 & 58 \\ 2 & 2 & 20 & 20 \\ 5 & 5 & 12 & 12 \end{bmatrix} UO = \begin{bmatrix} 1544 & 1544 & 578 & 1198 & 424 \\ 496 & 496 & 172 & 404 & 128 \\ 328 & 328 & 126 & 250 & 92 \end{bmatrix} \quad (31)$$

To evaluate the influence of defined failures on failure modes and prioritize failure, we can classify failure modes in the following ways [42]:

1. Critical, the pump does not realize the intended function (no flow on the pump outlet):
    - Symptom: no shaft rotation T_1.
    - Symptom: pump leakage T_2.

2. Major, pump malfunction:
   - Symptom: noisy operation T_3.
   - Symptom: flow below rated T_4.
   - Symptom: pressure below rated T_5.
3. Minor:
   - Symptom: unusual heat level T_6.
   - Symptom: shaft leaks T_7.

   The tests vectors for the abovementioned symptoms are presented in (Table 1)

Table 1. Test vector values.

	c_1	c_2	c_3	c_4	c_5	c_6	c_7	c_8	c_9	c_{10}	c_{11}	c_{12}
T_1	1	0	1	1	1	0	1	0	0	0	0	0
T_2	1	1	1	1	0	0	0	0	0	0	0	1
T_3	0	0	1	1	1	1	1	0	0	1	0	0
T_4	0	0	0	1	1	1	0	1	1	0	0	0
T_5	0	0	0	1	1	1	0	0	0	0	0	0
T_6	0	0	1	1	1	1	1	0	0	0	0	0
T_7	0	0	1	0	0	0	0	0	1	0	0	0

Discussion.

The results presented in the above section allow for the classification of failures and their causes depending on the user's needs. Let us assume that, as in a typical FMEA, we are interested in prioritizing failures. Using the matrix *UF* Equation (28), we can determine the failures that occur most often for the functions performed by the pump components. Taking into account the end effects, we can classify the failures as follows:

1. Brittle fracture (f_{11}), fatigue fracture (f_{12}), yielding (f_{21}).
2. Abrrasive wear (f_{31}), adhesive wear (f_{32}), particles erosion (f_{41}), cavitaion erosion (f_{42}), chemical attack (f_{52}).
3. Pitting (f_{13}).
4. Seizing (f_{62}).
5. Aging (f_{71}), hardening (f_{72}).

Analysis of the matrix *OC* in Equation (29) shows that the dominant causes of failures are as follows:

1. Design (o_1) for shaft (c_3) and vane (c_5).
2. Maintenance/Fluid (o_4) for shaft (c_3), vane (c_5) and bearing (c_7).
3. Material/Manufacturing (o_2) for stator (c_4), shaft (c_3).
4. Assembly/Installation (o_3) for fastener (c_{12}).
5. Operation: Overload (o_5) for housing (c_1), cover (c_2), shaft (c_3), stator (c_4), vane (c_5), pin (c_{10}), and fastener (c_{12}).

The matrix *LC* in Equation (30) indicates that, for given failures, the pump component end effects are marginal (l_3) and minor (l_4).

We can use the matrix *LC* in Equation (31) prioritize causes of failures for given end effects. For example for end effects: *Catastrophic* (l_1) and *Critical* (l_2), these causes are as follows:

1. Design (o_1), Material/Manufacturing (o_2).
2. Assembly/Installation (o_3).
3. Operation: Overload (o_5).
4. Maintenance/Fluid (o_4).

In the similar way, we can obtain causes of failures for end effects *marginal* (l_3) and *minor* (l_4), which are

1. Design (o_1), Material/Manufacturing (o_2).

2. Maintenance/Fluid (o_4).
3. Assembly/Installation (o_3).
4. Operation: Overload (o_5).

Causes of failures for given component functions can be obtained from matrix *UO* in Equation (31) and they are:

1. Design (o_1), Material/Manufacturing (o_2).
2. Maintenance/Fluid (o_4).
3. Assembly/Installation (o_3).
4. Operation: Overload (o_5).

Matrix *UL* in Equation (31) indicates that potential failures make end effects marginal (l_3) and minor (l_4) for components performing functions (u_1), (u_2), and (u_3), but it has to be added that the analyzed failures are the primary ones that can be an initiating factor for other types of failures. The obtained results from Equation (29) agree with well-known quality management rules that nearly 80% of failures are caused at the conceptual and design stages. According to the presented method, maintenance/fluid is the major cause of failures for vane pumps, which is also confirmed in practice [29]. An additional confirmation factor is that the vane pumps are used as reference devices for standardized validation tests [43–45] for evaluating the quality of working fluids.

The Jaccard similarity index (Table 2) shows the influence of individual failures on failure modes. For critical failure modes, the dominant failures are brittle fracture (f_{11}), fatigue fracture (f_{12}), yielding (f_{21}), abrasive wear (f_{31}), adhesive wear (f_{32}), particles erosion (f_{41}), cavitation erosion (f_{42}), and chemical attack (f_{52}). For major failure modes, the dominant failure are abrasive wear (f_{31}) and chemical attack (f_{52}), while for minor failure modes, the following failures are dominant: yielding (f_{21}), abrasive wear (f_{31}), adhesive wear (f_{32}), particles erosion (f_{41}), cavitation erosion (f_{42}), and chemical attack (f_{52}).

The obtained results agree with the empirical data included in [29], which is a compendium of vane pump and motor failures and their causes. The main failures according to [29] for pump components that perform main functions (shaft, stator, and rotor) are fatigue rupture (shaft) and brittle rupture (rotor and stator). The Jaccard similarity index (Table 2) indicates the same failures. It has to be mentioned that other failures presented in the table such as abrasive wear, particle erosion, and cavitation erosion can also be initiators of fatigue crack. The data from the table also agree with the results from *UF* in Equation (28), which defines the failures for components performing main functions. Noisy pump operation according to [29] is mainly caused by oil aeration and/or cavitation. This leads to wear of port plates or vanes. The Jaccard similarity index (Table 2) for failure mode noisy operation shows also abrasive wear as a dominant failure.

Table 2. The Jaccard similarity index for test vectors and failures.

	f_{11}	f_{12}	f_{13}	f_{21}	f_{22}	f_{31}	f_{32}	f_{41}	f_{42}	f_{51}	f_{52}	f_{61}	f_{62}	f_{71}	f_{72}
T_1	0.33	0.42	0.33	0.33	0.00	0.42	0.42	0.42	0.42	0.00	0.42	0.00	0.25	0.00	0.00
T_2	0.42	0.42	0.17	0.42	0.00	0.33	0.33	0.33	0.33	0.08	0.33	0.08	0.08	0.00	0.00
T_3	0.33	0.42	0.33	0.33	0.00	0.50	0.42	0.42	0.42	0.00	0.50	0.00	0.33	0.00	0.00
T_4	0.17	0.25	0.17	0.17	0.08	0.42	0.33	0.25	0.25	0.08	0.33	0.00	0.25	0.08	0.08
T_5	0.17	0.25	0.17	0.17	0.00	0.25	0.25	0.25	0.25	0.00	0.25	0.00	0.17	0.00	0.00
T_6	0.25	0.42	0.33	0.25	0.00	0.42	0.42	0.42	0.42	0.00	0.42	0.00	0.33	0.00	0.00
T_7	0.08	0.08	0.08	0.08	0.08	0.17	0.08	0.08	0.08	0.00	0.08	0.00	0.08	0.08	0.08

4. Conclusions

The presented method uses the matrix analysis along with elements of statistics that allows for the classification of potential failures and their causes in relation to the end effects inducted by the failures and to component functions. It also allows for performing select elements of RCA analysis. The proposed practical sheet allows for preparing data and its initial classification before they are used at the analysis stage. This sheet can be extended and adapted to the existing needs. The application of a similarity index allows for the identification of contribution to a specified pump failure mode. The presented method allows for identification failures, which refers to critical, major, and minor failure modes. The obtained results can be input data for further analysis aimed at searching for root cause of failures and at defining preventative measures. The presented method is simple for algorithmization and can be extended using elements of a statistical tool or matrix analysis. The method can also be used to analyze hydraulics components as well as full hydraulic systems.

Author Contributions: Conceptualization, J.F.-D. and M.D.; methodology, J.F.-D.; software, H.M.; validation, M.D., J.F.-D., and H.M.; formal analysis, J.F.-D.; investigation, J.F.-D.; resources, J.F.-D.; data curation, M.D.; writing—original draft preparation, M.D.; writing—review and editing, H.M.; visualization, H.M.; supervision, M.D.; project administration, J.F.-D.; funding acquisition, H.M. All authors have read and agreed to the published version of the manuscript.

Funding: This research received no external funding.

Institutional Review Board Statement: Not applicable.

Informed Consent Statement: Not applicable.

Data Availability Statement: Not applicable.

Conflicts of Interest: The authors declare no conflict of interest.

Abbreviations

The following abbreviations are used in this manuscript:

MDPI	hlMultidisciplinary Digital Publishing Institute
DOAJ	Directory of open access journals
MIL-STD	US Defense Standard
FMEA	Failure Modes and Effect Analysis
FMEDA	Failure Modes, Effects, and Diagnostic Analysis
FMCA	Failure Modes, Effects, and Criticality Analysis
NASA	National Aeronautics and Space Administration
RCA	Root Cause Analysis
RPN	Risk Priority Number
ISO	International Organization for Standardization

References

1. Lee, Y.; Lee, G.; Yang, J.; Park, J.; Baek, D. Failure analysis of a hydraulic power system in the wind turbine. *Eng. Fail. Anal.* **2020**, *107*, 104218. [CrossRef]
2. Li, Y.; Coolen, F.P.A.; Zhu, C.; Tan, J. Reliability assessment of the hydraulic system of wind turbines based on load-sharing using survival signature. *Renew. Energy* **2020**, *153*, 766–776. [CrossRef]
3. Gao, P.; Yu, T.; Zhang, Y.; Wang, J.; Zhai, J. Vibration analysis and control technologies of hydraulic pipeline system in aircraft: A review. *Chin. J. Aeronaut.* **2020**, *34*, 83–114. [CrossRef]
4. Wang, G.; Song, Y.; Wang, J.; Chen, W.; Cao, Y.; Wang, J. Study on the Shifting Quality of the CVT Tractor under Hydraulic System Failure. *Appl. Sci.* **2020**, *10*, 681. [CrossRef]
5. Ma, Z.; Wang, S.; Shi, J.; Li, T.; Wang, X. Fault diagnosis of an intelligent hydraulic pump based on a nonlinear unknown input observer. *Chin. J. Aeronaut.* **2018**, *31*, 385–394. [CrossRef]
6. Li, T.; Wang, S.; Zio, E.; Shi, J.; Ma, Z. A numerical approach for predicting the remaining useful life of an aviation hydraulic pump based on monitoring abrasive debris generation. *Mech. Syst. Signal Process.* **2018**, *136*, 106519. [CrossRef]
7. Hast, D.; Findeisen, R.; Streif, S. Detection and isolation of parametric faults in hydraulic pumps using a set-based approach and quantitative–qualitative fault specifications. *Control Eng. Pract.* **2015**, *40*, 61–70. [CrossRef]
8. Lu, C.; Wang, S.; Wang, X. A multi-source information fusion fault diagnosis for aviation hydraulic pump based on the new evidence similarity distance. *Aerosp. Sci. Technol.* **2017**, *71*, 392–401. [CrossRef]
9. Tague, N.R. *The Quality Toolbox*, 2nd ed.; ASQ Quality Press: Milwaukee, WI, USA, 2005.
10. Carlson, C.S. *Effective FMEAs: Achieving Safe, Reliable, and Economical Products and Processes Using Failure Mode and Effects Analysis*; John Wiley & Sons, Inc.: Hoboken, NJ, USA, 2012.

11. Byington, C.S.; Roemer, M.J.; Galie, T. Prognostic enhancements to diagnostic systems for improved condition-based maintenance [military aircraft]. In Proceedings of the IEEE Aerospace Conference, Big Sky, MT, USA, 9–16 March 2002.
12. Tidriri, K.; Chatti, N.; Verron, S.; Tiplica, T. Bridging data-driven and model-based approaches for process fault diagnosis and health monitoring: A review of researches and future challenges. *Annu. Rev. Control* **2016**. [CrossRef]
13. Djeziri, M.A.; Benmoussa, S.; Zio, E. Review on Health Indices Extraction and Trend Modeling for Remaining Useful Life Estimation. In *Artificial Intelligence Techniques for a Scalable Energy Transition*; Sayed-Mouchaweh, M., Ed.; Springer: Berlin, Germany, 2020.
14. MIL-P 1629. *USA Military Standard, Procedure for Performing a Failure Mode, Effects and Criticality Analysis (MIL-P 1629, USA)*; Military Specifications and Standards: Washington, DC, USA, 1949.
15. J1739 200901. *Potential Failure Mode and Effects Analysis in Design (Design FMEA), Potential Failure Mode and Effects Analysis in Manufacturing and Assembly Processes (Process FMEA)*; SAE International J1739 200901: Warrendale, PA, USA, 2009.
16. Grebe, J.C.; Goble, W.M. *FMEDA—Accurate Product Failure Metrics, FMEDA Development Paper*; Revision 1.1, 19 February 2007; FMEDA: Exida, PA, USA, 2007.
17. IEC 61508. *Functional Safety of Electrical/Electronic/Programmable Electronic Safety-Related Systems*; International Electrotechnical Commission, IEC 61508: Geneva, Switzerland, 2010.
18. Globe, W.M.; Bukowski, J.V. Development of a Mechanical Component Failure Database. In Proceedings of the annual Reliability and Maintainability Symposium, Orlando, FL, USA, 22–25 January 2007.
19. Wirth, R.; Berthold, B.; Kramer, A.; Peter, G. Knowledge-based Support Analysis for the Analysis of Failure Modes and Effects. *Eng. Appl. Artif. Intell.* **1996**, *9*, 219–229. [CrossRef]
20. Liu, H.C.; Liu, L.; Liu, N. Risk evaluation approaches in failure mode and effects analysis: A literature review. *Expert Syst. Appl.* **2013**, *40*, 828–838. [CrossRef]
21. Liu, H.-C. *FMEA Using Uncertainty Theories and MCDM Methods*; Springer Science+Business Media: Singapore, 2016.
22. Liu, H.C.; Liu, L.; Bian, Q.H.; Lin, Q.L.; Dong, N.; Xu, P.C. Failure mode and effects analysis using fuzzy evidential reasoning approach and grey theory. *Expert Syst Appl.* **2011**, *38*, 4403–4415. [CrossRef]
23. Bahrebar, S.; Blaabjerg, F.; Wang, H.; Vafamand, N.; Khooban, M.H.; Rastayesh, S.; Zhou, D. A Novel Type-2 Fuzzy Logic for Improved Risk Analysis of Proton Exchange Membrane Fuel Cells in Marine Power Systems Application. *Energies* **2018**, *11*, 721. [CrossRef]
24. Liu, S.; Guo, X.; Zhang, L. An Improved Assessment Method for FMEA for a Shipboard Integrated Electric Propulsion System Using Fuzzy Logic and DEMATEL Theory. *Energies* **2019**, *12*, 3162. [CrossRef]
25. Lv, Y.; Liu, Y.; Jing, W.; Wozniak, M.; Damasevicius, R.; Scherer, R.; Wei, W. Quality Control of the Continuous Hot Pressing Process of Medium Density Fiberboard Using Fuzzy Failure Mode and Effects Analysis. *Appl. Sci.* **2020**, *10*, 4627. [CrossRef]
26. Filo, G.; Fabis-Domagala, J.; Domagala, M.; Lisowski, E.; Momeni, H. Knowledge-based Support Analysis for the Analysis of Failure Modes and Effects. *MATEC Web Conf.* **2018**, *183*, 03009. [CrossRef]
27. *AIAG & VDA FMEA Handbook*; Automotive Industry Action Group (AIAG): Southfield, MI, USA, 2019.
28. Watton, J. *Modelling, Monitoring and Diagnostic Techniques for Fluid Power Systems*; Springer: London, UK, 2007.
29. *Catalogue HY29-0022/UK, 2M, 01/2012, IRO*; Parker Hannifin Corporation, Parker Hannifin Limited: Warwick, UK, 2014.
30. Barbour, G.L. Failure Modes and Effects Analysis by Matrix Method. In Proceedings of the Annual Reliability Symposium, Philadelphia, PA, USA, 18–20 January 1977; pp. 114–119.
31. Stock, M.E.; Stone, R.B.; Tumer, I.Y. Going back in time to improve design: The elemental function-failure design method. In Proceedings of the DETC 03 ASME 2003 Design Engineering Technical Conference Computers and Information in Engineering Conference, Chicago, IL, USA, 2–6 September 2003.
32. Arunajadai, S.G.; Uder, S.J.; Stone, R.B.; Tumer, I.Y. Failure Mode Identification through Clustering Analysis. *Qual. Reliab. Eng. Int.* **2004**, *20*, 511–526. [CrossRef]
33. Wilson, P.F.; Dell, L.D.; Anderson, G.F. *Root Cause Analysis: A Tool for Total Quality Management*; ASQ Quality Press: Milwaukee, WI, USA, 1993.
34. Jaccard, P. Etude comparative de la distribution florale dans une portion des Alpes et du Jura. *Bull. Soc. Vaudoise Des Sci. Nat.* **1901**, *37*. [CrossRef]
35. McDermott, R.E.; Mikulak, R.J.; Beauregard, M.R. *The Basics of FMEA*, 2nd ed.; CRC Press: New York, NY, USA, 2009.
36. Bloch, H.P.; Geitner, F.K. *Machinery Failure Analysis and Troubleshooting*; Gulf Professional Publishing: Huston, TX, USA, 1999; Volume 2.
37. Stachowiak, G.W. *Wear-Materials, Mechanism and Practice*; John Wiley & Sons Ltd.: Chichester, UK, 2005.
38. Kato, K. Classification of wear mechanisms/models. *J. Eng. Tribol.* **2002**, *216*. [CrossRef]
39. Donald, J.; Wulpi, L. *Understanding How Components Fail*, 3rd ed.; Miller, B., Ed.; ASM International: Materials Park, OH, USA, 2013.
40. Hossain, F.A.; Scutti, J.J. Four Fundamental Root Causes of Failure: Case Histories, Failure Analysis: A Foundation for Diagnostics and Prognostics Development. In Proceedings of the 53rd Meeting of Society for Machinery Failure Prevention Technology, Virginia Beach, VA, USA, 19–22 April 1999.
41. Flitney, R. *Seals and Sealing Handbook*, 6th ed.; Butterworth-Heinemann: Oxford, UK, 2014.
42. *Bulletin HY29-0106/UK, XM, 08/2012*; Parker Hannifin Manufacturing France SAS: Vierzon Cedex, France, 2012.

43. ASTM D7043-17. *Standard Test Method for Indicating Wear Characteristics of Non-Petroleum and Petroleum Hydraulic Fluids in a Constant Volume Vane Pump*; ASTM International: West Conshohocken, PA, USA, 2017.
44. DIN 51389-1. *Determination of Lubricants; Mechanical Testing of Hydraulic Fluids in the Vane-Cell-Pump; General Working Principles*; DIN Deutsches Institut für Normung e. V.: Berlin, Germany, 1982.
45. ISO 20763:2004. *Petroleum and Related Products—Determination of Anti-Wear Properties of Hydraulic Fluids—Vane Pump Method*; International Organization for Standardization: Geneva, Switzerland, 2004.

energies

MDPI

Article

Pressure Loss Reduction in an Innovative Directional Poppet Control Valve

Grzegorz Filo [1,*], Edward Lisowski [1] and Janusz Rajda [2]

1 Faculty of Mechanical Engineering, Cracow University of Technology, Jana Pawła II 37, 31-864 Cracow, Poland; lisowski@mech.pk.edu.pl

2 PONAR Wadowice, Wojska Polskiego 29, 34-100 Wadowice, Poland; janusz.rajda@ponar-wadowice.pl

* Correspondence: filo@mech.pk.edu.pl; Tel.: +48-12-374-33-35

Received: 31 May 2020; Accepted: 12 June 2020; Published: 17 June 2020

Abstract: This article presents the results of computational fluid dynamics (CFD) analysis of an innovative directional control valve consisting of four poppet seat valves and two electromagnets enclosed inside a single body. The valve has a unique design, allowing the use of any poppet valve configuration. Both normally opened (NO) and normally closed (NC) seat valves can be applied. The combination of four universal valve seats and two electromagnets gives a wide range of flow path configurations. This significantly increases the possibility of practical applications. However, due to the significant miniaturization of the valve body and the requirement to obtain necessary connections between flow paths, multiple geometrically complex channels had to be made inside the body. Hence, the main purpose of work was to shape the geometry of the flow channels in such a way as to minimize pressure losses. During the CFD analyses velocity distribution in flow channels and pressure distribution on the walls were determined. The results were used to obtain pressure loss as a function of flow rate, which was then verified by means of laboratory experiments conducted on a test bench.

Keywords: CFD analysis; directional control valve; poppet valve; flow resistance; pressure loss reduction

1. Introduction

Hydraulic drive and control systems are widely used in both stationary and mobile devices. Most hydraulic drives implement a fixed work cycle, usually involves obtaining the required direction of the cylinder or hydraulic motor movement. Hydraulic control valves are commonly used to set the proper flow direction. They can be controlled in various ways, including the usage of a digital controller or a computer in case when the valve is equipped with electromagnets. In the most commonly used spool valves, the movable element in the form of a spool with appropriately made holes is displaced under the electromagnet force inside a sleeve, thus opening or closing individual flow channels. A standard one-solenoid spool valve provides two operating positions, the first one obtained by switching on the electromagnet and the other by means of a return spring. Application of two electromagnets allows three operating positions to be obtained. For example, cylinder movement in one direction, stop at the fixed position and return movement. This is sufficient for most standard applications, such as control of fluid flow direction in a system with a hydraulic cylinder or a hydraulic motor. However, in some cases spool valves have also some disadvantages, such as impossibility to ensure complete tightness, which requires the use of additional shut-off valves and relatively small number of possible flow paths, which may be a serious problem in more advanced systems.

Simulation research on hydraulic control valves is usually carried out using CFD methods. However, the efforts can be focused on various aspects, as flow force reduction, innovative design, flow channel geometry optimization or metering edge formation. Simic and Herakovic [1] used CFD methods to reduce the axial flow force acting on the slider in a small seat valve by means of spool and housing geometry optimization. Huovinen et al. [2] conducted both simulation and experimental research on a choke valve in a turbulent flow using two turbulence models, $k - \varepsilon$ and $k - \omega$, respectively. Improvement in characteristics of a proportional spool valve by means of geometrical modifications of the spool was the subject of several works including Simic et al. [3], Amirante et al. [4], Park [5], as well as Lisowski and Filo [6]. In all the mentioned cases the authors were mainly focused on design or modification of a spool notch geometry. Similarly, Woldemariam et al. performed a CFD-driven optimization of a micro turbine valve [7], while Gomez et al. carried out CFD analysis of a cartridge poppet valve [8].

Studies on the orifice flow by means of CFD method were carried out by Zhou et al. [9] and Pan et al. [10] based on a lumped parameter modelling technique. Zhou used a two-dimensional approach on an axisymmetric fluid model, pointing out that it is widely utilized in the research related to hydraulic applications due to its relatively low computational requirements and convenience to apply. Similarly, Sapra et al. [11] used the CFD method to obtain flow characteristics of a large cone element, Ferreira et al. studied a ball valve behaviour [12], Brunone and Morelli [13] tested automatic control valves (ACV), while Liao et al. [14] carried out analysis of a large flow directional valve hysteresis. Recent studies include In addition to the CFD method, analyzes of hydraulic spool valve characteristics were also performed using different methods as particle image velocimetry [15] and various simulation environments such as Matlab-Simulink [16–19], SimScape [20] and AMESim [21,22].

This article concerns the results of research on a poppet control valve in a seat housing. Compared to the spool valves, the proposed solution is characterized by higher tightness and greater possibility of flow paths configuring. This work is a continuation of previous research based on the CFD method, including studies on proportional spool valves [23,24] as well as a poppet switching valve [25]. The proposed seat valve system used in the considered solution allows four different positions and four combinations of flow path connections to be obtained. It contains four universal sockets designed for installation of seat valves and two electromagnets. The each electromagnet controls two valve poppets. In addition, the valve housing has been adapted for mounting on a default plate in the ISO 1440 standard [26]. Hence, it can be used interchangeably with any usual control valves or even may be combined with other valves by means of a sandwich type arrangement.

2. Working Principle of the Valve

A general view of the tested valve is shown in Figure 1, while its cross-section is presented in Figure 2.

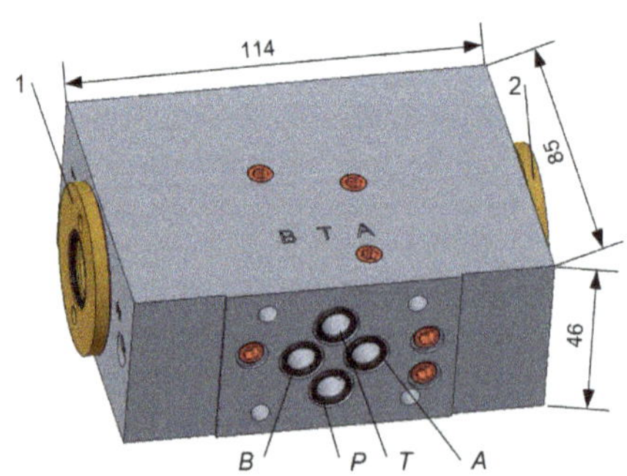

Figure 1. General view of the valve; 1, 2—electromagnet connectors, P, T, A, B—hydraulic connectors.

Four two-position seat valves are installed inside the valve block. Two electromagnets are mounted on both sides using (1) and (2) connecting ports (Figure 1). Thus, the each electromagnet controls opening or closing of two valves simultaneously. There are standard connection ports arranged in accordance with ISO 4401 at the inlet as well as at the outlet of the valve block. The individual connectors are designated as follows: *P*—supply port, *A*, *B*—flow lines, *T*—return port. The valve is designed for a flow rate up to $Q = 25\ \mathrm{dm^3\,min^{-1}}$ and a maximum operational pressure $p = 35$ MPa. It can be used to control flow to the receiver directly or as a pilot valve.

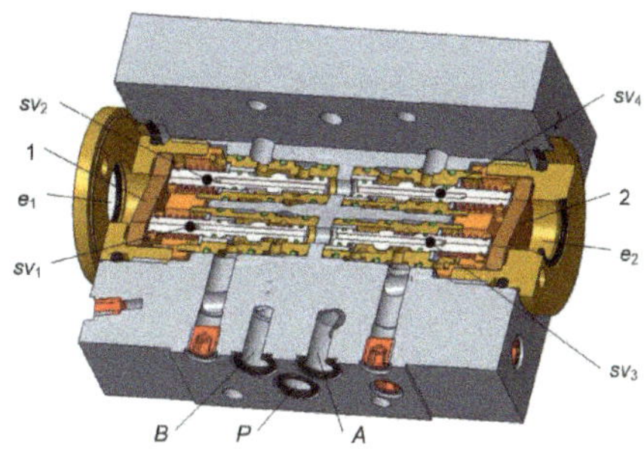

Figure 2. Valve cross-section; 1—platen, 2—cover; $sv_1 \ldots sv_4$—seat valve; *P*, *A*, *B*—hydraulic connectors.

In the default configuration, two normally opened (*NO*) poppet valves denoted sv_1, sv_2 and two normally closed (*NC*) valves, sv_3, sv_4 have been installed inside the block. The sv_1 and sv_2 valves have opened flow paths, when the e_1 electromagnet is not powered. Switching on the e_1 electromagnet causes pressure of the platen (1) on valve poppets, displacement of the poppets and thus closing flow. In contrast, the flow through sv_3 and sv_4 valves is cut-off and requires turning on power supply of e_2 electromagnet to open it by the platen (2) pressing on the poppets of sv_3 and sv_4 valves. The tested valve provides four positions as shown in the scheme (Figure 3).

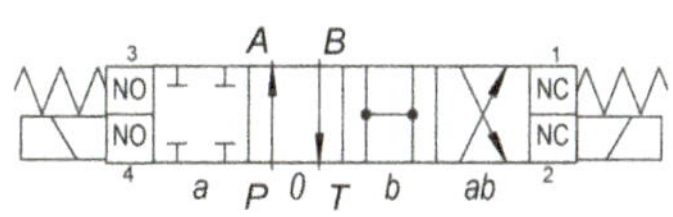

Figure 3. Connection diagram of a configuration with two *NO* and two *NC* seat valves.

The connection ports are denoted *P*, *A*, *B*, *T*, respectively, while the state of the electromagnets is described as follows: *a*—only left switched *on*, *b*—only right switched *on*, 0—both unpowered, *ab*—both powered. Since the *NO* and *NC* seat valves have identical mounting dimensions, they can be used interchangeably. This allows multiple different flow way configurations to be obtained. Table 1 shows several available arrangements with different numbers of *NC* and *NO* valves used.

Table 1. Available seat valve configurations.

Diagram	Number of NC	Number of NO
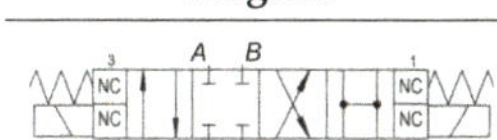	4	0
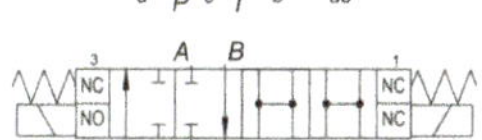	3	1
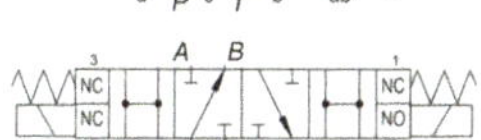	3	1
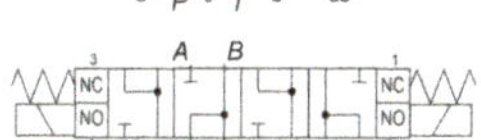	2	2

Design of *NO* and *NC* seat valves is shown in Figures 4 and 5, respectively. Both types consist of a body (1) in which a poppet (2) moves using the guide sleeve (3) under the force of an electromagnet and a return spring (4). The angle of the poppet head inclination is 30 °, which ensures the adequate tightness.

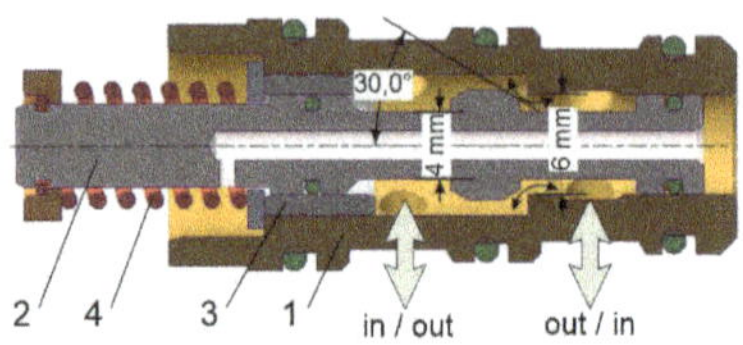

Figure 4. *NO* type seat valve; 1—body, 2—poppet, 3—sleeve, 4—spring.

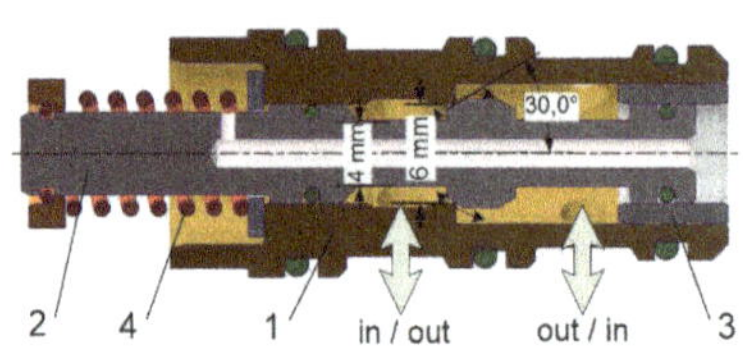

Figure 5. *NC* type seat valve; 1—body, 2—poppet, 3—sleeve, 4—spring.

3. Flow Analysis by Means of CFD Method

The analysis includes generating fluid model which comprises all the flow paths, creating discrete models, setting up parameters and carrying out simulations.

3.1. Geometrical Fluid Model

CFD analysis requires geometrical models of the fluid. The models correspond to geometry of the individual flow paths. The paths are separated geometrically, hence the analyses can be carried out independently of each other. The path models have been created in Creo Parametric system using *Boolean* operations for all considered fixed positions of seat valves.

A general view of the obtained fluid models is shown in Figure 6. Based on the valve scheme (Figure 3), the following flow paths can be distinguished: $P-A$, $P-B$, $B-T$ (Figure 7), as well as $A-T$ and $P-T$ (Figure 8). Each of the $P-A$, $P-B$ and $B-T$ paths contains one seat valve. In the $A-T$ path fluid flows through two seat valves, while in the case of the $P-T$ path, the flow occurs through all four seat valves in the parallel arrangement.

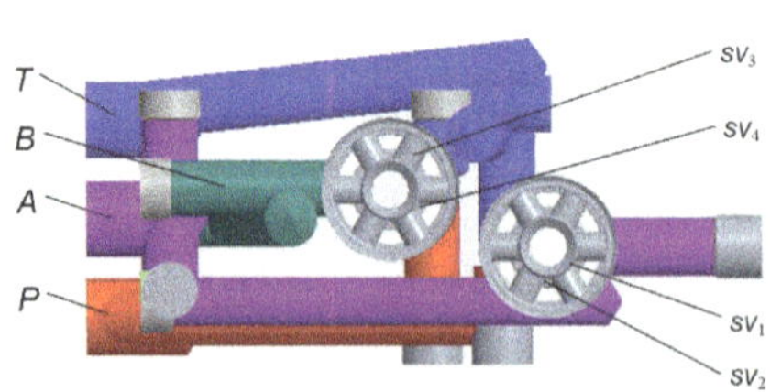

Figure 6. Fluid model (side view); *P*—supply channel, *A*, *B*—flow lines, *T*—return channel; $sv_1 \ldots sv_4$—seat valves.

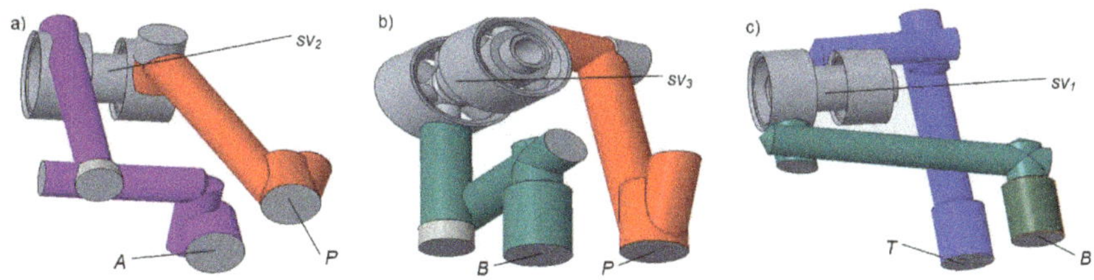

Figure 7. Fluid models of the flow lines: (**a**) $P - A$, (**b**) $P - B$, (**c**) $B - T$; sv_1, sv_2, sv_3—seat valves.

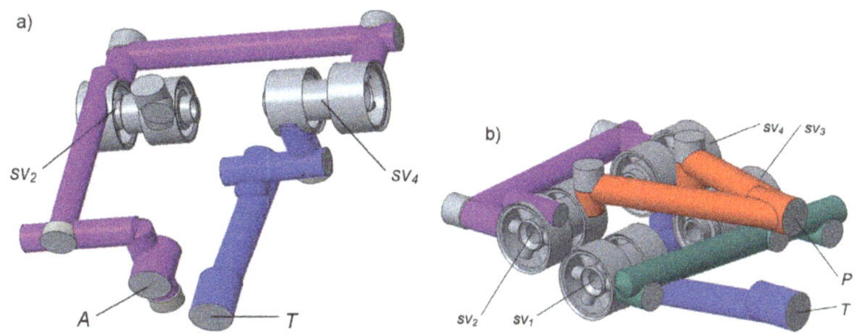

Figure 8. Fluid models of the flow lines: (**a**) $A - T$, (**b**) $P - T$; $sv_1, \ldots sv_4$—seat valves.

3.2. Discrete Model

The each created fluid model has one inlet surface and one outlet surface. Due to the complex geometry, the models cannot be simplified using axes or planes of symmetry. Therefore, the fluid flow should be treated as three-dimensional and turbulent.

All the models were meshed in the ANSYS/Fluent system using irregular elements, including tetrahedrons in the bulk flow and three layers of prisms at the boundaries. Number of generated cells varies from 250,000 for $P - B$ flow path (Figure 9a) to over 2,900,000 for $P - T$ (Figure 9b) due to its complexity. A pressure-velocity coupling was done by the *Pressure based* solver with the *Segregated* algorithm. Mesh parameters were adjusted on the base of the previous research results and according to general recommendations provided by ANSYS/Fluent. The following quality criteria of the mesh were achieved: minimal value of the orthogonal quality 0.70, skewness between 0.40 and 0.45 and the maximum aspect ratio close to 4.50. The maximum allowable absolute values of both mass and momentum residuals 10^{-3} were adopted as the convergence criteria.

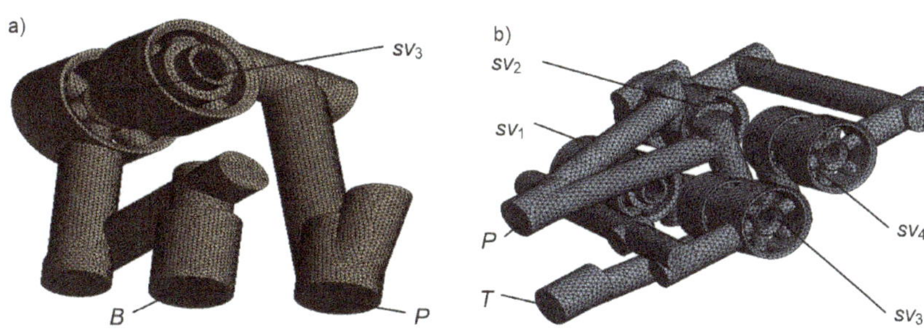

Figure 9. Meshed models: (**a**) $P - B$ line, (**b**) $P - T$ line; $sv_1, \ldots sv_4$—seat valves.

3.3. CFD Model Parameters

In order to carry out CFD simulation, first the kind of flow has to be determined. The Reynolds number calculated on the base of flow channel geometry, fluid viscosity and flow rate range varies from 10^3 to 10^4. Hence, the turbulent flow pattern was used. ANSYS/Fluent provides a wide set of turbulence models, including *Standard* $k-\omega$, *SST* $k-\omega$, *Standard* $k-\varepsilon$, *RNG* $k-\varepsilon$, *Realizable* $k-\varepsilon$ and others. Based on geometrical complexity of flow channels, relatively high Reynolds number value and a general assumption that the flow is considered over the entire cross section of the channels, the *Standard* $k-\varepsilon$ model was used in the research. This model was also used in similar research on the flow through control valves [8,14,23,27].

The main model parameters, including k—kinetic energy of the turbulence and ε—kinetic energy dissipation factor, can be calculated based on the transport equations Equations (1) and (2). The nature of turbulence is described using *Intensity* I and *Length scale* ℓ options, Equations (3) and (4), respectively.

$$\frac{\partial(\rho k)}{\partial t}+\frac{\partial(\rho k u_i)}{\partial x_i}=\frac{\partial}{\partial x_j}\left[\left(\mu+\frac{\mu_t}{\sigma_k}\right)\frac{\partial k}{\partial x_j}\right]+G_k+G_b-\rho\varepsilon-Y_M+s_k, \tag{1}$$

$$\frac{\partial(\rho \varepsilon)}{\partial t}+\frac{\partial(\rho \varepsilon u_i)}{\partial x_i}=\frac{\partial}{\partial x_j}\left[\left(\mu+\frac{\mu_t}{\sigma_k}\right)\frac{\partial \varepsilon}{\partial x_j}\right]+C_{1\varepsilon}\frac{\varepsilon}{k}\left(G_k+C_{3\varepsilon}G_b\right)-C_{2\varepsilon}\rho\frac{\varepsilon^2}{k}+s_\varepsilon, \tag{2}$$

$$I=0.16\cdot Re^{-0.125}, \tag{3}$$

$$\ell=0.07\cdot D_H, \tag{4}$$

where G_k is the increase in kinetic energy of turbulence, which is caused by gradient of average velocities, G_b and Y_M represent energy generated by the buoyancy phenomenon and the compressibility of fluid, respectively, Re is Reynolds number and D_H is relevant hydraulic diameter of the flow channel. The dimensionless constants were assigned using recommended values [28]: $s_k=1.00$, $s_\varepsilon=1.30$, $C_{1\varepsilon}=1.44$, $C_{2\varepsilon}=1.92$, $C_\mu=0.09$. Turbulent viscosity was calculated based on k and ε values:

$$\mu_t=\rho\cdot C_\mu\cdot k^2\cdot\varepsilon^{-1}, \tag{5}$$

where ρ is fluid density. The boundary conditions were defined using physical parameters of the fluid: average velocity in the inlet channel and pressure at the outlet. The velocity magnitude was set as normal to the boundary in the *Boundary Conditions/Velocity Specification Method* selection box. The outlet pressure was defined as a *Gauge Pressure* of value 0.1 MPa using the *Outlet condition* option. Table 2 summarizes physical parameters of the fluid and boundary conditions of the simulation model.

Table 2. Physical parameters and boundary conditions.

Fluid Dynamic Viscosity	Fluid Density	Fluid Temperature	Turbulence Intensity	Length Scale	Min Inlet Flowrate	Max Inlet Flowrate	Outlet Pressure
32.2 mPa s	870 $kg\,m^{-3}$	40.0 °C	5.1 %	0.42 mm	5.0 $dm^3\,min^{-1}$	25.0 $dm^3\,min^{-1}$	0.1 MPa

3.4. Results of CFD Simulations

Calculations were made for each of the distinguished flow paths ($P-T$, $P-A$, $P-B$, $A-T$ and $B-T$) with volumetric flow rate values $Q=5$, 10, 15, 20, 25 dm^3min^{-1}. Since there are seat valves in each flow path, and the valves are characterized by similar flow resistance values, the main differences occur due to the various length and geometry of both inlet and outlet flow channels. As an example, the results obtained for $P-A$ and $P-T$ flow paths in the form of velocity and pressure distributions are shown in Figures 10 and 11, respectively.

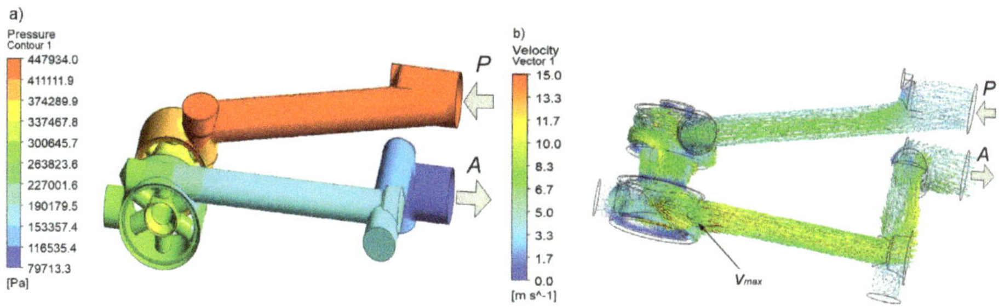

Figure 10. Simulation results obtained with the $P-A$ flow path at a flow rate $Q = 10\ \mathrm{dm^3 min^{-1}}$: (**a**) pressure distribution, (**b**) velocity distribution.

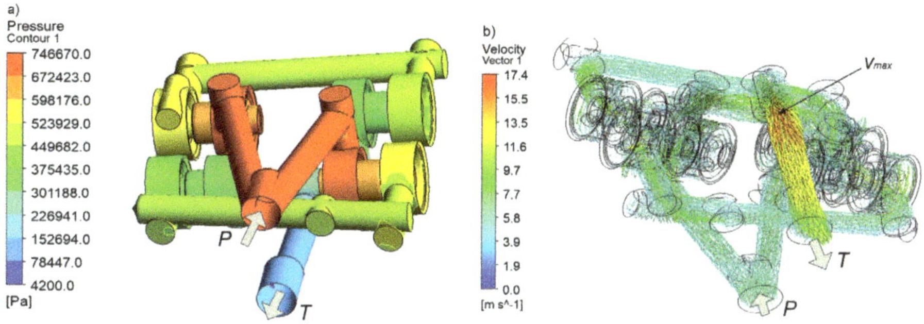

Figure 11. Simulation results obtained with the $P-T$ flow path at a flow rate $Q = 20\ \mathrm{dm^3 min^{-1}}$: (**a**) pressure distribution, (**b**) velocity distribution.

3.4.1. $P-A$ and $P-T$ Flow Paths

Pressure distribution on the walls of the $P-A$ flow path for $Q = 10\ \mathrm{dm^3 min^{-1}}$ is shown in Figure 10a. The inlet pressure is near to 0.45 MPa, and the highest pressure drop can be observed in the seat valve flow gaps, because of sudden step changes in cross-sectional areas of flow channels. This causes swirling and deflection of the fluid stream, as illustrated in Figure 10b, by means of velocity vectors. The maximum obtained fluid speed in the seat valve gap is 15.0 $\mathrm{m\,s^{-1}}$.

In the case of the $P-T$ flow path, the result obtained for $Q = 20\ \mathrm{dm^3 min^{-1}}$ is presented. In this case, despite the twice higher flow rate, the inlet pressure is less than 0.75 MPa, due to the fact, that pressure drop on two seat valves connected in series is compensated by two parallel flow paths. The maximum obtained fluid speed is 17.4 $\mathrm{m\,s^{-1}}$.

3.4.2. Determination of Flow Characteristics

The CFD analysis allowed flow resistance in all considered paths to be determined. Figure 12 shows a graph of the obtained pressure drop against flow rate. As it arises from the figure, the largest pressure drop occur in the $A-T$ path, which results from the largest length of flow channels and a serial flow through two seat valves. In contrast, the smallest drop occurs in the $P-T$ path because of a parallel flow through two pairs of seat valves.

The cross-sectional areas of the flow channels in the valve block are varied depending on the considered location. At the entrance flow area is determined by diameter of the connection port, then diameters of the channels and width of a throttling gap between the poppet and the seat valve body. The highest resistance occurs in the gap, where flow area is significantly decreased. Flow through the gap can be described by Equation (6):

$$Q = \mu_g\, A_g \sqrt{2\, \Delta p\, \rho^{-1}}. \tag{6}$$

With the known flow rate Q, fluid density ρ, gap area A_g and the determined value of discharge coefficient μ_g, a local pressure drop in the gap Δp can be calculated. However, in the case of the considered system, apart from the seat valve itself, pressure drop also occur in the flow channels. To estimate its influence, flow simulations of a singular seat valve were carried out. Figure 13 shows pressure (*a*) and velocity (*b*) distribution obtained at a flow rate $Q = 10\ \mathrm{dm^3 min^{-1}}$, respectively. The results indicate that change in the flow direction and the division of the fluid stream may increase value of the total pressure drop significantly.

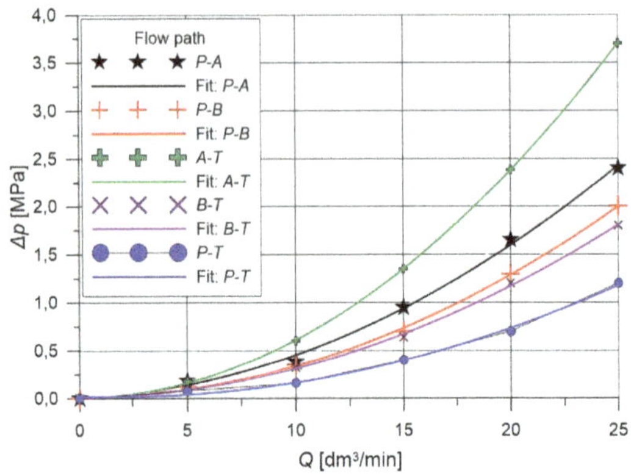

Figure 12. Pressure drop in individual flow paths (CFD analysis).

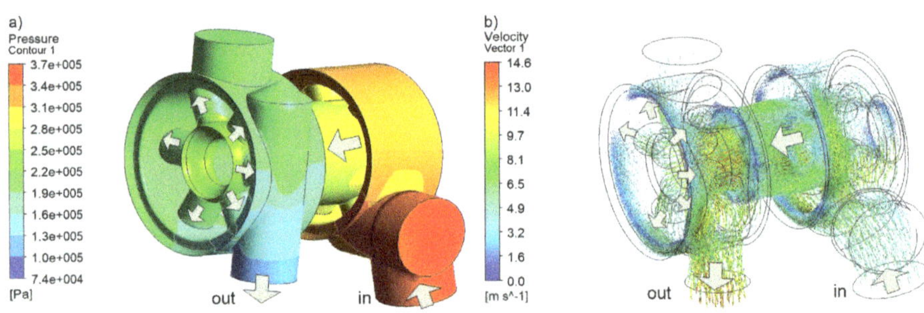

Figure 13. Simulation results obtained for a single seat valve at a flow rate $Q = 10\ \mathrm{dm^3 min^{-1}}$: (**a**) pressure distribution, (**b**) velocity distribution.

3.4.3. Discussion of Simulation Results

As part of the analysis of simulation results, pressure drop on the seat valve was compared with the one occurring in flow channels. Figure 14 shows the results obtained for considered flow paths with one seat valve: $P-A$, $P-B$ and $B-T$, respectively. As arises from the figure, the most significant flow resistance occurs on the seat valve itself, while the share of flow channels is between 5% ($P-B$) and 30% ($P-A$).

In contrast, Figure 15 presents the results obtained for $A-T$ and $P-T$ flow paths. The paths contain two seat valves connected in series. In the $A-T$ path, the share of flow resistance of the flow channels does not exceed 20%. On the other hand, flow resistance of the $P-T$ path is about 30% lower comparing to a single seat valve due to the flow through two parallel channels.

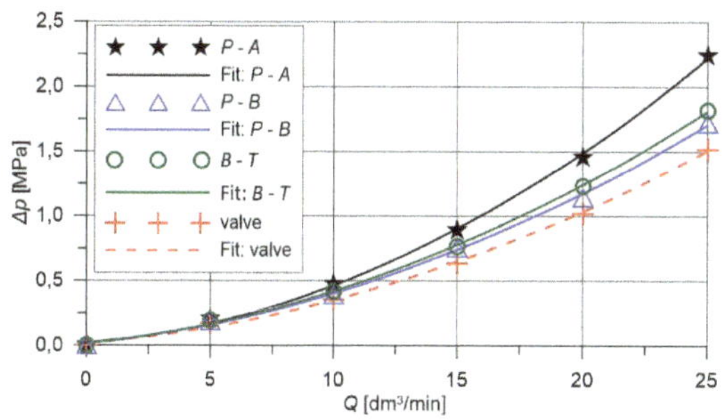

Figure 14. Comparison of pressure drop across a seat valve to $P-A$, $P-B$, $B-T$ flow paths.

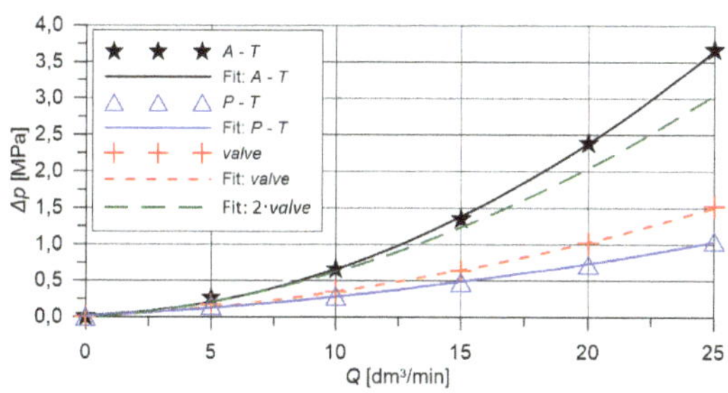

Figure 15. Comparison of pressure drop over the flow paths containing two seat valves connected in series; $A-T$ and $P-T$.

4. Laboratory Experiments

The results of CFD simulations were verified by means of laboratory experiments. The experiments were performed on a test bench, which was made according to the scheme shown in Figure 16. First a valve prototype was made (Figure 17a) and next the test bench was built (Figure 17b).

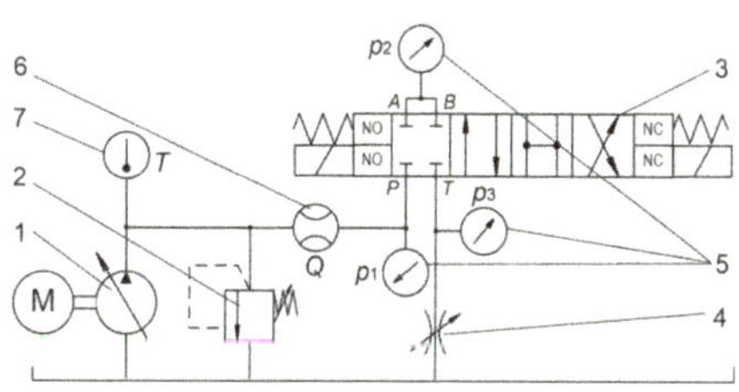

Figure 16. Test bench scheme: 1—pump, 2—relief valve, 3—tested 4/4UREZ6 valve, 4—throttle valve, 5—pressure transducers, 6—flow meter, 7—temperature gauge.

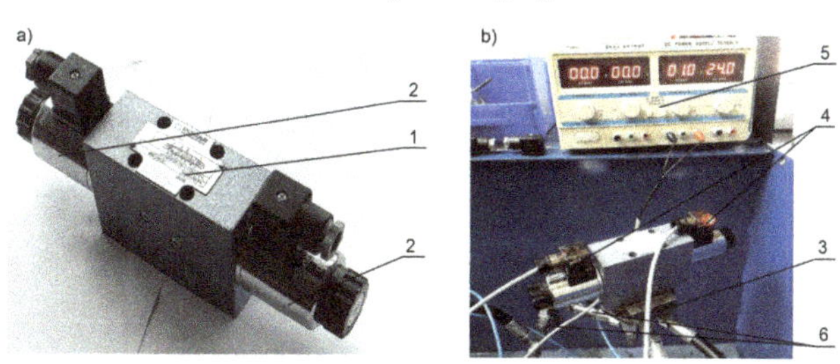

Figure 17. Laboratory equipment; (**a**) valve prototype, (**b**) test bench; 1—valve body, 2—electromagnet, 3—mounting plate, 4—power connectors, 5—power supply, 6—pressure transducers.

The test bench contained a variable flow rate pump, a flow meter and three pressure sensors with voltage transducers. A Kracht TM flow meter with a measuring range 0.1–80 $dm^3 min^{-1}$ and an accuracy class ±0.3% was used. Pressure sensors Trafag NAT100.0V had an accuracy class ±0.2% and a range 0–10 MPa. The data was acquired using a 16-bit DAQ card at a sampling rate of 100 Hz. The measurements were carried out for identical flow rates as CFD simulations, which allowed comparison of the results to be made. Figure 18a shows comparison of flow characteristics obtained for the $P-A$ flow path, while Figure 18b provides the analogous graphs for the $P-T$ line. The summary of all the results is presented in Table 3.

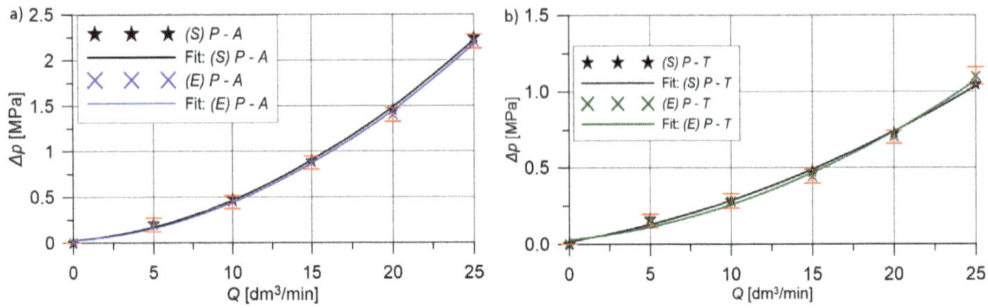

Figure 18. Comparison of simulation and experimental results of pressure drop; (**a**) $P-A$ flow path, (**b**) $P-T$ flow path; error bars—the measurement uncertainty.

Table 3. Comparison of simulation (S) and experimental (E) values of pressure drop Δp MPa.

Path Parameter	S/E	Q $dm^3 min^{-1}$					Average Diff.%
		5	10	15	20	25	
$P-A$ Δp (MPa)	S	0.21	0.47	0.89	1.46	2.24	4.71
	E	0.19	0.44	0.87	1.40	2.20	
$P-B$ Δp (MPa)	S	0.19	0.39	0.75	1.15	1.72	6.26
	E	0.18	0.37	0.70	1.20	1.85	
$B-T$ Δp (MPa)	S	0.20	0.42	0.76	1.23	1.82	5.13
	E	0.18	0.38	0.74	1.20	1.80	
$A-T$ Δp (MPa)	S	0.27	0.65	1.35	2.38	3.66	5.00
	E	0.25	0.60	1.35	2.30	3.40	
$P-T$ Δp (MPa)	S	0.16	0.29	0.48	0.73	1.05	5.04
	E	0.15	0.28	0.44	0.70	1.10	

It arises from the table, that the results obtained from CFD analysis do not differ from the laboratory experiments by more than 6.5%.

5. Conclusions

The article presents results of both CFD simulations and laboratory experiments of an innovative electromagnetically controlled hydraulic control valve adapted for installation in the ISO 4401 standard. The proposed solution relies on replacing the commonly used spool with four miniature seat poppet valves, two of which are normally closed and two normally opened in the default configuration. This results in an increase in the number of operating position from three to four and provides significantly higher tightness comparing to spool valves. Simulation model was built in ANSYS environment with the *Standard* $k-\varepsilon$ turbulence model applied. Simulations were made for the volumetric flow rate in the range of 5–25 $dm^3 min^{-1}$. The simulation results showed that the maximum pressure drop $\Delta p = 3.66$ MPa occurred on the $A-T$ flow path at the flow rate $Q = 25$ $dm^3 min^{-1}$. It has also been shown that the largest share in the pressure drop have seat valves, from 70% to

95% depending on the flow path. This proves that the flow channels are well profiled. Laboratory experiments were made for the same flow rate values as simulations. The each experiment consisted of setting a fixed flow rate value and measuring pressure drops at specific points of the flow paths. This allowed pressure drops to be calculated. High compliance of simulation results and experiments was obtained, the maximum difference did not exceed 6.5%. The following practical conclusions can be drawn from the research:

- the available four operating positions is of significant practical importance, especially in the context of energy saving. This is particularly important in systems where there is a need to cut off the flow in one position of an actuator and open it in the another one. In case when the actuator is locked in the working position under a pressure of $p = 20$ MPa and the flow rate of the relieved pump is 20 $dm^3\,min^{-1}$, the power consumption is instantaneously reduced by approximately 5 kW;
- the use of poppet seat valves ensures full tightness between the flow paths, which is difficult to achieve with spool valves. Assuming a typical operating pressure $p = 32$ MPa, the average power consumption can be reduced by 60 W due to avoiding the leaks;
- although the proposed solution control valve contains more elements than a typical spool valve, it is characterized by simpler geometry of flow channels and includes typical two-way cartridge valves which are easy to mount or replace;
- application of CAD methods together with CFD analysis is a useful and convenient tool in the search for new design solutions of hydraulic valves as well as introducing changes in the operating characteristics.

Author Contributions: Conceptualization, E.L. and J.R.; methodology, G.F.; software, E.L.; validation, G.F. and E.L.; formal analysis, J.R.; investigation, E.L., G.F. and J.R.; resources, J.R.; writing—original draft preparation, G.F.; writing—review and editing, E.L.; visualization, G.F.; supervision, E.L.; project administration, G.F.; funding acquisition, E.L. and G.F. All authors have read and agreed to the published version of the manuscript.

Funding: This research received no external funding.

Conflicts of Interest: The authors declare no conflict of interest.

Nomenclature

A_g	poppet gap area (m^2)
$C_{1\varepsilon}$, $C_{2\varepsilon}$, $C_{3\varepsilon}$	turbulence model constants (-)
D_H	hydraulic diameter (mm)
G_k, G_b, Y_M	energy components of turbulence model (J)
I	turbulence intensity (-)
k, ε	specific kinetic energy of turbulence and kinetic energy dissipation ($m^2\,s^{-2}$, $m^2\,s^{-3}$)
ℓ	turbulence length scale (mm)
p, Δp	pressure, pressure drop (MPa)
Q	volumetric flow rate ($dm^3\,min^{-1}$)
Re	Reynolds number (-)
s_k, s_ε	Prandtl numbers (-)
u_i	fluid velocity component ($m\,s^{-1}$)
μ	fluid dynamic viscosity (Pa s)
μ_g	poppet gap flow coefficient (-)
μ_t	eddy viscosity (-)
ρ	fluid density ($kg\,m^{-3}$)

References

1. Herakovic, N.; Duhovnik, J.; Simic, M. CFD simulation of flow force reduction in hydraulic valves. *Teh. Vjesn./Tech. Gaz.* **2015**, *22*, 453–463. [CrossRef]
2. Huovinen, M.; Kolehmainen, J.; Koponen, P.; Nissilä, T.; Saarenrinne, P. Experimental and numerical study of a choke valve in a turbulent flow. *Flow Meas. Instrum.* **2015**, *45*, 151–161. [CrossRef]

3. Šimic, M.; Debevec, M.; Herakovič, N. Modelling of Hydraulic Spool-Valves with Specially Designed Metering Edges. *Stroj. Vestn.-J. Mech. Eng.* **2014**, *60*, 77–83. [CrossRef]
4. Amirante, R.; Distaso, E.; Tamburrano, P. Sliding spool design for reducing the actuation forces in direct operated proportional directional valves: Experimental validation. *Energy Convers. Manag.* **2016**, *119*, 399–410. [CrossRef]
5. Park, S. Development of a proportional poppet-type water hydraulic valve. *Proc. Inst. Mech. Eng. Part C J. Mech. Eng. Sci.* **2009**, *223*, 2099–2107. [CrossRef]
6. Lisowski, E.; Filo, G. CFD analysis of the characteristics of a proportional flow control valve with an innovative opening shape. *Energy Convers. Manag.* **2016**, *123*, 15–28. [CrossRef]
7. Woldemariam, E.T.; Lemu, H.G.; Wang, G.G. CFD-Driven Valve Shape Optimization for Performance Improvement of a Micro Cross-Flow Turbine. *Energies* **2018**, *11*, 248. [CrossRef]
8. Gomez, I.; Gonzalez-Mancera, A.; Newell, B.; Garcia-Bravo, J. Analysis of the Design of a Poppet Valve by Transitory Simulation. *Energies* **2019**, *12*. [CrossRef]
9. Zhou, J.; Hu, J.; Jing, C. Lumped Parameter Modelling of Cavitating Orifice Flow in Hydraulic Systems. *Stroj. Vestn.-J. Mech. Eng.* **2016**, *62*, 373–380. [CrossRef]
10. Pan, X.; Wang, G.; Lu, Z. Flow field simulation and a flow model of servo-valve spool valve orifice. *Energy Convers. Manag.* **2011**, *52*, 3249–3256. [CrossRef]
11. Sapra, M.; Bajaj, M.; Kundu, S.; Sharma, B. Experimental and CFD investigation of 100 mm size cone flow elements. *Flow Meas. Instrum.* **2011**, *22*, 469–474. [CrossRef]
12. Ferreira, J.P.B.; Martins, N.M.; Covas, D.I. Ball Valve Behavior under Steady and Unsteady Conditions. *J. Hydraul. Eng.* **2018**, *144*, 04018005. [CrossRef]
13. Brunone, B.; Morelli, R. Automatic control valve-induced transients in an operative pipe system. *J. Hydraul. Eng.* **1999**, *125*, 534–542. [CrossRef]
14. Liao, Y.; Yuan, H.; Lian, Z.; Feng, J.; Guo, Y. Research and Analysis of the Hysteresis Characteristics of a Large Flow Directional Valve. *Stroj. Vestn.-J. Mech. Eng.* **2015**, *61*, 355–364. [CrossRef]
15. El-Adawy, M.; Heikal, M.R.; Aziz, A.R.; Siddiqui, M.I.; Munir, S. Characterization of the Inlet Port Flow under Steady-State Conditions Using PIV and POD. *Energies* **2017**, *10*, 1950. [CrossRef]
16. Rybarczyk, D.; Sedziak, D.; Owczarek, P.; Owczarkowski, A. Modelling of electrohydraulic drive with a valve controlled by synchronous motor. *Adv. Intell. Syst. Comput.* **2015**, *350*, 215–222.
17. Wos, P.; Dindorf, R. Practical parallel position-force controller for electro-hydraulic servo drive using on-line identification. *Int. J. Dyn. Control* **2016**, *4*, 52–58. [CrossRef]
18. Stump, P.M.; Keller, N.; Vacca, A. Energy Management of Low-Pressure Systems Utilizing Pump-Unloading Valve and Accumulator. *Energies* **2019**, *12*, 4423. [CrossRef]
19. Dindorf, R.; Wos, P. Force and position control of the integrated electro-hydraulic servo-drive. In Proceedings of the 20th International Carpathian Control Conference (ICCC), Krakow-Wieliczka, Poland, 26–29 May 2019. [CrossRef]
20. Tamburrano, P.; Plummer, A.R.; De Palma, P.; Distaso, E.; Amirante, R. A Novel Servovalve Pilot Stage Actuated by a Piezo-Electric Ring Bender (Part II): Design Model and Full Simulation. *Energies* **2020**, *13*, 2267. [CrossRef]
21. Lei, L.; Desheng, Z.; Jiyun, Z. Design and Research for the Water Low-pressure Large-flow Pilot-operated Solenoid Valve. *Stroj. Vestn.-J. Mech. Eng.* **2014**, *60*, 665–674. [CrossRef]
22. Jin, Z.; Hong, W.; You, T.; Su, Y.; Li, X.; Xie, F. Effect of Multi-Factor Coupling on the Movement Characteristics of the Hydraulic Variable Valve Actuation. *Energies* **2020**, *13*, 2870. [CrossRef]
23. Lisowski, E.; Filo, G. Analysis of a proportional control valve flow coefficient with the usage of a CFD method. *Flow Meas. Instrum.* **2017**, *53*, 269–278. [CrossRef]
24. Lisowski, E.; Filo, G.; Rajda, J. Analysis of flow forces in the initial phase of throttle gap opening in a proportional control valve. *Flow Meas. Instrum.* **2018**, *59*, 157–167. [CrossRef]
25. Filo, G.; Lisowski, E.; Rajda, J. Flow analysis of a switching valve with innovative poppet head geometry by means of CFD method. *Flow Meas. Instrum.* **2019**, *70*, 101643. [CrossRef]
26. ISO 4401:2005 Hydraulic Fluid Power—Four-Port Directional Control Valves—Mounting Surfaces. 2005. Available online: www.iso.org (accessed on 23 May 2020).

27. Lisowski, E.; Filo, G.; Rajda, J. Pressure compensation using flow forces in a multi-section proportional directional control valve. *Energy Convers. Manag.* **2015**, *103*, 1052–1064. [CrossRef]
28. *ANSYS-Fluent: Users Guide*, 13th ed.; ANSYS, Inc.: Canonsburg, PA, USA, 2011.

energies

MDPI

Article

Control of Optoelectronic Scanning and Tracking Seeker by Means the LQR Modified Method with the Input Signal Estimated Using of the Extended Kalman Filter

Daniel Gapiński [1,*] and Zbigniew Koruba [2]

[1] Faculty of OPBMR, War Studies University, al. gen. A. Chruściela "Montera" 103, 00-910 Warszawa-Rembertów, Poland
[2] Faculty of Mechatronics and Mechanical Engineering, Kielce University of Technology, Aleja Tysiąclecia Państwa Polskiego 7, 25-314 Kielce, Poland; zbikoruba@wp.pl
* Correspondence: tu_daniel_kielce@wp.pl

Abstract: The paper presents the concept of controlling the designed optoelectronic scanning and tracking seeker. The above device is intended for the so-called passive guidance of short-range anti-aircraft missiles to various types of air maneuvering targets. In the presented control method, the modified linear-quadratic regulator (LQR) and the estimation of input signals using the extended Kalman filter (EKF) were used. The LQR regulation utilizes linearization of the mathematical model of the above-mentioned seeker by means of the so-called Jacobians. What is more, in order to improve the stability of the seeker control, vector selection of signals received by the optoelectronic system was used, which also utilized EKF. The results of the research are presented in a graphical form. Numerical simulations were carried out on the basis of the author's own program developed in the programming language C++.

Keywords: simulation; mechatronics; control systems; guided missile; flight dynamics; LQR control; Kalman filter

Citation: Gapiński, D.; Koruba, Z. Control of Optoelectronic Scanning and Tracking Seeker by Means the LQR Modified Method with the Input Signal Estimated Using of the Extended Kalman Filter. *Energies* **2021**, *14*, 3109. https://doi.org/10.3390/en14113109

Academic Editors: Paolo Mercorelli and Andrea Mariscotti

Received: 22 March 2021
Accepted: 18 May 2021
Published: 26 May 2021

1. Introduction

One of the most important components of an anti-aircraft self-guiding infrared missile is the optoelectronic self-guiding seeker. This type of device is still the subject of intensive research in many scientific centers around the world [1–15]. The issue of this article refers to the publications [16,17] and is a continuation of the research conducted on the designed optoelectronic scanning and tracing seeker, presented in Figure 1.

The drive system of the designed scanning and tracking seeker is the rotor shown in Figure 1a. It is suspended in two rotating housings forming the so-called Cardan joint (Figure 1c). The rotor axis is the optical axis of the search and tracking system for a detected target. By means of the motors mounted in the individual housings (Figure 1b), control moments are applied to the rotating rotor, which makes it possible to change the position of its axis in space and thus to control the seeker. Figure 1d shows a 3D visualization of the complete seeker. Thanks to the 3D software, the mathematical and dynamics model and problem of moving parts are easier to solve [18]. Figure 2 shows the seeker set in the first operating mode in which the device scans the air space with the so called large angle of scanning $\beta = 1.92^{\circ}$.

Figure 3 shows the area of the airspace scanned by the seeker set in the first operating mode (the plane is scanned perpendicularly to the head axis).

Figure 4 shows the seeker set in the second operating mode, where the device scans the air space with a so-called small scanning angle $\beta = 0.28^{\circ}$.

Figure 5 shows the area of air space scanned by the seeker set in the second operating mode (the plane scanned is perpendicular to the seeker axis).

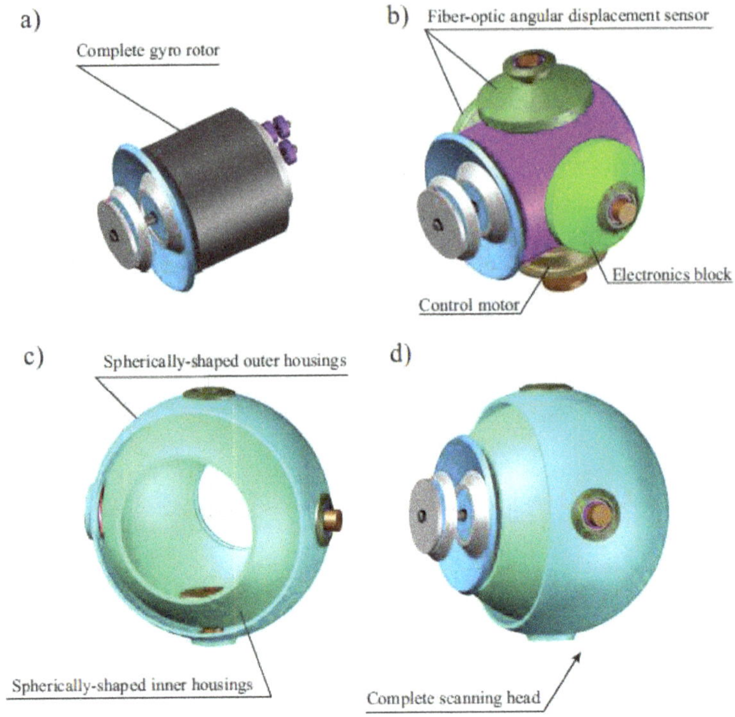

Figure 1. 3D view of the designed scanning and tracking seeker, where: (**a**) gyro rotor of scanning system; (**b**) sensors and control motors of the seeker; (**c**) spherically-shaped forming a cardan arrangement; (**d**) complete scanning head.

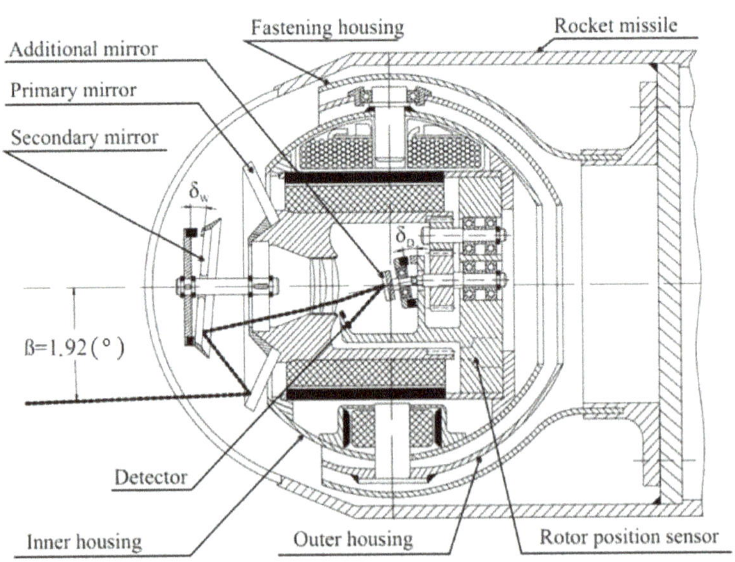

Figure 2. Scanning and tracking seeker set in the first operating mode.

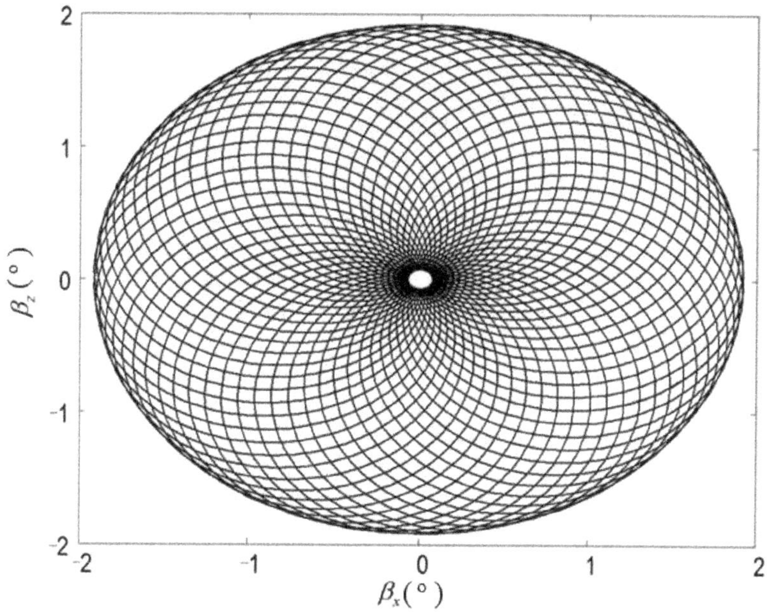

Figure 3. The area of scanned airspace in the first operating mode of the seeker.

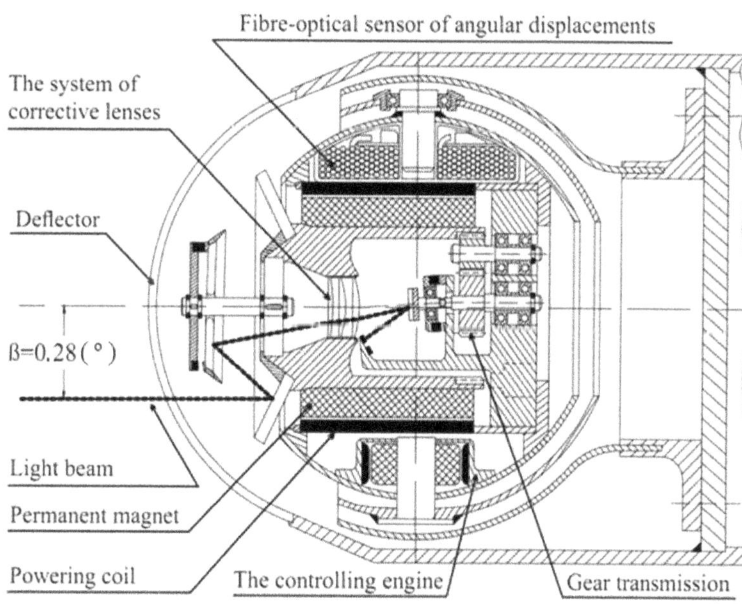

Figure 4. Scanning and tracking seeker set in the second operating mode.

The detailed principle of operation and innovation of the seeker is presented in [16,19]. At the present stage of research, a mathematical model of the dynamics of the presented device has been developed, various algorithms of control of the seeker's optical axis have been analysed in [16,20–23], and optimal operating parameters of the seeker have been determined while maintaining the stability conditions specified by the so-called Lapunov method [24]. In the course of the above-mentioned research, problems with precise control of the device axis in the so-called second operating mode of the seeker (Figure 4), in which the seeker tracks the previously detected air target with the small scanning angle (Figure 5).It should be noted that this type of solution for detection (space scanning) and tracking of the maneuvering air target is not described in the available literature. After a deeper analysis of the problem, it turned out that it is caused by too many pulses from infrared radiation emitted by the target that are received by the optoelectronic system. Too many detection pulses cause unfavorable overdriving of the seeker axis. It was, therefore, advisable to carry out additional filtering of signals received by the optoelectronic system. For this purpose, the so-called vector selection of signals received by the optoelectronic system was used, with a Kalman filter added [25–28]. Moreover, the so-called modified LQR control method was used to increase the precision of the seeker axis control. The results of this work are presented in subsequent points of this paper.

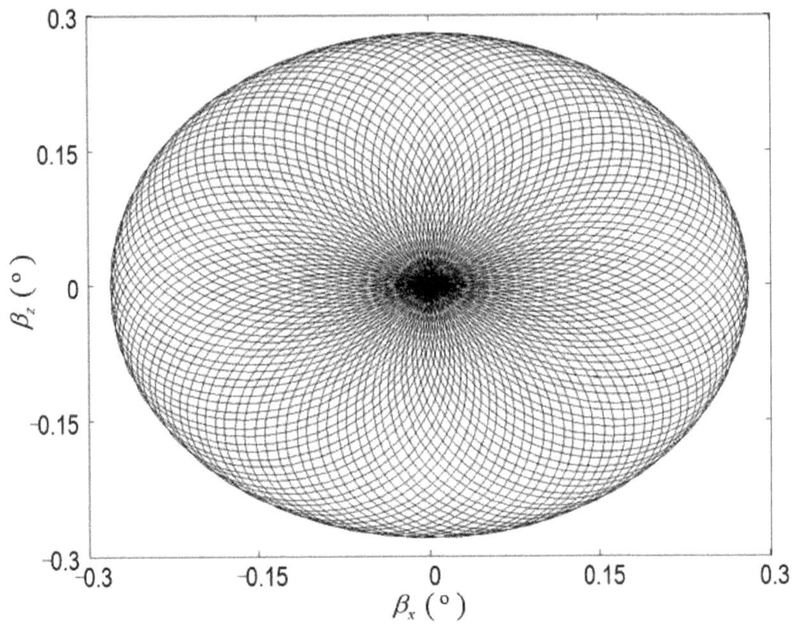

Figure 5. The area of scanned airspace in the second operating mode of the seeker.

2. Mathematical Model of the Scanning Seeker

Figure 6 shows the scanning seeker diagram together with the adopted coordinate systems and markings of individual angles of rotation of the respective systems in relation to each other. The origins of all coordinate systems are located at the intersection of the axis of rotation of the outer housing with the axis of rotation of the inner housing of the seeker. The movement of the seeker axis can be induced by moments of external forces M_Z and M_W forces generated by control motors or by moments of friction forces M_{TW} and M_{TZ} forces generated in the bearings of particular seeker housing as a result of angular displacement of the missile deck.

Angular movements of a missile are treated as external disturbances and are determined by the angular $\omega_{xp}, \omega_{yp}, \omega_{zp}$ velocities that cause the missile to rotate around the individual axes of the system x_P y_P z_P at the appropriate angles α_x α_y α_z. Angles ψ, ϑ are measured with fiber optic sensors (Figure 4) and angle φ is measured with the rotor position sensor (Figure 2).

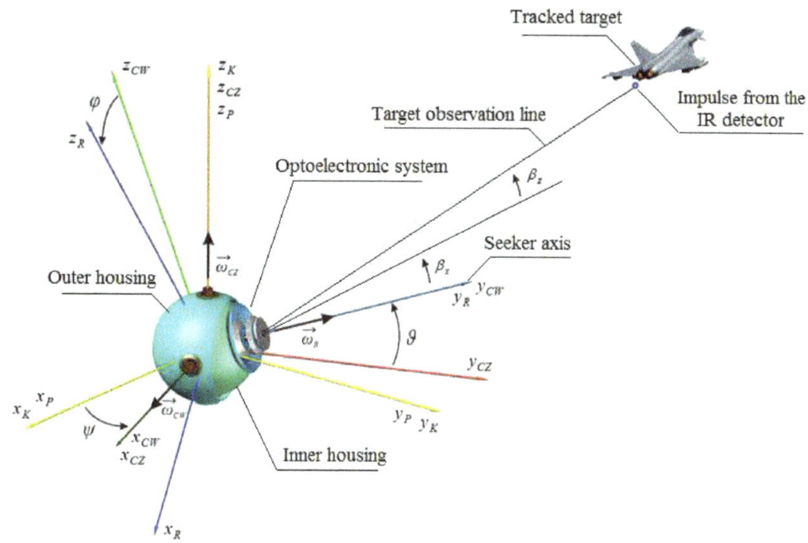

Figure 6. Seeker diagram with adopted coordinate systems.

The following coordinate systems have been introduced:

$x_K\ y_K\ z_K$—a coordinate system associated with the reference direction established in space;
$x_R\ y_R\ z_R$—a mobile coordinate system associated with the rotor;
$x_{CW}\ y_{CW}\ z_{CW}$—a mobile coordinate system associated with the inner housing;
$x_{CZ}\ y_{CZ}\ z_{CZ}$—a mobile coordinate system associated with the outer housing;
$x_P\ y_P\ z_P$—a mobile coordinate system associated with a missile;

The following marking of the angles of rotation has been adopted:

ψ-angle of rotation $x_{CZ}\ y_{CZ}\ z_{CZ}$ relative to $x_K\ y_K\ z_K$ around axis z_{CZ};
ϑ-angle of rotation $x_{CW}\ y_{CW}\ z_{CW}$ relative to $x_K\ y_K\ z_K$ around axis x_{CW};
φ-angle of rotation $x_R\ y_R\ z_R$ relative to $x_K\ y_K\ z_K$ around axis y_R;
α_x-angle of rotation $x_P\ y_P\ z_P$ relative to $x_K\ y_K\ z_K$ around axis x_P;
α_y-angle of rotation $x_P\ y_P\ z_P$ relative to $x_K\ y_K\ z_K$ around axis y_P;
α_z-angle of rotation $x_P\ y_P\ z_P$ relative to $x_K\ y_K\ z_K$ around axis z_P;

The following were assumed as given values:

$J_{x_{CZ}}, J_{y_{CZ}}, J_{z_{CZ}}$—moments of inertia of the complete outer housing,
$J_{x_{CW}}, J_{y_{CW}}, J_{z_{CW}}$—moments of inertia of the complete inner housing;
$J_{x_R}, J_{y_R}, J_{z_R}$—moments of inertia of the rotor;
$\vec{M}_Z$—the moment of forces of control motor's action on the outer housing;
$\vec{M}_W$—the moment of forces of control motor's action on the inner housing;
$\vec{\omega}_P(\omega_{x_P}, \omega_{y_P}, \omega_{z_P})$—angular velocity of the missile;
n-rotational speed of the rotor;
$\vec{M}_{TW}, \vec{M}_{TZ}$—the moments of the friction forces in the bearings of the inner and outer housing, hereby: $\vec{M}_{TW} = c_w\dot{\vartheta}$, $\vec{M}_{TZ} = c_z\dot{\psi}$, here: c_w is a coefficient of friction in the inner bowl bearing and c_z is a coefficient of friction in the outer bowl bearing.

Using the Lagrange II equation, the following gyroscope motion equations have been derived [29]:

$$\begin{gathered}(J_{x_{CW}} + J_{x_R})\ddot{\vartheta} + (J_{x_{CW}} + J_{x_R})\dot{\omega}_{x_{CZ}} - (J_{y_{CW}} - J_{z_{CW}} - J_{z_R})\omega_{y_{CW}}\omega_{z_{CW}} + \\ -J_{y_R} n \omega_{z_{CW}} = M_W - M_{TW}\end{gathered} \tag{1}$$

$$\begin{aligned}&\left[J_{z_{CZ}}+J_{z_{CW}}+J_{z_R}+\left(J_{y_{CW}}-J_{z_{CW}}-J_{z_R}\right)\sin^2\vartheta\right]\dot{\omega}_{z_{CZ}}+\\&+\tfrac{1}{2}\left(J_{y_{CW}}-J_{z_{CW}}-J_{z_R}\right)\sin 2\vartheta\left(\omega_{z_{CZ}}\dot{\vartheta}+\dot{\omega}_{y_{CZ}}\right)+\\&-\left[J_{z_{CW}}+J_{z_R}+\left(J_{y_{CW}}-J_{z_{CW}}-J_{z_R}\right)\sin^2\vartheta\right]\omega_{y_{CZ}}\dot{\vartheta}-\left(J_{z_{CW}}+J_{z_R}\right)\omega_{z_{CW}}\omega_{x_{CW}}\sin\vartheta+\\&+J_{y_{CW}}\omega_{y_{CW}}\omega_{x_{CW}}\cos\vartheta+J_{y_R}n\omega_{x_{CW}}\cos\vartheta-\left(J_{x_{CZ}}-J_{y_{CZ}}\right)\omega_{x_{CZ}}\omega_{y_{CZ}}+\\&-\left(J_{x_{CW}}+J_{x_R}\right)\omega_{x_{CW}}\omega_{y_{CZ}}=M_Z-M_{TZ}\end{aligned} \tag{2}$$

where the components of the angular velocity of the outer housing:

$$\begin{aligned}\omega_{x_{CZ}}&=\omega_{x_P}\cos\psi+\omega_{y_P}\sin\psi\\\omega_{y_{CZ}}&=-\omega_{x_P}\sin\psi+\omega_{y_P}\cos\psi\\\omega_{z_{CZ}}&=\dot{\psi}+\omega_{z_P}\end{aligned}$$

and the components of the angular velocity of the inner housing:

$$\begin{aligned}\omega_{x_{CW}}&=\omega_{x_{CZ}}+\dot{\vartheta}\\\omega_{y_{CW}}&=\omega_{y_{CZ}}\cos\vartheta+\omega_{z_{CZ}}\sin\vartheta\\\omega_{z_{CW}}&=-\omega_{y_{CZ}}\sin\vartheta+\omega_{z_{CZ}}\cos\vartheta\end{aligned}$$

Assuming that external kinematic impacts are negligible, we will obtain the following system of equations of motion of the seeker:

$$\left(J_{x_{CW}}+J_{x_R}\right)\ddot{\vartheta}-\frac{1}{2}\left(J_{y_{CW}}-J_{z_{CW}}-J_{z_R}\right)\dot{\psi}^2\sin 2\vartheta-J_{y_R}n\dot{\psi}\cos\vartheta+c_w\dot{\vartheta}=M_W \tag{3}$$

$$\begin{aligned}&\left[J_{z_{CZ}}+J_{z_{CW}}+J_{z_R}+\left(J_{y_{CW}}-J_{z_{CW}}-J_{z_R}\right)\sin^2\vartheta\right]\ddot{\psi}+\left(J_{y_{CW}}-J_{z_{CW}}-J_{z_R}\right)\dot{\psi}\dot{\vartheta}\sin 2\vartheta+\\&+J_{y_R}n\dot{\vartheta}\cos\vartheta+c_z\dot{\psi}=M_Z\end{aligned} \tag{4}$$

3. LQR Control of the Scanning Seeker

In this article, the authors proposed to control the seeker axis by means of a modified linear-quadratic regulator (LQR). This method can be used to determine such control that minimizes the integral quality indicator, given by the formula:

$$J=\int_0^\infty\left[\mathrm{x}^T\mathrm{Q}\mathrm{x}+\mathrm{u}^T\mathrm{R}\mathrm{u}\right]dt \tag{5}$$

where **Q** = matrix of state variable weights, **R** = matrix of control weights, **x**=state vector, **u** = $[M_W - M_{TW}]^T$ – control vector.

Q and **R** matrices are diagonal weight matrices that can be used to change the influence of particular state variables and controls on the presented quality criterion. The advantage of this method is that the entire state vector is the set point value, not just its selected values, as is the case with other controllers (e.g., PID) [30–33].

LQR regulation requires linearization and discretisation of state equations. Jacob's matrix—a matrix of successive partial derivatives—was used in the process of linearization.

To Equations (3) and (4), we introduce the signs:

$$J_{s1}=J_{x_{CW}}+J_{x_R},\ J_{s2}=J_{y_{CW}}-J_{z_{CW}}-J_{z_R},\ J_{s3}=J_{z_{CZ}}+J_{z_{CW}}+J_{z_R}$$

$$x_1=\vartheta,\ x_2=\dot{\vartheta},\ x_3=\psi,\ x_4=\dot{\psi} \tag{6}$$

thanks to which we get a nonlinear system of equations:

$$\dot{\mathrm{x}}=f(\mathrm{x}) \tag{7}$$

where:

$$\dot{\mathrm{x}}=\begin{bmatrix}\dot{x}_1 & \dot{x}_2 & \dot{x}_3 & \dot{x}_4\end{bmatrix}^T$$

$$\begin{bmatrix} \dot{x}_1 \\ \dot{x}_2 \\ \dot{x}_3 \\ \dot{x}_4 \end{bmatrix} = \begin{bmatrix} x_2 \\ \frac{1}{2}\frac{J_{s2}}{J_{s1}} x_4^2 \sin 2x_1 + \frac{J_{yR}}{J_{s1}} n x_4 \cos x_1 - \frac{c_w x_2}{J_{s1}} + \frac{M_W}{J_{s1}} \\ x_4 \\ -\frac{J_{s2} x_4 x_2 \sin 2x_1}{(J_{s3}+J_{s2}\sin^2 x_1)} - \frac{J_{yR} n x_2 \cos x_1}{(J_{s3}+J_{s2}\sin^2 x_1)} - \frac{c_z x_4}{(J_{s3}+J_{s2}\sin^2 x_1)} + \\ + \frac{M_Z}{(J_{s3}+J_{s2}\sin^2 x_1)} \end{bmatrix}.$$

Then, in the above system, the components dependent on the so-called owndynamics of the system (state variables) and the components dependent on external actions (control moments) will be separated, as shown below:

$$\begin{bmatrix} \dot{x}_1 \\ \dot{x}_2 \\ \dot{x}_3 \\ \dot{x}_4 \end{bmatrix} = \begin{bmatrix} f_1 \\ f_2 \\ f_3 \\ f_4 \end{bmatrix} + \begin{bmatrix} z_1 \\ z_2 \\ z_3 \\ z_4 \end{bmatrix} = \\ = \begin{bmatrix} x_2 \\ \frac{1}{2}\frac{J_{s2}}{J_{s1}} x_4^2 \sin 2x_1 + \frac{J_{yR}}{J_{s1}} n x_4 \cos x_1 \\ x_4 \\ -\frac{J_{s2} x_4 x_2 \sin 2x_1}{(J_{s3}+J_{s2}\sin^2 x_1)} - \frac{J_{yR} n x_2 \cos x_1}{(J_{s3}+J_{s2}\sin^2 x_1)} \end{bmatrix} + \\ + \begin{bmatrix} 0 \\ -\frac{c_w x_2}{J_{s1}} + \frac{M_W}{J_{s1}} \\ 0 \\ -\frac{c_z x_4}{(J_{s3}+J_{s2}\sin^2 x_1)} + \frac{M_Z}{(J_{s3}+J_{s2}\sin^2 x_1)} \end{bmatrix} \quad (8)$$

where:

f_i = a component dependent on the own dynamics of the system,
z_i = a component dependent on control and external interference:

The control law takes the form:

$$u = K(x_Z - x) \quad (9)$$

where x_Z is the matrix of set state variables, while the matrix of amplification is calculated from the dependency:

$$K = (R + B^T PB)^{-1} B^T PA \quad (10)$$

where **P** matrix is the solution to *Riccati* discrete equation [34,35]:

$$A^T P + PA - PBR^{-1} B^T P = 0 \quad (11)$$

Selection of LQR regulator settings consists in the determination of the **Q** and **R** weights matrices. The LQR algorithm does not have a universal method for selecting the above parameters and they are usually iteratively selected. In this paper, when selecting the initial values of **Q** and **R** matrices, the authors used the Bryson [36] rule, which suggests the selection of the following input parameters:

$$Q_{ii} = \frac{1}{x_{ii}^2} \quad (12)$$

$$R_{ii} = \frac{1}{u_{ii}^2} \quad (13)$$

where i-means another element of the state vector; x_{ii}—these are the maximum values for individual elements of the state vector **x**; u_{ii}—these are the maximum control moments.

The maximum operating parameters of the seeker were determined using the Lapunov method [24], and they are respectively:

$$\mathrm{u}_{11} = \mathrm{u}_{22} = 1.5\ (\mathrm{N}\cdot\mathrm{m}),\ \mathrm{x}_{11} = 0.5(^{\circ}),\ \mathrm{x}_{22} = 40\ (^{\circ}/\mathrm{s}),\ \mathrm{x}_{33} = 0.5(^{\circ}), x_{44} = 40(^{\circ}/s)$$

Q and R weights matrices:

$$\mathrm{Q} = \begin{bmatrix} \frac{1}{0.5^2} & 0 & 0 & 0 \\ 0 & \frac{1}{40^2} & 0 & 0 \\ 0 & 0 & \frac{1}{0.5^2} & 0 \\ 0 & 0 & 0 & \frac{1}{40^2} \end{bmatrix}$$

$$\mathrm{R} = \begin{bmatrix} \frac{1}{1.5^2} & 0 \\ 0 & \frac{1}{1.5^2} \end{bmatrix}$$

The P matrix was determined by numerically solving Riccati discrete equations according to the formula:

$$\mathrm{P}_{\mathrm{j}-1} = \mathrm{Q} + \mathrm{A}^{\mathrm{T}}(\mathrm{P}_{\mathrm{j}} - \mathrm{P}_{\mathrm{j}}\mathrm{B}(\mathrm{R} + \mathrm{B}^{\mathrm{T}}\mathrm{P}\mathrm{B})^{-1}\mathrm{B}^{\mathrm{T}}\mathrm{P})\mathrm{A} \tag{14}$$

Matrix $\mathbf{P}_{\mathrm{j}-1}$, according to the above formula, is calculated iteratively from the back. $\mathbf{P}_{\mathrm{j}}$ = $\mathbf{Q}$ is assumed as the input value. Jacobians were used to determine the state matrix **A** and control matrix **B** [37]. The individual elements of the matrix **A** are given according to the dependency:

$$\mathbf{A} = \begin{bmatrix} \frac{\partial f_1}{\partial x_1} & \frac{\partial f_1}{\partial x_2} & \frac{\partial f_1}{\partial x_3} & \frac{\partial f_1}{\partial x_4} \\ \frac{\partial f_2}{\partial x_1} & \frac{\partial f_2}{\partial x_2} & \frac{\partial f_2}{\partial x_3} & \frac{\partial f_2}{\partial x_4} \\ \frac{\partial f_3}{\partial x_1} & \frac{\partial f_3}{\partial x_2} & \frac{\partial f_3}{\partial x_3} & \frac{\partial f_3}{\partial x_4} \\ \frac{\partial f_4}{\partial x_1} & \frac{\partial f_4}{\partial x_2} & \frac{\partial f_4}{\partial x_3} & \frac{\partial f_4}{\partial x_4} \end{bmatrix} \tag{15}$$

After calculating the partial derivatives, further elements of the matrix **A** were determined, represented by the following equations:

$$\frac{\partial f_1}{\partial x_1} = 0,\ \frac{\partial f_1}{\partial x_2} = 1,\ \frac{\partial f_1}{\partial x_3} = 0,\ \frac{\partial f_1}{\partial x_4} = 0 \tag{15a}$$

$$\frac{\partial f_2}{\partial x_1} = \frac{J_{s2}x_4^2\cos 2x_1 - J_{y_R}nx_4\sin x_1}{J_{s1}},\ \frac{\partial f_2}{\partial x_2} = 0,\ \frac{\partial f_2}{\partial x_3} = 0 \tag{15b}$$

$$\frac{\partial f_2}{\partial x_4} = \frac{J_{s2}x_4\sin 2x_1 + J_{y_R}n\cos x_1}{J_{s1}} \tag{15c}$$

$$\frac{\partial f_3}{\partial x_1} = 0,\ \frac{\partial f_3}{\partial x_2} = 0,\ \frac{\partial f_3}{\partial x_3} = 0,\ \frac{\partial f_3}{\partial x_4} = 1 \tag{15d}$$

$$\begin{aligned} \frac{\partial f_4}{\partial x_1} = & \frac{-2J_{s2}x_2x_4\cos 2x_1 + J_{y_R}nx_2\sin x_1}{(J_{s3}+J_{s2}\sin^2 x_1)} + \\ & + \frac{J_{s2}^2x_2x_4 2\sin x_1\cos x_1\sin 2x_1 + J_{s2}J_{y_R}nx_2\cos^2 x_1 2\sin x_1}{(J_{s3}+J_{s2}\sin^2 x_1)^2} \end{aligned} \tag{15e}$$

$$\frac{\partial f_4}{\partial x_2} = \frac{-J_{s2}x_4\sin 2x_1 - J_{y_R}n\cos x_1}{(J_{s3} + J_{s2}\sin^2 x_1)} \tag{15f}$$

$$\frac{\partial f_4}{\partial x_3} = 0 \tag{15g}$$

$$\frac{\partial f_4}{\partial x_4} = \frac{-J_{s2}x_2\sin 2x_1}{(J_{s3} + J_{s2}\sin^2 x_1)} \tag{15h}$$

The individual elements of the matrix **B** are given according to the dependency:

$$\mathbf{B} = \begin{bmatrix} \frac{\partial z_1}{\partial u_1} & \frac{\partial z_1}{\partial u_2} \\ \frac{\partial z_1}{\partial u_1} & \frac{\partial z_1}{\partial u_2} \\ \frac{\partial z_1}{\partial u_1} & \frac{\partial z_1}{\partial u_2} \\ \frac{\partial z_1}{\partial u_1} & \frac{\partial z_1}{\partial u_2} \end{bmatrix} \tag{16}$$

After calculating the partial derivatives, the individual elements of the control matrix **B** were obtained, represented by the following equations:

$$\frac{\partial z_1}{\partial u_1} = 0, \ \frac{\partial z_1}{\partial u_2} = 0, \tag{16a}$$

$$\frac{\partial z_2}{\partial u_1} = 0, \ \frac{\partial z_2}{\partial u_2} = \frac{1}{J_{s1}}, \tag{16b}$$

$$\frac{\partial z_3}{\partial u_1} = 0, \frac{\partial z_3}{\partial u_2} = 0, \tag{16c}$$

$$\frac{\partial z_4}{\partial u_1} = \frac{1}{(J_{s3} + J_{s2}\sin^2 x_1)}, \ \frac{\partial z_4}{\partial u_2} = 0. \tag{16d}$$

4. Vector Filtration of Control Signals by Means of the Extended Kalman Filter

The accurate angle measurement of the detected object has a significant impact on the accuracy of its tracking [38–42]. In order to correctly determine the angular position of the detected object, it should be determined the law of airspace scanning by the optoelectronic system.

The law of airspace scanning by the optoelectronic system of the seeker is presented in the paper [19]. On its basis, linear equations describing the model of the scanning process were derived:

$$\beta_X(t) = a\tan(\tan(\beta(t)) \cdot \cos(a\sin(z_{zp}(t)/\sqrt{x_{zp}(t)^2 + z_{zp}(t)^2})) \tag{17}$$

$$\beta_Z(t) = a\tan(\tan(\beta(t)) \cdot \sin(a\sin(z_{zp}(t)/\sqrt{x_{zp}(t)^2 + z_{zp}(t)^2})) \tag{18}$$

where:

$\beta_X(t)$, $\beta_Z(t)$-angular coordinates of the detected target relative to the axis of the scanning seeker;

$\beta(t)$-resultant angle of deflection of the light beam from the optical axis;

x_{zp} , z_{zp}-the components of the position of the light beam on the plane of the original mirror.

Angular coordinates of the detected target β_X , β_Z are measured with respect to the optical axis of the seeker. These coordinates are the position desired to control the axis of the seeker so that it tracks the detected target.

Due to the high scanning density, especially in the second operating mode of the seeker (see Figure 5), there is a large number of pulses received from the infrared detector, which causes an unfavorable overdriving of the seeker axis. For the reasons mentioned above, it was necessary to apply appropriate filtration. Although a large number of control signals do not cause losing track of the target, it has a negative effect on the precision of the control. The method of filtering signals received by the optoelectronic system of the seeker presented in the paper is divided into two stages:

- selecting the maximum signal (pulse),
- performing additional filtration of the determined maximum signals using the EKF.

Figure 7 shows a diagram of filtering the signals received by the optoelectronic system of the seeker.

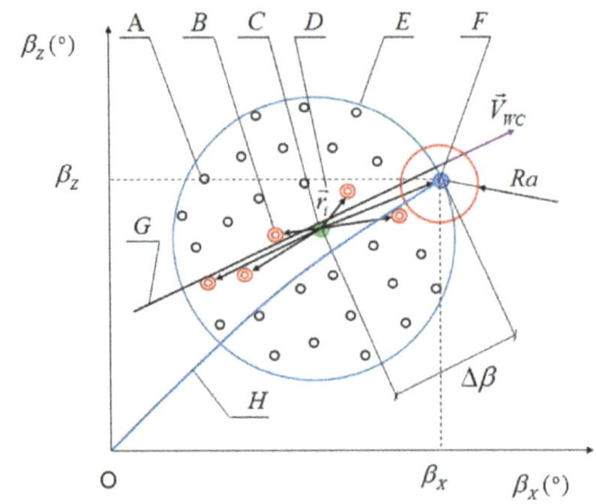

Figure 7. Diagram of filtering signals received by the optoelectronic system of the seeker. O, current position of the seeker axis; A, infrared detector pulses; B, selected maximum pulses; C, currently set angular position for controlling the seeker axis; D, status vectors (one of the Kalman filter criteria); E, selection area (one of the Kalman filter criteria); F, next, set angle position for controlling the seeker axis (result of filtration operation); G, target flight trajectory; H, set trajectory for seeker axis movement; Ra, corrective lens system visual field radius; $\Delta\beta$, variable coefficient depending on the speed vector value of the detected V_{WC} target.

Based on the series of pulses from the infrared detector, only those for which the voltage value is the highest, i.e., theoretically the closest to the source of infrared radiation, are taken into account. These are the so-called maximum pulses marked in Figure 7 with the symbol B. At the next stage of selection, the Kalman filter was used, in which a variable coefficient $\Delta\beta$ was adopted as one of the quality criteria (see Figure 7), depending on the value of the V_{WC} velocity vector. The coefficient $\Delta\beta$ varies from 1.5 *Ra* to 6 *Ra*, where *Ra* is the radius of the visual field of the seeker corrective lens system. The algorithm according to which the Kalman filter works is divided into two stages. The first stage is called prediction and the second stage is called correction. During prediction, the velocity vector of the detected target is estimated based on the previous coordinates of the detected target [43–45].

Estimated values of the direction and orientation of the target velocity vector are additional quality criteria for filtering those maximum signals whose vectors have the opposite direction and orientation compared to the V_{WC} vector. Vectors of measurement signals are marked with the "r_i" symbol in Figure 7.

The prediction of the direction and orientation values of the target velocity vector is based on the matrix of coordinates of the detected target:

$$\beta = \begin{bmatrix} \beta_X(t_0) & \beta_Z(t_0) \\ \beta_X(t_1) & \beta_Z(t_1) \\ \beta_X(t_2) & \beta_Z(t_2) \\ \cdots & \cdots \\ \beta_X(t_i) & \beta_Z(t_i) \end{bmatrix} \quad (19)$$

where i = number of target detection pulses.

Equations describing the estimated velocity vector of a detected air target:

$$v_{WC}(t) = \frac{\sqrt{(tg(\beta_X(t_i) - \beta_X(t_i - \Delta t)))^2 + (tg(\beta_Z(t_i) - \beta_Z(t_i - \Delta t)))^2}}{t_i - \Delta t} \quad (20)$$

$$\gamma_{WC}(t) = \frac{\arccos(tg(\beta_{X_{SR}})}{\sqrt{(tg(\beta_{X_{SR}} - \beta_X(t_i)))^2 + (tg(\beta_{Z_{SR}} - \beta_{Z_K}(t_i)))^2}} \tag{21}$$

$$\beta_{X_{SR}} = \frac{\left(\sum_1^i \beta_{X_i}\right)}{i} \tag{22}$$

$$\beta_{Z_{SR}} = \frac{\left(\sum_1^i \beta_{Z_i}\right)}{i} \tag{23}$$

where: ν_{WC} = estimated value of the target velocity vector, γ_{WC} = estimated direction of the target velocity vector, t_i = time of measurement of consecutive pulses from the infrared detector, Δt = reverse time interval (in numerical simulations, it is the time of about 10 detection pulses).

In the next stage of filtration, called correction, the final control signal is determined (Figure 7, point E), for which the value of the determined "r_i" vector is greater than or equal to the quality $\Delta\beta$ coefficient. Signals selected in this way β_X, β_Z have been used to control the seeker axis and thereby track the detected air target, as described in the next chapter of the paper.

5. Results

The studies were carried out for different air situations. Numerical simulations were carried out on the basis of the author's own program developed in the C++ language.

5.1. Scanning Seeker Parameters

Moments of rotor inertia:

$$J_{x_R} = 0.00114143\ (\text{kg}\cdot\text{m}^2);\ J_{y_R} = 0.00157911\ (\text{kg}\cdot\text{m}^2);\ J_{z_R} = 0.00158234\ (\text{kg}\cdot\text{m}^2)$$

Moments of inertia of the complete inner housing:

$$J_{x_{CW}} = 0.0016663\ (\text{kg}\cdot\text{m}^2);\ J_{y_{CW}} = 0.0011666\ (\text{kg}\cdot\text{m}^2);\ J_{z_{CW}} = 0.0011463\ (\text{kg}\cdot\text{m}^2)$$

Moments of inertia of the complete outer housing:

$$J_{x_{CZ}} = 0.0003383\ (\text{kg}\cdot\text{m}^2);\ J_{y_{CZ}} = 0.0002213\ (\text{kg}\cdot\text{m}^2);\ J_{z_{CZ}} = 0.0002583\ (\text{kg}\cdot\text{m}^2)$$

Rotational speed of the rotor:

$n = 600(\text{rad/s})$ (The speed and torque of the motor depend on the strength of the magnetic field generated by the energized windings of the motor, which depends on the current through them-may slightly differ from the fixed value [46]).

The coefficient of friction in the inner housing bearing:

$c_w = 0.05\ (\text{N}\cdot\text{m}\cdot\text{s})$

The coefficient of friction in the outer housing bearing:

$c_z = 0.05\ (\text{N}\cdot\text{m}\cdot\text{s})$

5.2. Results of the Simulation

Figure 8 shows a computer simulation image of the tracking of an air target moving at a speed of 350 m/s, located at a distance of 1600 m from the firing position, without the use of a filtration of pulses received by the optoelectronic system of the seeker.

Figure 9 shows a computer simulation image of tracking the same air target but using the signal filtering presented in Chapter 4. In both cases, the seeker axis is controlled by the method described in Chapter 3, using a modified linear-quadratic regulator (LQR).

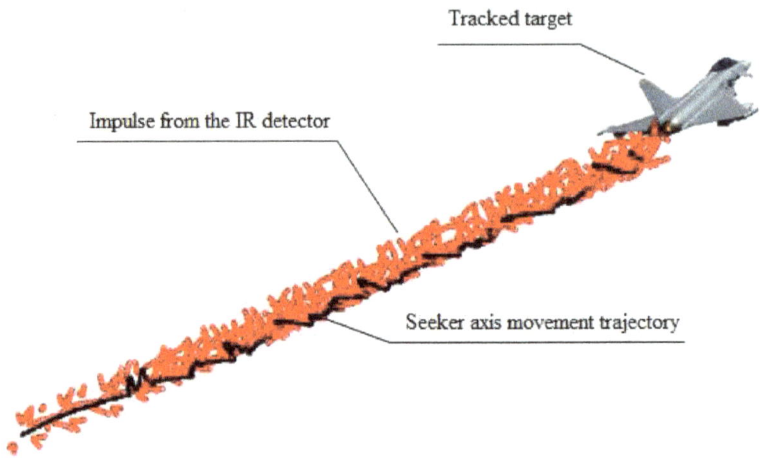

Figure 8. Tracking of a detected air target without signal filtering.

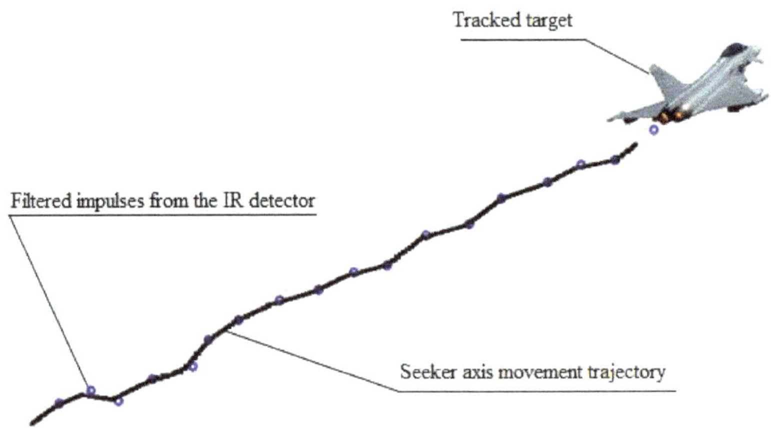

Figure 9. Tracking a detected air target using signal filtering.

For a better comparison of the simulations shown above, Figure 10 shows the set trajectory T_Z and the trajectory T_R pursued by the seeker axis when tracking a detected air target without signal filtering.

Figure 11 shows the same trajectories after signal filtering.

Figure 12 shows a computer simulation image of the seeker axis control in the airspace search phase and in the phase of tracking the detected target. Target speed: 250 m/s, target distance from fire station: 1100 m.

Description of the markings used in Figures 12–14:

A, scanning lines; B, trajectory of seeker axis motion in the programmatic airspace search phase; C, phase of seeker axis shifting to the detected target; D, tracking of the detected target.

Figure 13 shows the differences between the set trajectory and the trajectory pursued by the seeker axis when controlled with the use of the PID method, while Figure 14 shows the differences between the set trajectory and the trajectory pursued by the seeker axis when controlled with the use of modified LQR method.

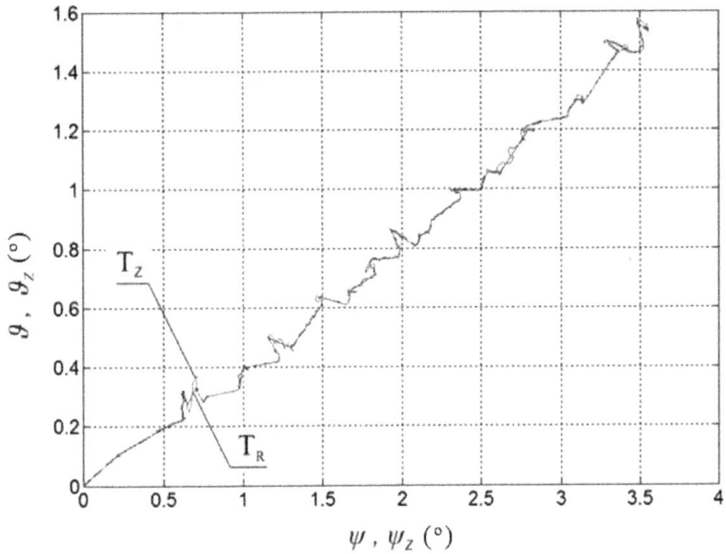

Figure 10. T_Z set trajectory and pursued T_R trajectory of the seeker axis without filtering the signals received from the infrared detector.

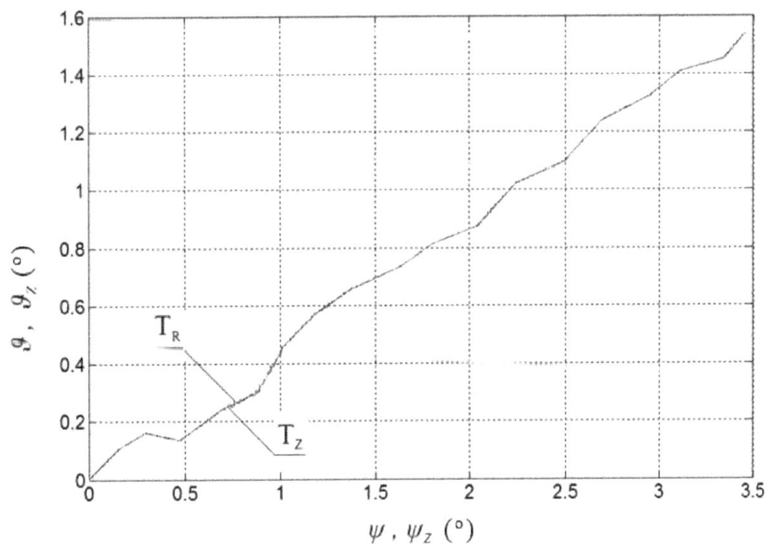

Figure 11. T_Z set trajectory and pursued T_R trajectory of the seeker axis with the filtration of signals received from the infrared detector.

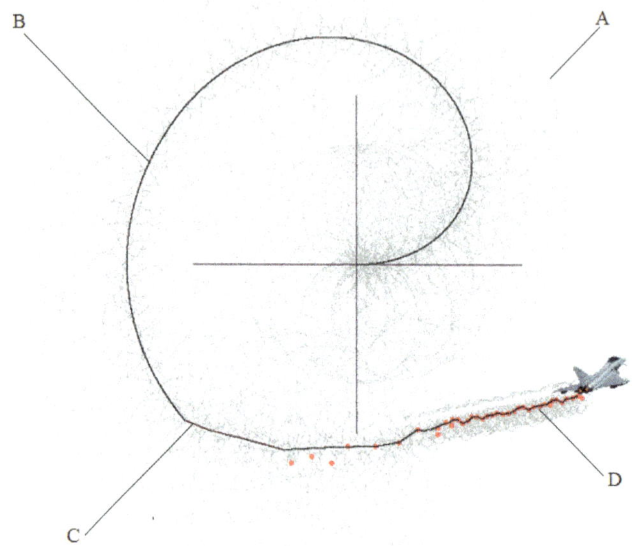

Figure 12. The process of modified LQR control of the seeker axis in the airspace search phase and in the phase of tracking a detected target.

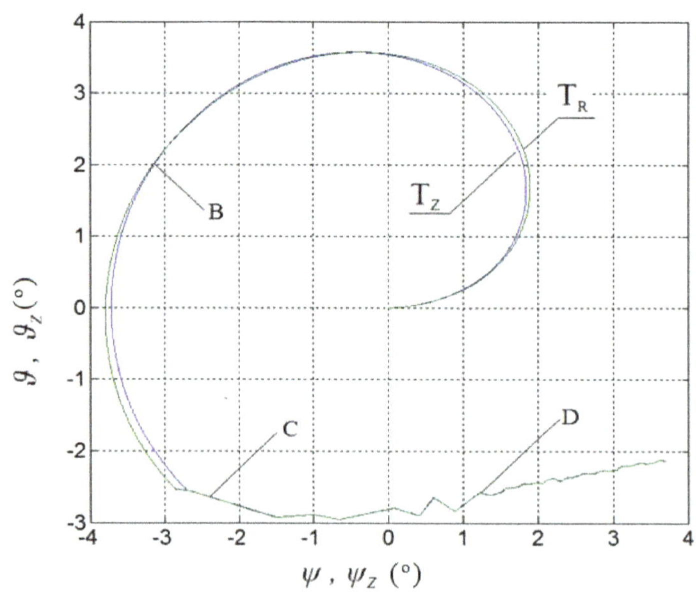

Figure 13. T_Z set trajectory and pursued T_R trajectory of the seeker axis in the airspace search phase and in the detected target tracking phase–PID control.

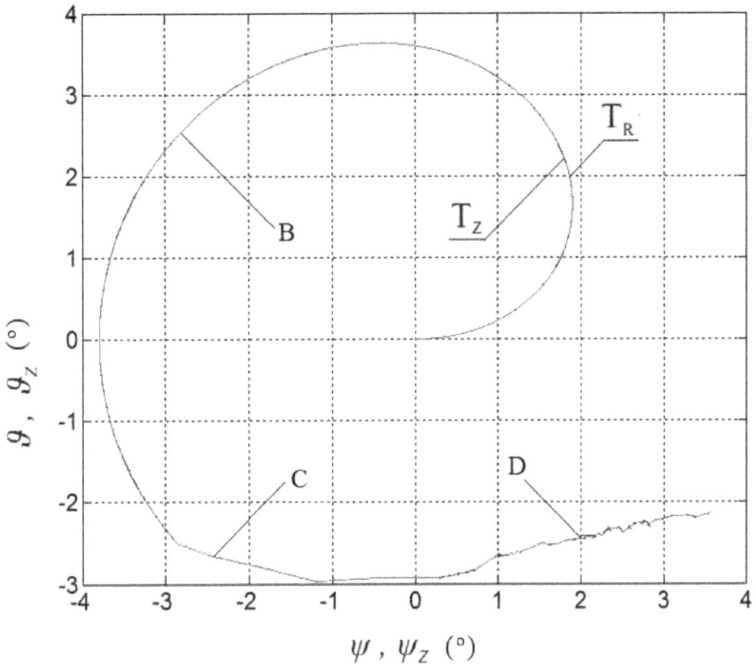

Figure 14. T_Z set trajectory and pursued T_R trajectory of the seeker axis in the airspace search phase and in the detected target tracking phase–modifiedLQR control.

6. Conclusions

The paper presents the application ofmodified LQR control and the estimation of input signals using Kalman filter for the process of detection and tracking of air targets.

LQR regulation uses linearization of the mathematical model of the tested scanning seeker with the use of the so-called Jacobians, while in order to improve the stability of the seeker's operation, vector selection of signals received by the optoelectronic system, which utilizes, among others, an extended Kalman filter, was used.

Computer simulations have shown that tracking of the maneuvering air target by the seeker being studied, using a Jacobian in a closed-loop control, is more precise than using the classical PID control method. The results also confirm the effectiveness of the developed method of filtering the signals received by the optoelectronic system of the presented seeker. After applying the vector selection of signals and Kalman's linear filter, we can clearly see a significant improvement in the stability of the trajectory of seeker axis motion.

In further research, statistical results will be presented and analyzed, which will be compared with the results obtained in this article. Moreover, in the future, it is planned to conduct research on the use of a more powerful filter, the "unscented Kalman filter", which was widely discussed in articles [47–51].

Author Contributions: Conceptualization, D.G. and Z.K.; methodology, D.G. and Z.K; software, D.G.; validation, D.G. and Z.K.; formal analysis, D.G and Z.K; investigation, D.G.; resources, D.G. and Z.K.; data curation, D.G. and Z.K; writing—original draft preparation, D.G.; writing—review and editing, D.G. and Z.K.; visualization, D.G.; supervision, D.G. All authors have read and agreed to the published version of the manuscript.

Funding: This research received no external funding.

Institutional Review Board Statement: Not applicable.

Informed Consent Statement: Not applicable.

Data Availability Statement: Not applicable.

Conflicts of Interest: The authors declare no conflict of interest.

References

1. LFK. Lenkflugköerper System GmbH, Infrared Seeker Head. U.S. Patent Number US20120248238A1, 4 October 2012.
2. Jenkins, J.; Diehl BGT Defence GmbH & Co. Verfahren Zum Steuern Eines Lenkflugkörper und Suchkopf für Einen Lenkflugkörper. Deutsches Patent—Und Markenamt DE102010055493A1, 15 December 2012.
3. Han, X.J.; Bi, Q.S.; Ji, P.; Kurths, J. Fast-slow analysis for parametrically and externally excited systems with wo slow rationally related excitation frequencies. *Phys. Rev. E* **2015**, *92*. [CrossRef]
4. Taylor, B.; Schaub, M.; Jenkins, D. Projectile Guidance System Including a Compact Semi-Active Laser Seeker. U.S. Patent Number US8207481B2, 26 June 2012.
5. Raytheon Company. Optical System for Aide Field of View Staring Infrared Sensor Having Improved Optical Symmetry. European Patent Number EP1618358B1, 24 March 2010.
6. Dharmveer, S.; Kurode, S.R.; Parkhi, P.; Tamhane, B. Robust control for seekers canloop using sliding modes. In Proceedings of the 12th International Conference on Control, Automation and Systems (ICCAS), Jeju Island, Korea, 17–21 October 2012; pp. 49–54.
7. Zarchan, P. *Tactical and Strategic Missile Guidance*, 7th ed.; American Institute of Aeronautics and Astronautics: Reston, VA, USA, 2019; ISBN 978-1-62410-584-5.
8. Sakarya, D.U. Optical design of dual-mode seeker for long-wave infrared and mid-wave infrared seeker in missile application. In Proceedings of the Conference: Optical Modeling and Performance Predictions X, San Diego, CA, USA, 22–23 August 2018; p. 10743. [CrossRef]
9. Li, B.; Dong, C.; Yu, J.; Zhang, Q.; Zhoua, H.; Liua, R. Mechanical behavior and microstructural evolution of Ni-based single crystal alloys under shock loading. *RSC Adv.* **2018**, *8*, 22127–22135. [CrossRef]
10. Stefanski, K. The analysis of the anti-aircraft missile flight controlled with gyroscopic system. *AIP Conf. Proc.* **2019**, *2077*, 020054. [CrossRef]
11. Hu, Q.; Ji, H.B.; Zhang, Y.Q. Tracking of maneuvering non-ellipsoidal extended target with varying number of sub-objects. *Mech. Syst. Signal Process.* **2018**, *99*, 262–284. [CrossRef]
12. Krzysztofik, I. Analysis of gyroscope systems liding mode control in a target seeking and tracking process. *ZAMM J. Appl. Math. Mech. Z. Angew. Math. Mech.* **2018**, *99*. [CrossRef]
13. Moein, Z.; Gholami, H.; Norouzi, H.; Soltani, M.; Dehghan, S.; Haeri-Hamedani, M.; Eslamian, S.; Singh, V.P.; Ghane, M.; Ostad-Ali-Askari, K. Principleand Conducting Manners of Creating of Passive Defense Components as a Critical Artery. *Int. J. Constr. Res. Civ. Eng.* **2019**, *5*, 8–15. [CrossRef]
14. Cheng, C.; Gao, M.; Cheng, X.; Fang, D. New-generation hybrid guidance system based on infrared and millimeter waves. *IEEE Aerosp. Electron. Syst. Mag.* **2018**, *33*, 34–44. [CrossRef]
15. Wu, S.; Li, X.; Rong, W.; Zhou, J.; Tian, X.; Deng, Z.; Sheng, Z. Profile detection and surface fitting based on the key optical components of the seeker. In Proceedings of the 2018 IEEE International Conference on Mechatronics and Automation (ICMA), Changchun, China, 5–8 August 2018; pp. 1699–1704. [CrossRef]
16. Gapiński, D.; Koruba, Z.; Krzysztofik, I. The model of dynamics and control of modified optical scanning seeker in anti-aircraft rocket missile. *Mech. Syst. Signal Process.* **2014**, *45*, 433–447. [CrossRef]
17. Gapiński, D.; Krzysztofik, I.; Koruba, Z. Multi-channel, passive short-range anti-aircraft defence system. *Mech. Syst. Signal Process.* **2018**, *98*, 802–815. [CrossRef]
18. Łaski, P.; Takosoglu, J.; Błasiak, S. Design of a 3-DOF tripod electro-pneumatic parallel manipulator. *Robot. Auton. Syst.* **2015**, *72*, 59–70. [CrossRef]
19. Gapiński, D. Optical Scanning Coordinator. Patent PL 199721 B1, 31 October 2008.
20. Gapiński, D.; Krzysztofik, I.; Koruba, Z. Analysis of the dynamics and control of the modified optical target seeker used in anti-aircraft rocket missiles. *J. Theor. Appl. Mech.* **2014**, *52*, 629–639.
21. Gapiński, D.; Stefański, K. A control of modified optical scanning and tracking head to detection and tracking air targets. *Solid State Phenom.* **2014**, *210*, 145–155. [CrossRef]
22. Gapiński, D.; Krzysztofik, I. The process of tracking an air target by the designed scanning and tracking seeker. In Proceedings of the 15th International Carpathian Control Conference (ICCC), Velke Karlovice, Czech Republic, 28–30 May 2014; pp. 129–134. [CrossRef]
23. Gapiński, D.; Szmidt, P. The control process of a scanning and tracking IR seeker using inverse dynamics. In Proceedings of the 3rd International Conference "Engineering Mechanics 2017", Svratka, Czech Republic, 15–18 May 2017; pp. 326–329, ISBN 978-80-214-5497-2.
24. Gapiński, D.; Krzysztofik, I. Koruba, Stability of the designed scanning seeker in an anti-aircraft missile. *Probl. Mechatron. Armament Aviat. Saf. Eng.* **2015**, *6*, 56–70.
25. Kim, P. *Kalman Filter for Beginners with MATLAB Examples*; A-JIN Publishing: Seoul, Korea, 2011; ISBN 978-1463648350.

26. Lei, Y.; Xia, D.; Erazo, K.; Nagarajaiah, S. A novel unscented Kalman filter for recursive state-input-system identification of nonlinear systems. *Mech. Syst. Signal Process.* **2019**, *127*, 120–135. [CrossRef]
27. Zghal, M.; Mevel, L.; Moral, P.D. Modal parameter estimation using interacting Kalman filter. *Mech. Syst. Signal Process.* **2014**, *47*, 139–150. [CrossRef]
28. Li, S.E.; Li, G.; Yu, J.; Liu, C.; Cheng, B.; Wang, J.; Li, K. Kalman filter-based tracking of moving objects using linear ultrasonic sensor array for road vehicles. *Mech. Syst. Signal Process.* **2018**, *98*, 173–189. [CrossRef]
29. Awrejcewicz, J.; Koruba, Z. Classical mechanics. In *Applied Mechanics and Mechatronics: Advances in Mechanics and Mathematics*; Monograph; Springer: Heidelberg, Germany, 2012; p. 30.
30. Zhuang, M.; Atherton, D.P. Automatic tuning of optimum PID controllers. *IEE Proc. D Control Theory Appl.* **1993**, *140*, 216–224. [CrossRef]
31. Marquez, H.J. *Nonlinear Control Systems: Analysis and Design*; Wiley: Hoboken, NJ, USA, 2003; ISBN 978-0-471-42799-5.
32. Slotine, J.-J.E.; Li, W. *Applied Nonlinear Control*; Prentice Hall: Hoboken, NJ, USA, 1991.
33. Stefański, K.; Grzyb, M.; Nocoń, Ł. The analysis of homing of aerial guided bomb on the ground target by means of special method of control. In Proceedings of the 2014 15th International Carpathian Control Conference (ICCC), Velke Karlovice, Czech Republic, 28–30 May 2014; pp. 551–556. [CrossRef]
34. Arnold, W.F., III. *Numerical Solution of Algebraic Matrix Riccati Equations*; Report number NWCTP6521; Naval Weapons Center China Lake: Ridge Crest, CA, USA, 1984.
35. Zhou, Y. Convergence of the discrete-time riccati equation to its maximal solution. In Proceedings of the 35th IEEE Conferenceon Decisionand Control, Kobe, Japan, 13 December 1996; pp. 2665–2670. [CrossRef]
36. Bryson, A.E.; Ho, Y.C. Applied Optimal Control: Optimization, Estimation, and Control. *IEEE Trans. Syst. Man Cybern.* **1979**, *9*, 366–367. [CrossRef]
37. Souza, L.C.G.; Bigot, P. An adaptive method with weight matrix as a function of the state to design the rotatory flexible system control law. *Mech. Syst. Signal Process.* **2016**, *79*, 132–140. [CrossRef]
38. Zheng, Y.; Chen, H.; Zhou, Z. Angle Measurement of Objects outside the Linear Field of View of a Strap down Semi-Active Laser Seeker. *Sensors* **2018**, *18*, 1673. [CrossRef]
39. Titterton, D.H. *Military Laser Technology and Systems*; Artech House: Norwood, MA, USA, 2015; ISBN 9781608077793.
40. Allen, R.; Volpi, J.; Roemerman, S. COTS lens and detector characterization for low cost miniature SAL seekers. In Proceedings of the AIAA Guidance, Navigation, and Control Conference, Boston, MA, USA, 19–22 August 2013. [CrossRef]
41. Cheng, X.; Yang, Y.; Hao, Q. Analysis of the Effects of Thermal Environment on Optical Systems for Navigation Guidance and Control in Supersonic Aircraft Based on Empirical Equations. *Sensors* **2016**, *16*, 1717. [CrossRef] [PubMed]
42. Bużantowicz, W.; Pietrasieński, J. Dual-control missile guidance: A simulation study. *J. Theor. Appl. Mech.* **2018**, *56*, 727–739. [CrossRef]
43. Zhu, M.; Chen, H.; Xiong, G. A model predictive speed tracking control approach for autonomous ground vehicles. *Mech. Syst. Signal Process.* **2017**, *87*, 138–152. [CrossRef]
44. Guo, H.; Cao, D.; Chen, H.; Sun, Z.; Hu, Y. Model predictive path following control for autonomous cars considering a measurable disturbance: Implementation, testing, andverification. *Mech. Syst. Signal Process.* **2019**, *118*, 41–60. [CrossRef]
45. Zhang, Q.; Wang, Q.; Li, G. Nonlinear modeling and predictive functional control of Hammerstein system with application to the turn table servo system. *Mech. Syst. Signal Process.* **2016**, *72–73*, 383–394. [CrossRef]
46. Sobczynski, D. A concept of a power electronic converter for a BLDC motor drive system in aviation. *Aviation* **2015**, *19*, 36–39. [CrossRef]
47. Wei, Z.; Zhao, J.; He, H.; Ding, G.; Cui, H.; Liu, L. Future smart battery and management: Advanced sensing from external to embedded multi-dimensional measurement. *J. Power Sources* **2021**, *489*, 229462. [CrossRef]
48. Wei, Z.; He, H.; Pou, J.; Tsui, K.-L.; Quan, Z.; Li, Y. Signal-Disturbance Interfacing Elimination for Unbiased Model Parameter Identification of Lithium-Ion Battery. *IEEE Trans. Ind. Inform.* **2020**, *1*. [CrossRef]
49. Schimmack, M.; Haus, B.; Mercorelli, P. An Extended Kalman Filteras an Observer in a Control Structure for Health Monitoring of a Metal-Polymer Hybrid Soft Actuator. *IEEE/ASME Trans. Mechatron.* **2018**, *23*, 1477–1487. [CrossRef]
50. Schimmack, M.; Mercorelli, P.; Maiwald, M. Combining Kalman filter and RLS-algorithm to improve a textile based sensor system in the presence of linear time-varying parameters. In Proceedings of the 2015 17th International Conference on E-health Networking, Application & Services (Health Com), Boston, MA, USA, 14–17 October 2015; pp. 507–510. [CrossRef]
51. Chen, L.; Mercorelli, P.; Liu, S. A Kalman estimator for detecting repetitive disturbances. In Proceedings of the American Control Conference 3, Chicago, IL, USA, 1–3 July 2005; pp. 1631–1636. [CrossRef]

energies

MDPI

Article

Exhaust Noise Reduction by Application of Expanded Collecting System in Pneumatic Tools and Machines

Dominik Gryboś [1,*], Jacek S. Leszczyński [1], Dorota Czopek [2] and Jerzy Wiciak [2]

[1] Department of Thermal and Fluid Flow Machines, Faculty of Energy and Fuels, AGH University of Science and Technology, Al. Mickiewicza 30, 30-059 Krakow, Poland; jale@agh.edu.pl

[2] Department of Mechanics and Vibroacoustics, Faculty of Mechanical Engineering and Robotics, AGH University of Science and Technology, Al. Mickiewicza 30, 30-059 Krakow, Poland; dorota.czopek@agh.edu.pl (D.C.); jerzy.wiciak@agh.edu.pl (J.W.)

* Correspondence: dgrybos@agh.edu.pl

Abstract: In this paper, we demonstrate how to reduce the noise level of expanded air from pneumatic tools. Instead of a muffler, we propose the expanded collecting system, where the air expands through the pneumatic tube and expansion collector. We have elaborated a mathematical model which illustrates the dynamics of the air flow, as well as the acoustic pressure at the end of the tube. The computational results were compared with experimental data to check the air dynamics and sound pressure. Moreover, the study presents the methodology of noise measurement generated in a pneumatic screwdriver in a quiet back room and on a window-fitting stand in a production hall. In addition, we have performed noise measurements for the pneumatic screwdriver and the pneumatic screwdriver on an industrial scale. These measurements prove the noise reduction of the pneumatic tools when the expanded collecting system is used. When the expanded collecting system was applied to the screwdriver, the measured Sound Pressure Level (SPL) decreased from 87 to 80 dB(A).

Keywords: pneumatics; hybrid systems; fluid power; air expansion; noise reduction; SPL measurement

Citation: Gryboś, D.; Leszczyński, J.S.; Czopek, D.; Wiciak, J. Exhaust Noise Reduction by Application of Expanded Collecting System in Pneumatic Tools and Machines. *Energies* **2021**, *14*, 1592. https://doi.org/10.3390/en14061592

Academic Editor: Francesco Castellani

Received: 20 January 2021
Accepted: 8 March 2021
Published: 13 March 2021

1. Introduction

Manufacturing plants very often use compressed air for production lines and machines. All pneumatic machines and tools tend to produce noise, which is related to the discharge of high-pressure air through the exhaust port [1]. The unfavorable effect of air leaks from the installation may also increase the noise level [2]. In addition to the exhaust noise, mechanical noise, particularly from percussive tools, may also be present.. Aerodynamic noise from the turbulent air flow through the working elements is generally the dominant noise source. The sudden exhaust process of the air expanding from the typical gauge pressure of 6-8 bar to the level of atmospheric pressure generates transient aerodynamic noise [3]. The unstable exhaust noise in the pneumatic system can be additionally subdivided, based on the duration criterion, into intermittent (from a few to several seconds) and impulse (from several dozen to several hundred milliseconds) [4].

The generated aerodynamic noise, resulting from the transient chocked air flow with large pressure differences, is caused by the formation of air shock waves due to the discontinuity at the source–tube interface [3,5]. The transient exhaust air expansion can be divided into two regions—sonic and subsonic—with the boundary between them expressed in the value of the critical pressure of the choked flow. The majority of exhaust time is spent in the sonic region, which generates a high-amplitude roaring sound. The transient sonic flow has a greater pressure difference and a greater flow rate [5]. According to Lighthills's theory of aerodynamic noise, the airflow through the throttle is made up of acoustic monopoles induced by a change in the mass flow rate, dipoles showing the effect of constant boundaries on the base field, and turbulence quadrupoles [6,7]. The generated impulse sound can be distinguished by an initial sudden increase due to the large pressure difference and large flux [3], where the value can range from 96 to 120 dBA [8,9], or even up to 130 dBA [10]. Then, there is a gentle decrease due to the smooth reduction in the jet and the pressure difference.

The maximum noise level [id=R2, comment=Linguistic corrections] for an 8-h working day should not exceed the threshold value of 85 dB, defined in Directive 2003/10/EC [11]. There is an additional concept of a threshold value of 80 dB, which is a specific alarm signal of currents dangerously approaching the maximum permissible limit. Moreover, the maximum A-sound level must not exceed 115 dBA and the peak C-sound level must not exceed 135 dBC [11]. It is believed that work with exposure to noise exceeding 80 dBA does not expose the employee to health damage if it is properly organized: frequent breaks at work, limitations on working time, and the use of personal hearing protectors. From the perspective of the technological process, frequent interruptions have negative consequences, i.e., they reduce work efficiency. According to standard [12], when the noise level during an 8-h working day is exceeded by 1 dB, the working time in this environment is reduced by about 100 min. At the level of 90 dB, the employee should not work in such an environment for more than 2.5 h. The impact of noise on human health and condition can be analyzed in terms of direct impact—on the middle and inner ear—and indirect impact—on the nervous system, psyche and internal organs. The harmful effects of noise on the human body are manifested by temporary or permanent hearing impairment, increased fatigability, decreased learning efficiency, difficulties focusing, disorientation, irritability, decreased precision of movements, pain and dizziness, increased blood pressure, and abnormal heart rhythms. The scale of these systemic changes depends on the duration, frequency, intensity and nature of noise, as well as on the psychophysical condition of a person.

The reduction in noise is essential, especially in pneumatics industries, since it contributes to the comfort of the workplace and reduces environmental sound pollution. The large number of sources of impulse noise in manufacturing plants significantly increases the acoustic background. Generally, there are two different approaches to noise attenuation: passive and control techniques [13–15]. Currently, the most popular method of noise reduction in pneumatics is the use of various types of mufflers. The four main types of mufflers are: expansion chamber, sintered bronze silencer, porous-diffusion and perforated panel [16]. Commercial pneumatic silencers were tested by Daggerhart and Berger [17]. They showed a reduction in the muzzle noise down to 85–95 dB. The best noise suppression was found with the porous plastic muffler, which reduced noise from 120 to 90 dBA at a pressure of 8 bar [17]. Zhao et. al analysed the two-level exhaust process through expansion chamber mufflers with sound-absorbing stainless steel fiber and a switch valve from pneumatic presses PFC/B [10]. The exhaust air is first expanded into the muffler chamber at a pressure of 1.5–2 bar. With the decrease in the pressure difference, the noise generation was reduced from 115 to 82 dB. [10]. However, expansion chamber mufflers with a switch valve make the air exhaust process three times longer. Li et al. [5] focused on the sintered bronze silencer based on porous materials. They presented a mathematical model of the exhaust air process through porous materials based on Ergun's equation. Moreover, they showed that the sintered bronze silencer reduces the transient exhaust noise by 15 dBA at 1–10 kHz. In contrast, in intermittent exhaust-noise perforated silencers reduce noise to levels below 85 dBA [16]. Another solution, apart from pneumatic silencers, are appropriately shaped air outlet nozzles. Ivanov [9] achieved noise reductions of up to 82 dBA by appropriately shaping the nozzle and increasing the orifice number. A completely different method is presented by Li and Zhao [3]. The authors created a valve opening control algorithm, which causes a linear variation in the pressure difference during the transient exhaust air process. They managed to reduce the average noise emission by 2–4 dBA and the peak noise emission by up to 11 dB. The methods described in the above-mentioned papers, despite various attempts to reduce noise, fail to bring the noise level below 80 dB. They also do not account for the fact that, in production plants, the sources of aerodynamic noise are distributed and cause a high acoustic background.

In this paper, we present an alternative method that can reduce the noise of exhaust air from pneumatic tools and devices. According to the original method [18] of collecting the exhaust air from pneumatic tools and pneumatic machines with tubes, we are able to soundproof and spread the sources of pneumatic noise. We named the technology the expanded collecting system. To present this idea, a mathematical model of exhaust air expansion, air flow through the tube and sound generation was created. Moreover, the technology was tested on a technical and industrial scale on a pneumatic screwdriver, where Sound Pressure Level (SPL) was measured in a plant's production hall.

2. Mathematical Model

In order to illustrate the physical phenomena contributing to the generation of pneumatic noise, such as exhaust air expansion, air flow through tubes, and sound generation from monopole and quadrapol sources, a mathematical model was created using the methods described in [10,19,20]. Figure 1 shows a discretized scheme of a tube with the air flow from the source at point A to the outlet at point B. The tube of length l is divided into n control volumes, each Δx long, where the observables for each control volumes are the mass flow and the pressure, with indices i and j, respectively.

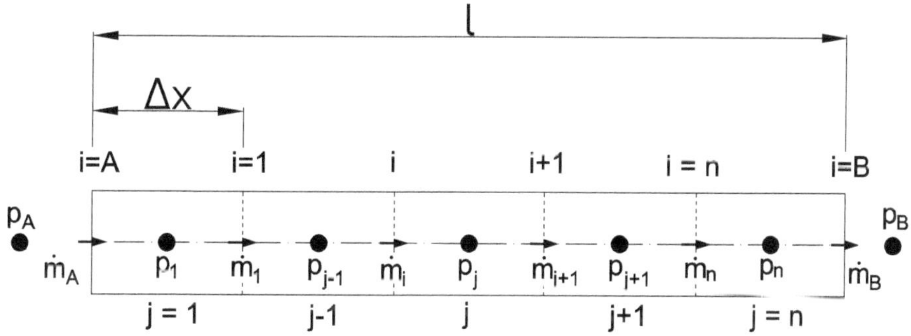

Figure 1. 1-D pneumatic transmission line model [19].

Numerical meshes of the mass flow and the pressure are shifted relative to each other so that pressure p_j is recorded at the midpoint of the control volume, for which index j is assigned, and mass flow $\dot{m}_i$ at its edges, for which index i is assigned.

Using the method of line [19,20], we assume that every observable and parameter is calculated in every space point as being dependent on time only. A pneumatic line of length l is parted into n segments of length $\Delta x = \frac{l}{n}$. Coordinate x is discretized by two staggered grids for the pressure and the mass flow separately, respectively, the j and i indexes. Therefore, control volumes for the pressure and the mass flow are shifted. For the pressure grid for index $j = 1, \ldots, n$ the following parameters are defined

- position of pressure nodes:

$$x_j = \frac{\Delta x}{2} + (j-1)\Delta x \tag{1}$$

- pressure:

$$p_j = p(t, x_j) \tag{2}$$

- mid cross section of each control volume:

$$A_j = A(x_j) = \frac{\pi d_j^2}{4} \tag{3}$$

- humid air density calculated on each control volume: Correction for comment 5 Reviewer 1. We unified format of formulas and figures

$$\rho_j = \rho(t, x_j) = \frac{p_j}{R_a T}\left(1 + \left(\frac{R_a}{R_v} - 1\right)\phi\frac{p_s}{p_0}\right) \tag{4}$$

where: air constant R_a, vapour constant R_v, temperature T, ambient pressure p_0, relative humidity ϕ, saturation pressure p_s for humid air temperature. For ϕ = 1 the product $\phi p_s(T)$ equals the pressure of vapor contained in the humid air p_v. For the mass flow grid for $i = 1, \ldots, n-1$ the following parameters are defined

- Position of mass flow nodes

$$x_i = i\Delta x \tag{5}$$

- Mass flow

$$\dot{m}_i = \dot{m}(t, x_i) \tag{6}$$

- Cross-section of each control volume edge

$$A_i = A(x_i) = \frac{\pi d_i^2}{4} \tag{7}$$

- Speed of sound: Correction for comment 5 Reviewer 1. We unified format of formulas and figures

$$a_i = \sqrt{\kappa\frac{p_j + p_{j-1}}{\rho_j + \rho_{j-1}}} \tag{8}$$

- Air velocity: Correction for comment 5 Reviewer 1. We unified format of formulas and figures

$$w_i = \frac{2\dot{m}_i}{A_i(\rho_j + \rho_{j-1})} \tag{9}$$

2.1. The Dynamics of Air Expansion at the Boundaries

The process of air discharge from the chamber is associated with a change in mass over time, due to the mass through the channel at different pressures. Due to the speed of the exhaust air, expansion process and the small air flows, it was assumed that air expansion is adiabatic [21].

$$\frac{dm_i}{dt} = \dot{m}_i \quad \text{for} \quad i = \{A, B\} \tag{10}$$

The mass flow between the volume under pressure p_{j-1} and the volume under pressure p_j is defined by Equation (11) [22,23]. If the pressure ratio for the left side of the tube $\frac{p_j}{p_{j-1}}$, for $i = A$, we have $j-1 = A$ and $j = 1$, is lower than $\zeta(\frac{2}{\kappa+1})^{\frac{\kappa}{\kappa-1}}$ the air exhaust flow is sonic. On the other hand, if the pressure ratio on the right side $\frac{p_j}{p_{j-1}}$, for $i = B$, we obtained $j-1 = n$ and $j = B$, is lower than $\zeta(\frac{2}{\kappa+1})^{\frac{\kappa}{\kappa-1}}$ the air exhaust flow is also sonic. While the pressure ratio $\frac{p_j}{p_{j-1}}$ is greater than $\zeta(\frac{2}{\kappa+1})^{\frac{\kappa}{\kappa-1}}$, the air exhaust flow is subsonic

$$\dot{m}_i = \begin{cases} A\sqrt{\frac{2\kappa}{\kappa-1}\rho_{j-1}p_{j-1}\left(\left(\frac{p_j}{Zp_{j-1}}\right)^{\frac{2}{\kappa}} - \left(\frac{p_j}{Zp_{j-1}}\right)^{\frac{\kappa+1}{\kappa}}\right)} & \text{for } \frac{p_j}{p_{j-1}} > \zeta(\frac{2}{\kappa+1})^{\frac{\kappa}{\kappa-1}} \\ A\sqrt{\rho_{j-1}p_{j-1}(\frac{2}{\kappa+1})^{\frac{\kappa+1}{\kappa-1}}} & \text{for } \frac{p_j}{p_{j-1}} \le \zeta(\frac{2}{\kappa+1})^{\frac{\kappa}{\kappa-1}} \end{cases} \tag{11}$$

Equation (11) is simultaneously valid for boundaries $i = A$ and $i = B$. Index j takes the following values: $j = 1, j-1 = A$ for $i = A$ and $j = B, j-1 = n$ for $i = B$. Moreover, the scaling factor Z and factor ζ caused by the non-ideal air mass flow due to the sharp-edged orifice [23] is formed as

$$Z = \frac{1-\zeta}{1-\beta^{2\zeta}}\left(\frac{p_{j-1}}{p_j}\right)^{2\zeta} + \frac{\zeta-\beta^{2\zeta}}{1-\beta^{2\zeta}} \tag{12}$$

2.2. Pneumatic Transmission Line Model

The 1-D pneumatic transmission line model (Figure 1) is given by way of a system of ordinary differential equations (ODE) based on the continuity equation and the equation of motion of air flow [21]

$$\frac{dp_j}{dt} = -\frac{p_j}{\rho_j A_j}\frac{\dot{m}_{i+1}-\dot{m}_i}{\Delta x} \tag{13}$$

$$\frac{d\dot{m}_i}{dt} = -A_i\frac{p_{j-1}-p_j}{\Delta x} - \frac{\lambda_i}{(\rho_{j-1}+\rho_j)A_i d_i}\dot{m}_i|\dot{m}_i| \tag{14}$$

where pressure $p_j = p(t, x_j)$ and mass flow $\dot{m}_i = \dot{m}(t, x_i)$ are dependent on time t and coordinate x, calculated according to Figure 1, A is the pneumatic line cross-section, ρ_j density of the moist air, ν dynamic viscosity and d_i the pneumatic line diameter. The Darcy–Weisbach equation is used to model the friction phenomenon in air-flow through a pneumatic line [19]. The Darcy friction factor λ_i, included in the Darcy–Weisbach equation, is a function of Reynolds number Re_i, kinematic viscosity ν, tube diameter d_i and tube roughness ϵ. However, as claimed in [24,25], the error in determining the Darcy friction factor λ_i for the compressible flow with the Mach number less than 0.6 ($\mathrm{Ma} = \frac{w_i}{a_i} < 0.6$) is not greater than 3 % and can be used successfully in such an application. In the case of air flows in a pneumatic, this condition is fulfilled [19]. The formulae proposed by [26] were used to determine the Darcy friction factor λ_i

$$\lambda_i = \left(\frac{64}{\mathrm{Re}_i}\right)^b_i \left(0.75\ln\frac{\mathrm{Re}_i}{5.37}\right)^{2(b_i-1)c_i} \left(0.88\ln 3.41\frac{d_i}{\epsilon}\right)^{2(b_i-1)(1-c_i)} \tag{15}$$

$$b_i = \frac{1}{1+\left(\frac{\mathrm{Re}_i}{2712}\right)^{8.4}} \quad \text{– semi-empirical constant} \tag{16}$$

$$c_i = \frac{1}{1+\left(\frac{\mathrm{Re}_i}{150\frac{d_i}{\epsilon}}\right)^{1.8}} \quad \text{– semi-empirical constant} \tag{17}$$

2.3. Acoustic Model

Due to the mass flow, the impulse noise is mainly composed of monopole and quadrupole sources. The acoustic pressure i.e., the difference between the instantaneous pressure at a certain point in the disturbed medium and its average value under equilibrium conditions, from a monopole source at distance s from this source, is [10]

$$p' = \frac{1}{4\pi s}\frac{d\dot{m}_B}{dt} \tag{18}$$

where $\dot{m}_B$ is the outlet air flow at the end of the pneumatic line (see Figure 1 and Equation (11)).

Furthermore, the acoustic pressure from the quadrapole source at distance s from the pneumatic line end is [10]

$$p'' = \frac{d_B(p_n-p_B)^2}{s}\sqrt{\frac{K}{p_n^2+(p_n-p_B)^2}} \tag{19}$$

where: proportional constant $K = 8 \times 10^{-6}$, exhaust line diameter d_B, pressure at the end of the pneumatic line p_n, ambient pressure p_B (see Figure 1). Sound Pressure Levels (SPL) from the monopole $L_{p'}$ and quadrapole $L_{p''}$ sources [10] are

$$L_{p'} = 20\log\frac{p'}{p_{ref}} \text{ and } L_{p''} = 20\log\frac{p''}{p_{ref}} \tag{20}$$

respectively, where the reference pressure p_{ref} = 20 μPa. Total Sound Pressure Level (SPL) L_p [8] combined from the monopole and quadrapole source is

$$L_p = 10\log\left(10^{\frac{Lp'}{10}} + 10^{\frac{Lp''}{10}}\right) \tag{21}$$

Depending on the mass flow velocity and frequency, one of the above-mentioned sources of noise may dominate. The complete acoustic model represents a practical tool to calculate the pneumatic noise generation from the expansion of exhaust air. The mathematical model was implemented in the MATLAB environment and the ODE system was solved using the Runge Kutty Felhberg procedure. The following assumptions were made for the mathematical model: air is treated as an ideal gas, the flow is isothermal and the flow velocity is less than 0.6 Ma, which allows for the application of the Darcy–Weisbach formula for air and, based on the 1-D flow model, the average air velocity in the tube profile is considered.

3. Experimental Setup

The idea of the expanded collecting (EC) system is to capture exhaust air from dispersed outlets of pneumatic tools and machines in production area and throw it outside the working environment(see Figure 2) [18].

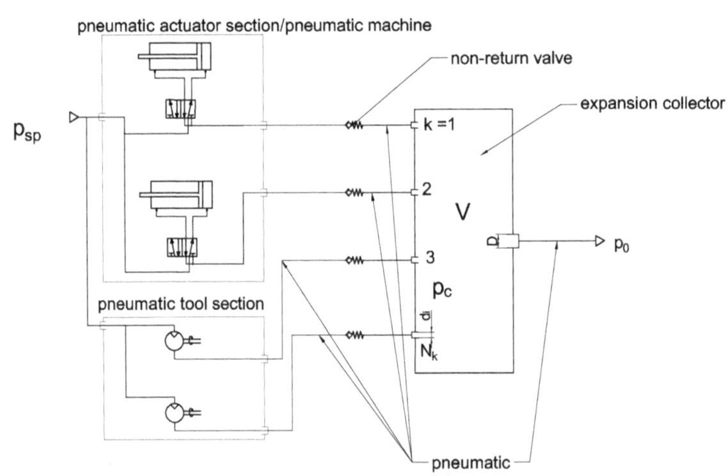

Figure 2. Outline of the expanded collecting system.

The existing pneumatic mufflers mounted at the outlet of pneumatic tools and machines are replaced by the pneumatic tubes, with cross-section areas no smaller than the cross-sectional areas of the exhaust air outlets, and non-return valves. The other ends of the pneumatic tubes are connected to the expanded collector with a volume defined by the formula

$$V = \frac{p_{sp}}{p_c}\frac{1}{\sum_{k=1}^{N_k} t_k}\sum_{k=1}^{N_k}\left(\dot{V}_k t_k^2\right) \tag{22}$$

where supply pressure p_p, pressure in expansion collector p_c, volume flow from pneumatic tool or machine $\dot{V}_k$, operating time t_k, number of collector inlet N_k and index of collector inlets k.Then, through one outlet pneumatic tube with the cross-sectional area, no smaller than the sum of the cross-sectional areas of the inlet pneumatic tubes to the expansion collector, the exhaust air is thrown outside the working environment; for example, outside the room, building, etc. Proper selection of the cross-sectional areas of the pneumatic tubes and the volume of the expanded collector is crucial in order not to disturb the operation of pneumatic tools and machines. The solution reduces the pneumatic noise generated during the dispersed expansion of the air through pneumatic mufflers. Then, the noise from part of or the entire production hall is collected in one place (expansion collector), which makes it easier to neutralize.

In order to determine the sound pressure level (SPL) to which the technical operator is exposed, two tests were performed in a partial isolated workshop adjacent to the office of the maintenance department and on window-fitting stand (Figure 3) Measurements of the sound pressure level were made using Svan 959 Class 1 sound-level meter by SVANTEK, with a preamplifier by SVANTEK, type SV12L and a microphone by G.R.A.S, type 40AE. Before the start of the measurements and after the completion of the measurements, the meter was calibrated using a Brüel & Kjær type 4231 calibrator. The measurements were made in the broadband range, taking into account the correction characteristics (A and Z). The sound spectrum in one-third octave bands and the following basic acoustic parameters were recorded

- $L_{Z_{eq}}$ (dB)—the equivalent sound pressure level;
- $L_{A_{eq}}$ (dBA)—the equivalent A-weighted sound pressure levell

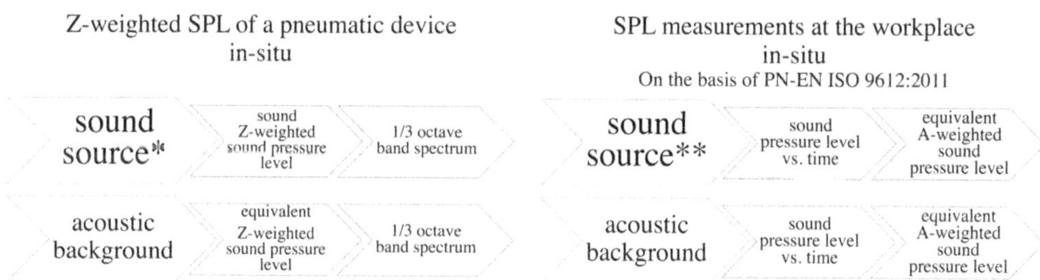

Figure 3. Schematic diagram of the measurement procedure for an acoustic test. Measurement was carried out for the following sound sources: * The hand-held pneumatic screwdriver with standard factory configuration (with muffler), with modified outlet port and with the expanded collecting system. ** The hand-held pneumatic screwdriver with standard factory configuration (with muffler) and with the expanded collecting system.

A hand-held pneumatic screwdriver SHINANO SI-1166-8A was used for both tests, whose parameters are presented in Table 1. The screwdriver was supplied with compressed air directly from the plant's compressed air supply system, with an operating gauge pressure of approximately 8.5–9.0 bar.

Table 1. Technical parameters of the hand-held pneumatic screwdriver SHINANO SI-1166-8A.

Parameter	Value
Maximum rotational speed	2000 rpm
Maximum torque	9 Nm
Supply gauge pressure	8.5–9.0 bar
Air volume flow	200 dm^3/min
Supply inlet diameter	1/4′
Exhaust outlet diameter	3/8′

The first test was to measure the noise level generated by a hand-held pneumatic screwdriver in a partial isolated workshop adjacent to the office of the maintenance department in order to reduce the impact of high acoustic background in the production hall (Figure 3a). The measurement was performed for the pneumatic screwdriver in three variants of its outlet, as shown in Figure 4.

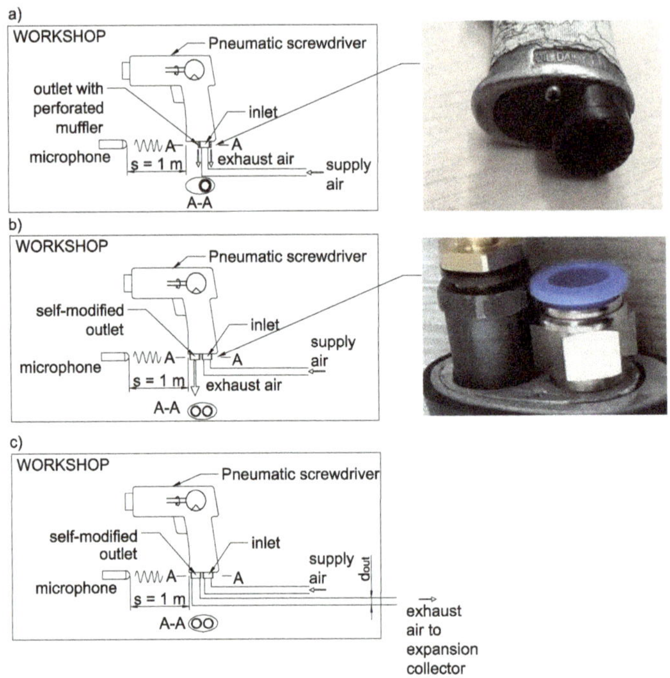

Figure 4. The hand-held pneumatic screwdriver used for measurements: (**a**) at standard factory configuration(with pneumatic muffler); (**b**) with self-modified outlet port (**c**) with the expanded collecting system.

The first variant, shown in Figure 4a, is a standard factory configuration of the hand-held pneumatic screwdriver with the outlet equipped with perforated pneumatic muffler in the form of air exhaust channels on the circumference of the inlet nozzle (photo in Figure 4a). Then, in variant 2, the outlet from the pneumatic screwdriver was modified to a separate outlet port, as shown in Figure 4b. In the last variant (Figure 4c), a pneumatic tube from the expanded collecting system was connected to the modified outlet port of the hand-held pneumatic screwdriver from variant 2. The pneumatic tube with an outlet diameter d = 12 mm and length l = 10 m was used, while the other end of it was placed outside the workshop. In all three variants, the in situ measurements were made at a distance of one meter from the screwdriver in a direction perpendicular to the direction of the air outlet.

Then, the test of the technical scale was performed on the production hall in the section of window fitting stands, Section 2, stand no. 1 (see Figure 5a).

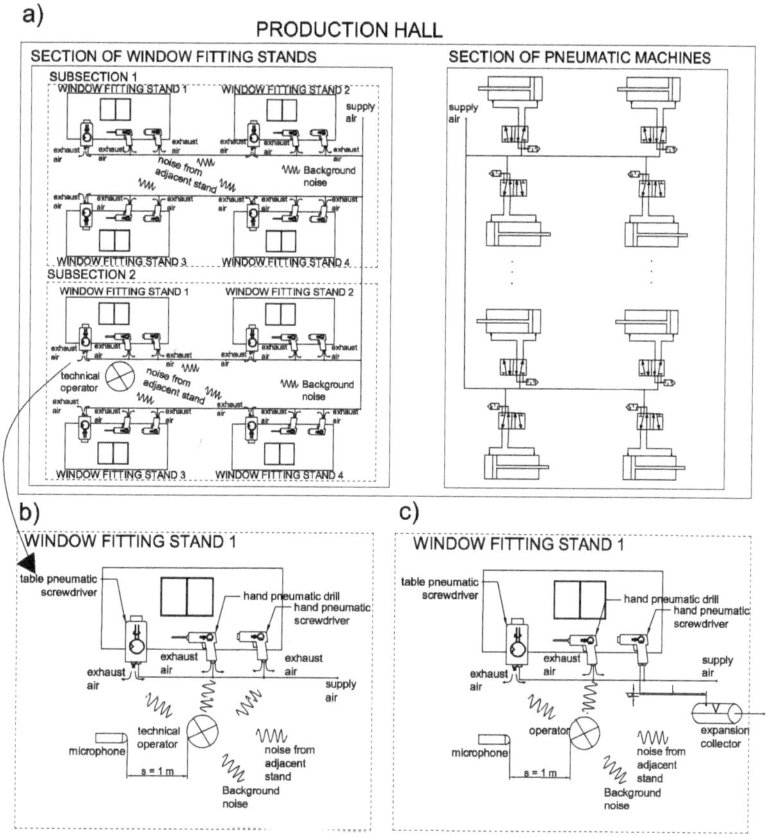

Figure 5. The outline of: (**a**) general production hall; (**b**) the window fitting stand no 1, Section 2; (**c**) the window-fitting stand no 1, Section 2 with the expanded collecting system.

Figure 5b shows an outline of the window-fitting stand, where noise level measurements took place. The technical operator has three pneumatic tools at his disposal: table pneumatic screwdriver, hand-held pneumatic drill and hand-held pneumatic screwdriver. During his work, he is exposed to noise coming directly from those three used devices, stands in the immediate vicinity in the subsection and the acoustic background of the entire plant. Moreover, the SPL measurements were carried out at the window-fitting stand equipped with the hand-held pneumatic screwdriver in two variants: standard factory configuration (with the pneumatic muffler) and with the expanded collecting system. The EC system was used by the attaching pneumatic tube (d = 12 mm, l = 8 m) to the outlet port of the hand-held pneumatic screwdriver and connected to the expanded collector with volume V = 15 dm^3. The pneumatic tube from the expanded collector was placed out of the subsection, according to the outline shown in Figure 5c. The noise level measurement was carried out on the basis of standard PN-EN ISO 9612:2011.

4. Results

4.1. Mathematical Model Validation

Therefore, to confirm the usefulness of the mathematical model, this should be validated. For this purpose, the results of a computer simulation in the MATLAB of the model of the exhaust air flow through a pneumatic tube, given by Equation (11), were compared with experimental data from the literature [19] (Figure 6). Next, a full validation of the mathematical model of air expansion, tube airflow, and sound generation was performed with the results of our own experimental results. The validation of the transmission line model (see Equation (11)) was performed by comparing the pressure values at the beginning of the tube p_1 and at the end p_n (see Figure 1). For this purpose, the conditions of the experiment presented in [19] were used. A pneumatic tube of length l = 5 m and inner diameter d = 5.7 mm was connected to a compressed air supply of gauge pressure p_A = 6 bar. The other end of the tube was connected to a closed air tank of volume 0.1 l.

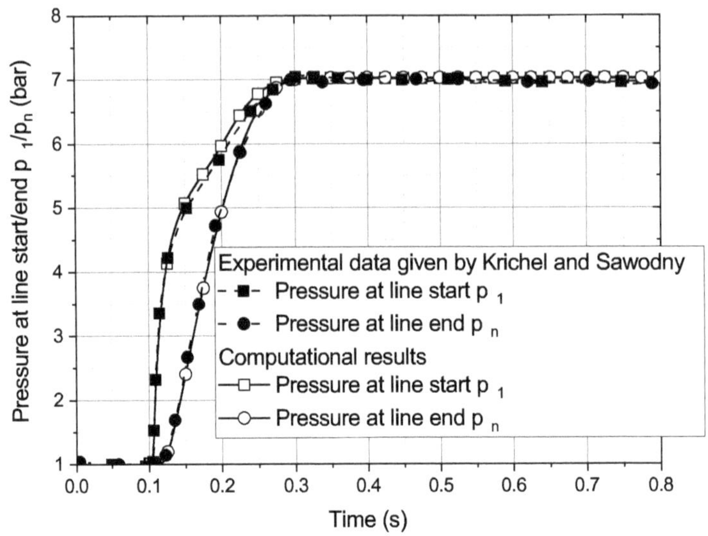

Figure 6. Validation of the mathematical model of the pneumatic transmission line with experimental data included in [19].

High compliance of the mathematical model with experimental data [19] was obtained as a result. The maximum error between the results of the simulation and the experiment is 4.5% for pressure p_1 and 3.6% for pressure p_n. The non-linear nature of the air flow may be observed on the pressure curve p_1. This results from the propagation of the shock wave due to the discontinuity at the source–tube interface at the initial moment t_0 through the sharp-edged orifice.

Then, the entire mathematical model was validated, taking the formulas responsible for sound generation into account, as in Equations (18)–(21), with the data obtained in the experimental measurements. The experiment on the laboratory-scale was carried out in the workshop, with background noise levels of 49 dB. The measurements consisted of measuring the sound generated by the hand-held pneumatic screwdriver, with parameters included in Table 1, with a free air outflow through the outlet (according to Figure 4b). The sound-level pressure was measured during idle operation of the tool for 20 s at a distance of 1 m. For the first 10 s, the tool worked in the intermittent mode with a interval of 2 s and duty cycle 50%, while the remaining time was in the continuous mode. The resulting time domain signal was then transformed into the frequency domain using the Fast Fourier Transformate (FFT) and 1/3 octave-band filter. As the experimental results are included in the range of 1/3 octave frequencies band from 100 Hz to 16 kHz, the results from the computer simulation are also limited to the same frequency range. Then, the signal processed in this way was compared with the sound pressure levels obtained in the experimental measurements using the Z filter (see Section 3). The Figure 7 shows a comparison of the pneumatic noise generated by the free flow of air through the outlet.

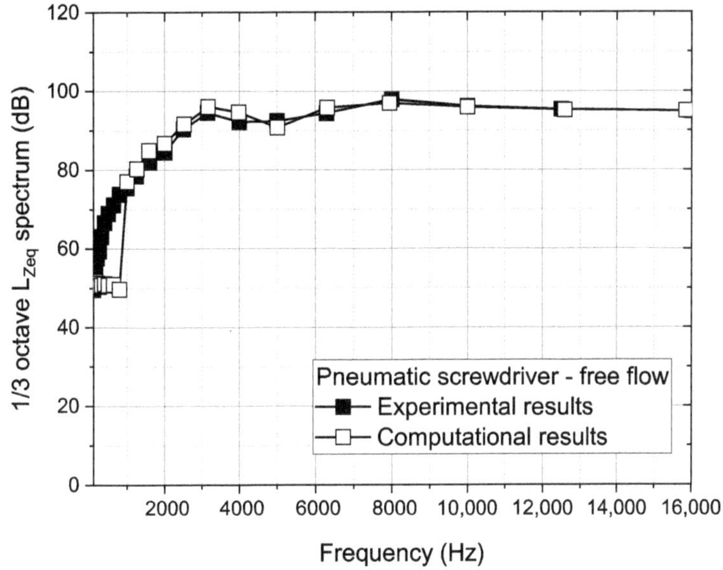

Figure 7. Comparison of the 1/3 octave band equivalent of A-weighted sound-pressure-level average spectrum of the noise of a hand-held pneumatic screwdriver obtained from the computer simulation and the experiment.

A high convergence of the computer simulation data values with the measurement data is obtained for the one-third octave frequency bands from 1 to 16 kHz, for which most of the differences ranged from 1 to 2.4 dB. The maximum difference was 2.9 dB for the frequency 1600 Hz. Similar values of the computer simulation data, juxtaposed with the measurement data, were also obtained for frequencies 100 and 200 Hz. In the frequency range from 315 to 630 Hz, the model begins to diverge from the measured data. The discrepancy of the mathematical model for low frequencies may result from unidentified acoustic sources in the real system and not the smooth combination of two sound-generation formulas.

4.2. Industrial Test

The developed theoretical models were verified for two operating modes of the device: for an idle mode in the workshop adjacent to the office of the maintenance department, and for standard operations on the window-fitting stand in the production hall. The in situ measurements were carried out to determine the background sound-pressure level and the sound-pressure level for a screwdriver damped with a factory pneumatic muffler and a tube collector. The background noise level in the office of the maintenance department was 49 dBA (Table 2).

Table 2. The acoustic background level of the production hall.

No. of Measurement	L_{Aeq} (dBA)	L_{Zeq} (dB)
1	49.3	80.9
2	48.7	80.1
3	49.9	81.4

A-weighted equivalent sound-pressure level $L_{A_{eq}}$ of the idle device was determined by twenty-second measurements at a distance of 1 meter. As already mentioned, two cases of screwdriver operation were considered: standard factory configuration (see Figure 4a) and that with an expansion collecting system (see Figure 4c). A-weighted equivalent sound level $L_{A_{eq}}$ for the first case was 96 dBA and for the second case it was 78 dB (Figure 8).

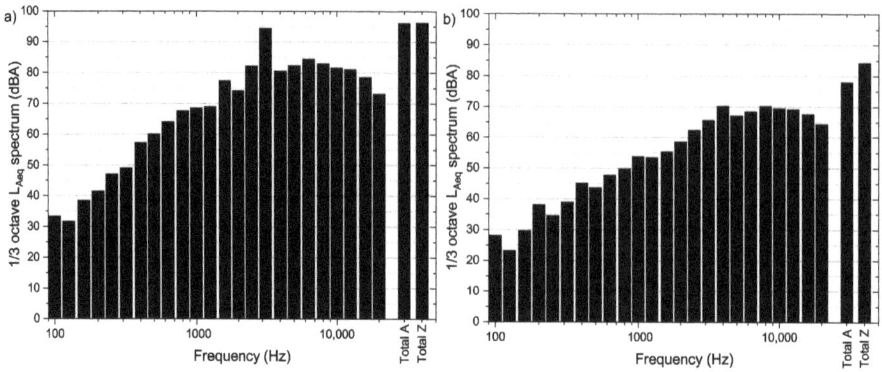

Figure 8. The 1/3 octave band equivalent of A-weighted sound-pressure level average spectrum of the noise of a hand-held pneumatic screwdriver: (**a**) at standard factory configuration(with pneumatic muffler); (**b**) with the expanded collecting system.

It is worth noting that the distance between the measured signals ($L_{A_{eq}}$ of screwdriver operation and the background level) is 47 and 29 dBA, respectively. Such a large difference between the measured acoustic signals makes it possible to observe the actual decrease in the noise level generated by the screwdriver. The reduction of the noise level of the screwdriver was about 18 dBA, i.e., the acoustic pressure at a distance of 1 meter from the device decreased by eight times. This means that the modification of pneumatic devices at all positions in the production hall, consisting of the expanded collecting system, should significantly reduce the overall noise level at workplaces.

The second part of the experimental research was conducted at the window-fitting work station in the production hall. In the first stage, the background sound-pressure level measurement was performed over short intervals between production cycles at the test stand. The average SPL in the production hall was calculated based on three measurements, and was 77.9 dBA (Table 3). Such a high level of acoustic background results from the fact that normal work was carried out on the neighbouring stands during the in situ tests.

Table 3. The acoustic background level of the production hall.

No. of Measurement	L_{Aeq} (dBA)	L_{Zeq} (dB)
1	78.0	80.4
2	78.6	80.8
3	77.2	79.6

Measurements were made for production cycles lasting from 3 to 5 min. Two cases of screwdriver operation were considered: at standard factory configuration(with pneumatic muffler) and with expanded collecting system. The equivalent sound level $L_{A_{eq}}$ for the first production cycle (see Figure 5b) is 82.8 dBA, and (see Figure 5c) is 81.1 dB for the second production cycle. The time-frequency characteristics of the A-weighted sound-pressure level of one production cycle of the window fitting are shown in Figure 9.

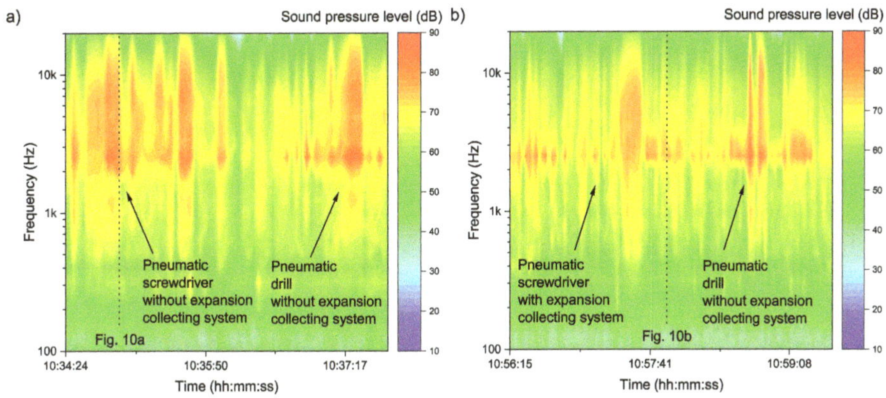

Figure 9. Spectrogram of a sample window fitting cycle: (**a**) at standard factory configuration(with pneumatic muffler); (**b**) with the expanded collecting system.

A comparison between Figure 9a and Figure 9b shows that the highest sound levels are recorded for frequencies above 1 kHz, with the maximum values occurring for one-third octave bands with center frequencies from 2.5 to 12 kHz. In that frequency range, the greatest decrease in the sound-pressure level can also be observed after the use of the expanded collecting system. On the spectrogram shown in Figure 9b, we can also observe an example from a fragment of the process for which no significant decrease in the noise level was noticed (just before 10:59). This is the moment when an unmuted pneumatic drill works—the new solution was not used.

The 1/3 octave band equivalents of A-weighted sound-pressure level spectrum of the noise of a pneumatic device standard factory configuration and with expansion collecting system are shown on Figure 10, respectively.

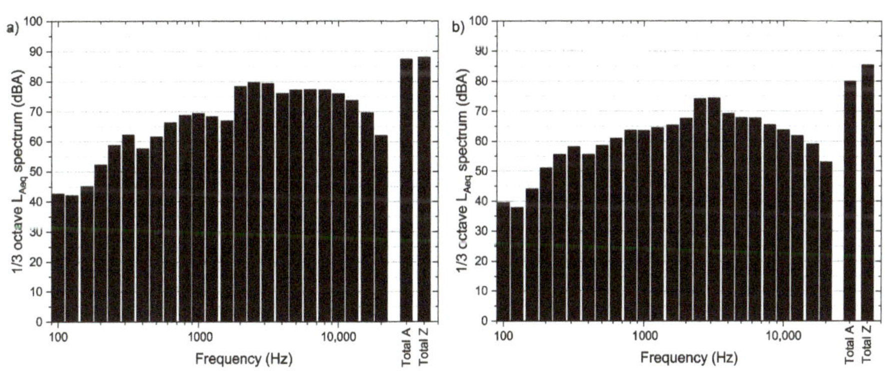

Figure 10. The 1/3 octave band equivalent of A-weighted sound pressure level spectrum of the noise of a hand–held pneumatic screwdriver: (**a**) at standard factory configuration(with pneumatic muffler); (**b**) with expanded collecting system.

Sound spectrum (Figure 10b) shows a significant decrease in the sound pressure level, especially for high frequencies. Based on the analysis of the presented measurement results, it can be seen that the total A SPL value for the situation with the expanded collecting system is 79.9 dBA, while at standard factory configuration(with pneumatic muffler), it is 87.3 dBA. After the expanded collecting system was used, the noise generated on the stand while the operator used a hand-held pneumatic screwdriver was almost reduced to the background noise level in the production hall. Therefore, it was not possible to achieve a lower value.

The results of the tests showed that the noise level at the workplace is very high, almost 90 dBA, despite the use of factory-made pneumatic mufflers. The measurements confirmed the high potential of the proposed solution, which reduces noise level from pneumatic tools and machines in production hall. However, due to the high acoustic background of the hall and the noise generated directly from the neighboring stands, the average noise level at the stand was reduced from 82.8 to 81.1 dBA. It is worth noting that a 1 dBA change in noise level is just a noticeable difference. Therefore, the expanded collection system should be applied to at least one subsection of window-fitting stands to achieve a significant noticeable difference.

Figure 11 shows the proposed design of the expanded collecting system for one subsection in the production hall. At each stand, there is a local expansion collector to which all tools are connected through a pneumatic tube. The local expansion collector is then linked to the main expansion collector. It should be noted that, at a given time, only one pneumatic tool is used at a given stand; therefore, the diameters of the tubes to and from local collectors may have the same internal diameter—in this case, d = 12 mm. According to the Equation (22), the minimum volumes of local and main collectors are: 15 dm^3 and 60 dm^3, respectively. As such, the expanded collecting system will remove the pneumatic noise in the showed subsection, leaving only noise coming from other sections of the production hall.

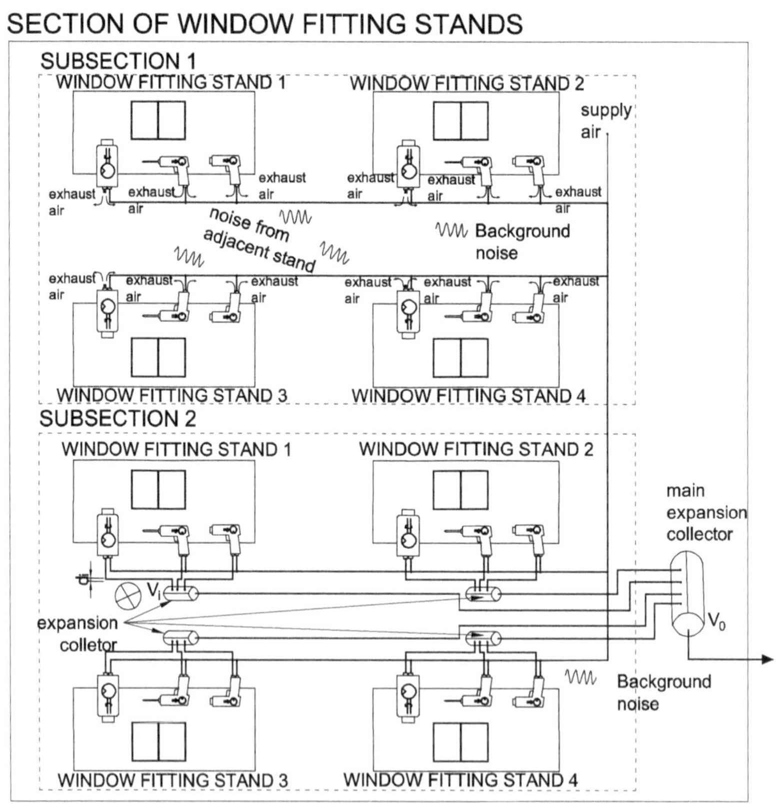

Figure 11. Proposed modification of the subsection no 2 of window-fitting stands by applied the EC system.

5. Conclusions

The article presents a new, consistent method of reducing pneumatic noise from machines and tools in production halls. The expanded collecting system captures exhaust air from pneumatic tools and machines outlets scattered throughout the production hall in one main collector. Then, the collected exhaust air in one place can be neutralized or thrown outside the production hall and reducing the noise generated as a result of air expansion in the working environment.This solution has been tested on the industrial scale.

To illustrate how our approach works, we built a mathematical model of the entire process of sound generation during the exhaust air expansion. The pneumatic line model is necessary for assessment of the dynamics of expanding air. We obtain changes in mass flow and pressure over time and at the end of pipes. Such observables are sensitive to the line topology (i.e., the diameter and length of pipes) and variables of initial conditions, such as, for example, pressure in expansion collector. Finally, through the use the acoustic theory and observables determined above, one may assess the noise level over time.Then, the model is validated with the measurement results of a pneumatic screwdriver without a muffler. The analytical results show a very good correlation with the measurement data at a high-frequency sound, from 1 to 16 kHz, and also for low frequencies of 100 and 200 Hz. The discrepancy of the mathematical model for low frequencies from 200 Hz to 1 kHz may result from unidentified acoustic sources in the real system, and the not-smooth combination of two sound generation formulas.

Then, measurement of the noise reduction in the hand-held pneumatic screwdriver in a partially insulated workshop was carried out. The results showed an approximately 8-fold reduction in the SPL (from 96 do 78 dBA) of the pneumatic screwdriver with the expanded collecting system compared to a standard configuration with a factory muffler. The expanded collecting system is able to reduce noise by from 1.5 to even 4 times compared to current solutions, which are capable of reducing pneumatic noise to 82–90 dBA. It should be emphasized that the tests were conducted in a separate workshop, but still located in the production hall.

Then, the industrial test was carried out in the production hall (see Figure 5) at a very high background noise of 77.9 dBA, caused by the large number of machines and pneumatic tools. Moreover, at one window-fitting stand, the operator is exposed to an average noise level of 82.8 dBA, temporarily exceeding 91.8 dBA. Thus, the noise level at the window-fitting stand exceeded 80 dBA, which, according to the Occupational Safety and Health Administration, is considered dangerous to human health. On the other hand, when the operator used a pneumatic screwdriver with the expanded collecting system, the noise level was reduced to 79.9 dBA, which was almost the background noise level. Because it is impossible to separate the background noise from the stand in the production hall, further noise reduction is presented. It would be necessary to apply the solution to subsequent tools and stands, as shown in Figure 11, to obtained further noise reduction.Due to the measured, very high level of acoustic background in the hall, of 77.9 dBA, noise measurements were carried out on one window-fitting stand. The aim was to apply the EC system to one pneumatic screwdriver and determine the effect on noise generated at the stand. The conducted noise measurements of the hand-held pneumatic screwdriver in an industrial environment confirm the possibility of reducing the pneumatic noise at a workplace on a full technical scale. When the expansion-collecting system is applied to the screwdriver, the measured SPL decreases from 87 to 80 dBA. It should be noted that the EC system was installed for only one pneumatic tool, so the implementation of the proposed system for the entire subsection would radically change the conditions of a working environment. Further, the EC system can also be used for other subsections or for other pneumatic machines in production hall.

Author Contributions: Conceptualization, J.S.L. and J.W.; methodology, D.C., J.S.L. and J.W.; software, D.G.; validation, D.G. and D.C.; formal analysis, D.G., D.C., J.S.L. and J.W.; investigation, D.G. and D.C.; resources, D.G. and D.C.; writing—original draft preparation, D.G. and D.C.; writing—

review and editing, J.S.L. and J.W.; visualization, D.G. All authors have read and agreed to the published version of the manuscript.

Funding: This research was funded by research subvention supported by the Polish Ministry of Science and Higher Education grant number 16.16.210.476 and 16.16.130.942.

Institutional Review Board Statement: Not applicable.

Informed Consent Statement: Not applicable.

Data Availability Statement: Not applicable.

Acknowledgments: This research was supported in part by PLGrid Infrastructure.

Conflicts of Interest: The authors declare no conflict of interest.

References

1. Dindorf, R.; Wos, P.; Pawelec, K. Automatic device for indirect measurement of leakage flow rate in compressed air pipeline. *IOP Conf. Ser. Mater. Sci. Eng.* **2017**, *233*, 159–170. [CrossRef]
2. Dindorf, R.; Takosoglu, J.; Wos, P. *Development of Pneumatic Control Systems*, m89 ed.; Wydawnictwo Politechniki Świętokrzyskiej: Kielce, Poland, 2017.
3. Li, J.; Zhao, S. Optimization of valve opening process for the suppression of impulse exhaust noise. *J. Sound Vib.* **2017**, *389*, 24–40. [CrossRef]
4. Raghunathan, S.; Kim, H.D.; Setoguchi, T. Impulse noise and its control. *Prog. Aerosp. Sci.* **1998**, *34*, 1–44. [CrossRef]
5. Li, J.; Zhao, S.; Ishihara, K. Study on acoustical properties of sintered bronze porous material for transient exhaust noise of pneumatic system. *J. Sound Vib.* **2013**, *332*, 2721–2734. [CrossRef]
6. Lighthill, M.J. On sound generated aerodynamically I. General theory. *Proc. R. Soc. Lond. Ser. A. Math. Phys. Sci.* **1952**, *211*, 564–587. [CrossRef]
7. Lighthill, M.J. On sound generated aerodynamically II. Turbulence as a source of sound. *Proc. R. Soc. Lond. Ser. A Math. Phys. Sci.* **1954**, *222*, 1–32. [CrossRef]
8. Shi, H.; Zhao, S. Prediction of radiation characteristic of intermittent exhaust noise generated via pneumatic value. *Noise Control Eng. J.* **2009**, *57*, 157–168. [CrossRef]
9. Ivanov, Y.V. Reducing the aerodynamic noise of pneumatic nozzles in the pneumatic mechanisms of forges and presses. *Metallurgist* **2011**, *55*, 139–142. [CrossRef]
10. Zhao, S.; Wang, J.; Wang, J.; He, Y. Expansion-chamber muffler for impulse noise of pneumatic frictional clutch and brake in mechanical presses. *Appl. Acoust.* **2006**, *67*, 580–594. [CrossRef]
11. Official Journal of the European Union. *Directive 2003/10/EC of the European Parliament and of Council of 6 February 2003 on the Minimum Health and Safety Requirements Regarding the Exposure of Workers to the Risks Arising from Physical Agents (Noise)*; European Union: Luxembourg, 2003; pp. 38–44.
12. ISO 1999:2013. *Acoustics-Estimation of Noise-Induced Hearing Loss*; International Organization for Standardization: Geneva, Switzerland, 2006.
13. Czopek, D.; Korbiel, T.; Kukulski, B.; Małecki, P.; Pawlik, P.; Stępień, B.; Wszołek, T.; Wszołek, W. *New Methods of Signal Processing in the Selected Vibroacoustics Problems*; Monograph AGH University of Science and Technology: Cracow, Poland, 2016.
14. Kozupa, M.; Wiciak, J. Comparison of passive and active methods for minimization of sound radiation by vibrating clamped plate. *Acta Phys. Pol. A* **2011**, *119*, 1013–1017. [CrossRef]
15. Kozień, M.S.; Wiciak, J. Reduction of structural noise inside crane cage by piezoelectric actuators—FEM simulation. *Arch. Acoust.* **2008**, *33*, 643–652. [CrossRef]
16. Li, J.; Zhao, S.; Ishihara, K.; Shi, H. Numerical and Experimental Studies on Aerodynamic Characteristics of Pneumatic Exhaust with Perforated Panel Muffler. *Adv. Mater. Res.* **2011**, *294*, 2125–2129. [CrossRef]
17. Daggerhart, J.A.; Berger, E. Evaluation of Mufflers To Reduce Punch Press Air Exhaust Noise. *Noise Control Eng.* **1975**, *4*, 120–123. [CrossRef]
18. Leszczynski, J.S.; Grybos, D.; Borek, M. *Układ Do Kolektorowania Powietrza Odpadowego z Sekcji Instalacji Pneumatycznej*; PL432516A1; Polish Patent Office: Warsaw, Poland, 2020.
19. Krichel, S.V.; Sawodny, O. Non-linear friction modelling and simulation of long pneumatic transmission lines. *Math. Comput. Model. Dyn. Syst.* **2014**, *20*, 23–44. [CrossRef]
20. Patankar, S.V. *Numerical Heat Transfer and Fluid Flow*; Hemisphere Publishing Corporation: New York, NY, USA, 1980.
21. Beater, P. *Pneumatic Drives*; Springer: Berlin/Heidelberg, Germany, 2007; p. 319, [CrossRef]
22. Cengel, Y.A.; Boles, M.A. *Thermodynamics. An Engineering Approach*, 5th ed.; Mc.Graw–Hill Education: New York, NY, USA, 2015.
23. Leszczynski, J.S.; Grybos, D. Compensation for the complexity and over-scaling in industrial pneumatic systems by the accumulation and reuse of exhaust air. *Appl. Energy* **2019**, *239*, 1130–1141. [CrossRef]
24. Idelchik, I.E. Handbook of Hydraulic Resistance, 2nd Edition. *J. Press. Vessel. Technol.* **1987**, *109*, 260–261. [CrossRef]

25. Voronin, S.F. Effect of contraction on the friction coefficient in a turbulent gas flow. *Inzh. Fiz. Z.* **1959**, *23*, 81–85.
26. Bellos, V.; Nalbantis, I.; Tsakiris, G. Friction Modeling of Flood Flow Simulations. *J. Hydraul. Eng.* **2018**, *144*, 04018073. [CrossRef]

energies

Article

Study on the Sensitivity of a Gyroscope System Homing a Quadcopter onto a Moving Ground Target under the Action of External Disturbance

Izabela Krzysztofik * and **Zbigniew Koruba**

Faculty of Mechatronics and Mechanical Engineering, Kielce University of Technology, Aleja Tysiąclecia Państwa Polskiego 7, 25-314 Kielce, Poland; ksmzko@tu.kielce.pl
* Correspondence: pssik@tu.kielce.pl

Abstract: This paper investigates the sensitivity (resistance) of a quadcopter on-board gyroscope system for the observation and tracking of a moving ground target to changing parameters of its regulator under interference conditions. It was shown that the gain in matrix elements is most sensitive, and even their slightest deviation from optimal values can lead to reduced target tracking efficiency and even loss of control system stability. Furthermore, the authors studied the energy expenditure at various gyroscope system control parameter values, while homing a quadcopter onto a ground target. A Matlab/Simulink environment was used to conduct simulations of the controlled gyroscope system dynamics. Selected test results are shown in graphic form.

Keywords: gyroscope system; sensitivity; optimal regulator; quadcopter

Citation: Krzysztofik, I.; Koruba, Z. Study on the Sensitivity of a Gyroscope System Homing a Quadcopter onto a Moving Ground Target under the Action of External Disturbance. *Energies* **2021**, *14*, 1696. https://doi.org/10.3390/en14061696

Academic Editor: Francesco Castellani

Received: 31 January 2021
Accepted: 16 March 2021
Published: 18 March 2021

1. Introduction

Currently, one of the most important elements in the equipment of a quadcopter unmanned aerial vehicle (QUAV) is its observation and tracking head. It is used to automatically search and track ground targets, both moving and stationary. Its objective is to determine the position of target line of sight (TLOS) [1,2]. A Gyroscope System (GS) was suggested as a device to control and stabilize TLOS. A relevant issue in terms of such devices is their control under conditions of disturbance induced by the QUAV maneuvering deck.

Previous studies on the dynamics of unmanned aerial vehicles, including a quadcopter, indicated that external disturbance acting on them led to significant errors in the tracking, laser illumination and homing onto both a moving and stationary target [3,4]. In particular, the control system onboard the aforementioned aerial vehicles, did not provide sufficient resistance to vibrations [5–7]. A head with controlled gyroscope (its drive) parameters selected optimally for homing precision, should be chosen in order to minimize the aforementioned error.

Such parameters can change in the course of gyroscopic system operation, and the system's sensitivity to their modification should be tested. This is mainly about determining the scope of change within which the tracking and illumination of a ground target is still sufficiently accurate.

This paper discusses an example of a quadcopter equipped with an EFP (Explosively Formed Projectile) shaped charge that can attack tanks or armored vehicles from the upper ceiling, i.e., from an altitude of several dozen meters, in which an observation and tracking head scans the surface of the Earth from onboard the drone, searching for an object emitting infrared radiation. Upon detecting a target, the QUAV enters the self-homing phase using the proportional navigation method. The target can be simultaneously illuminated with a laser beam enabling it to be attacked, with other external means of precise striking (antitank missile or a bomb homing onto a reflected laser beam). It is also possible for an ultralight drone to land directly on a tank in order to emit signals that can be intercepted by the heads of other means of attack, e.g., rocket missiles on ground launchers or fixed under the wings of flying vehicles (helicopters, aircraft or unmanned aerial vehicles).

In the source literature, there are many examples of the widespread use of drones in various areas of today's life, such as in [8,9]. It should be noted that the proposed system in this paper differs significantly from the above-mentioned works and the those discussed in [10,11], mainly due to the use of an original gyro guidance head. Also, many algorithms and onboard systems have already been developed for quadcopter stabilization and motion control. Most of them use sensors to measure the position of the copter, such as gyroscopes, accelerometers or IMU/MEMS units [12,13] but lack information on conducted studies involving the use of a mechatronic controller gyroscope onboard a quadcopter. It should be stressed that one of the major advantages of such a gyroscope is its resistance to vibrations and random interference, including noise. Furthermore, the authors propose gyroscope system controls, optimal in terms of precision and energy expenditure, ensuring the most stable and effective homing of the drone onto a target.

The parts of this publication have been organized as follows: Section 2 discusses a method for the determination of optimal parameters for the controlled GS. Section 3 includes the results of an analysis covering the simulation tests of the gyroscope system sensitivity to changing its regulator parameters under the conditions of interference, when illuminating a ground target with a QUAV onboard laser. It also contains the results of simulation tests regarding GS control energy expenditure within the process of homing a quadcopter onto a moving ground target. Section 4 summarizes the study and presents the final conclusions.

2. Determining Optimal Parameters for a Controlled Gyroscope System

A linearized model of a controlled gyroscope system is expressed in the following form [14–16]:

$$\frac{d\boldsymbol{x}_g}{dt} = \boldsymbol{A}_g \boldsymbol{x}_g - \boldsymbol{B}_g \boldsymbol{u}_g, \tag{1}$$

where:

$\boldsymbol{x}_g = \begin{bmatrix} \vartheta_g & \dot{\vartheta}_g & \psi_g & \dot{\psi}_g \end{bmatrix}^T$—state vector, $\boldsymbol{u}_g = \begin{bmatrix} M_b & M_c \end{bmatrix}^T$—control vector,

$\boldsymbol{A}_g = \begin{bmatrix} 0 & 1 & 0 & 0 \\ 0 & -b_b & 0 & -1 \\ 0 & 0 & 0 & 1 \\ 0 & 1 & 0 & -b_c \end{bmatrix}$—state matrix, $\boldsymbol{B}_g = \begin{bmatrix} 0 & 0 \\ c_b & 0 \\ 0 & 0 \\ 0 & c_c \end{bmatrix}$—control matrix,

$b_b = \frac{\eta_b}{J_{gk}\Omega}$, $b_c = \frac{\eta_c}{J_{gk}\Omega}$, $c_b = c_c = \frac{1}{J_{gk}\Omega^2}$, $\Omega = \frac{J_{go} n_g}{J_{gk}}$,

ϑ_g, ψ_g—angles defining the position of the GS axis in space,

M_b, M_c—control moments,

η_b, η_c—damping coefficients in GS frame suspension bearings,

J_{go}—moment of inertia of a GS rotor relative to the longitudinal axis,

J_{gk}—moment of inertia of a GS rotor relative to the transverse axis,

n_g—rotary speed of the GS rotor.

In order to provide the controlled gyroscope, described by Equation (1), with the stability and shortest decay time of transition to a set value, let us introduce optimal control in the form:

$$\boldsymbol{u}_g = -\boldsymbol{K}_g \boldsymbol{x}_g, \tag{2}$$

where:

$\boldsymbol{K}_g = \begin{bmatrix} k_{11} & k_{12} & k_{13} & k_{14} \\ k_{21} & k_{22} & k_{23} & k_{24} \end{bmatrix}$—gain matrix.

After taking into account the above assumptions using the LQR method and analytical solution of the Riccati equation, individual elements of the gain matrix $\boldsymbol{K}_g$ for the analysed gyroscope system satisfy the following relationship [14]:

$$k_{11} = k_{23} = \overline{k}_b,\ k_{12} = k_{14} = k_{22} = k_{24} = \overline{h}_g,\ k_{21} = -k_{13} = \overline{k}_c. \tag{3}$$

After substituting gain coefficient (3) to (2), correction controls will be expressed in the following form:

$$M_b = -\overline{k}_b\overline{\vartheta}_g + \overline{k}_c\psi_g - \overline{h}_g\frac{d\vartheta_g}{dt}, \tag{4}$$

$$M_c = -\overline{k}_c\overline{\vartheta}_g - \overline{k}_b\psi_g - \overline{h}_g\frac{d\psi_g}{dt}, \tag{5}$$

where:

$$\overline{k}_b = \frac{k_b}{J_{gk}\Omega^2},\ \overline{k}_c = \frac{k_c}{J_{gk}\Omega^2},\ \overline{h}_g = \frac{h_g}{J_{gk}\Omega}. \tag{6}$$

Therefore, the open-system gyroscope system (1), taking into account (4), i.e., after substituting M_b and M_c in the control vector $\boldsymbol{u}_g$, is reduced to a new form (closed-system):

$$\frac{d\boldsymbol{x}_g}{dt} = \boldsymbol{A}_g^*\boldsymbol{x}_g, \tag{7}$$

where:

$$\boldsymbol{A}_g^* = \begin{bmatrix} 0 & 1 & 0 & 0 \\ -\overline{k}_b & -\overline{h}_g - b_b & +\overline{k}_c & 1 \\ 0 & 0 & 0 & 1 \\ -\overline{k}_c & -1 & -\overline{k}_b & -\overline{h}_g - b_c \end{bmatrix}.$$

Further, let us assume that friction in the gyroscope suspension bearings is negligible i.e., $b_b = b_c = 0$. For such a described gyroscope system, let us additionally look for such parameters and inter-relations, which guarantee the shortest transient process damping time. In this case let us also use a modified method of the Golubientsev optimization method, which consists in ensuring the fastest disappearance of transient processes that appear after switching on the control, or a sudden operation of a disturbance. The algorithm of this method is presented in [14].

Using the Hurwitz stability criterion and the modified Golubientsev optimization method [14,17], we obtain the following system of equations and inequalities:

$$\overline{k}_b > 0,\ \overline{k}_c > 0,\ \overline{h}_g > 0, \tag{8}$$

$$2\overline{k}_b - \frac{1}{2}\overline{h}_g^2 + 1 > 0, \tag{9}$$

$$\overline{k}_c = \frac{1}{2}\overline{h}_g, \tag{10}$$

$$\frac{1}{16}\overline{h}_g^4 + \frac{1}{4}\overline{h}_g^2 - \frac{1}{2}\overline{h}_g^2\overline{k}_b - \overline{h}_g\overline{k}_c + \overline{k}_b^2 + \overline{k}_c^2 > 0. \tag{11}$$

Taking into account the maximization condition for the absolute trace value of matrix $\boldsymbol{A}_g^*$

$$\left|Tr\boldsymbol{A}_g^*\right| \to max, \tag{12}$$

Using inequality (9) we obtain the following value of the damping coefficient:

$$\overline{h}_g = \sqrt{2 + 4\overline{k}_b}. \tag{13}$$

Substituting (13) to the Equation (10) we get:

$$\overline{k}_c = \frac{1}{2}\sqrt{2 + 4\overline{k}_b}. \tag{14}$$

Whereas, after taking into account (6) we get:

$$h_g = \sqrt{2J_{go}^2 n_g^2 + 4J_{gk}k_b}, \tag{15}$$

$$k_c = \frac{1}{2}\frac{J_{go}^2 n_g^2}{J_{gk}}\sqrt{2J_{go}^2 n_g^2 + 4J_{gk}k_b}. \tag{16}$$

Thus, coefficients $\overline{h}_g$ and $\overline{k}_c$ are explicitly defined as the gyroscope parameter functions J_{go}, J_{gk} and n_g of the coefficient $\overline{k}_b$, which should satisfy the stability criteria as well as the technical restrictions arising from the strength of the gyroscope structure itself.

The obtained relationships can be used for controlling the gyroscope under conditions of changing specific rotation angular velocity (e.g., in certain drones with a limited and unstable power source). In such a case, one should measure the $n_g(t)$ values on an ongoing basis and updated the $\overline{h}_g$ and $\overline{k}_c$ regulator coefficient values in accordance with the relationships (15) and (16). Coefficient $\overline{k}_b$ is set through software. This enables adaptive gyroscope control. The adaptive control algorithm described in [14] should be used for cases of numerous other parameters of the gyroscope changing over time. Figures 1 and 2 graphically characterize the relationships between individual parameters of the gyroscope and its regulator. In order to obtain the aforementioned relationships, it was assumed that $J_{gk} = J_{go}/2$.

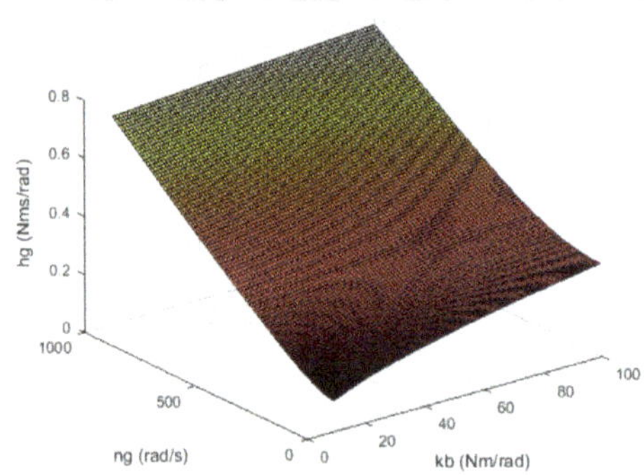

Figure 1. Graph of optimal inter-relations between the regulator damping coefficient h_g, angular velocity n_g and the gain coefficient k_b.

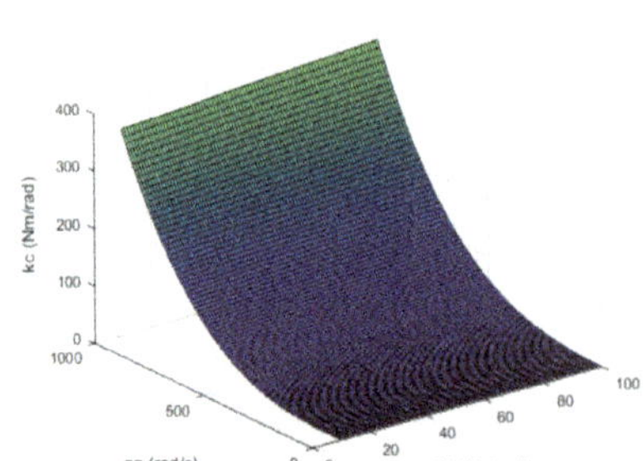

Figure 2. Graph of optimal inter-relations between the regulator gain coefficient k_c, angular velocity n_g and the gain coefficient k_b.

Therefore, if the system of terms (8)–(11) is satisfied, the gyroscope system transient process induced by external disturbance (1) is damped first.

3. Test Results

This section presents the results and analysis of the tests covering the dynamics of the controlled gyroscope system with the following parameters [2,18]:

$$J_{x_1} = 2.5{\cdot}10^{-5}\ \text{kgm}^2,\ J_{y_1} = J_{x_1},\ J_{z_1} = J_{x_1},$$

$$J_{x_2} = 5{\cdot}10^{-5}\ \text{kgm}^2,\ J_{y_2} = J_{x_2},\ J_{z_2} = J_{x_2},$$

$$J_{x_3} = J_{go} = 5{\cdot}10^{-4}\ \text{kgm}^2,\ J_{y_3} = J_{gk} = 2.5{\cdot}10^{-4}\ \text{kgm}^2,$$

$$J_{z_3} = J_{x_3},\ m_2 = 0.1\ \text{kg},\ m_3 = 0.14\ \text{kg},\ m_g = m_2 + m_3,$$

$$n_g = 600\ \text{rad/s};\ \eta_b = \eta_c = 0.05\ \text{Nm/s}.$$

The kinematic excitations acting on the gyroscope system were assumed to be harmonic forms with an amplitude equal to 0.5 rad/s and frequency equal to 15 rad/s.

The simulation testing was conducted in the Matlab/Simulink environment (Version R2020a, MathWorks, Natick, MA, USA), with an integration step of $dt = 0.00001$ s [19].

3.1. Test Results Regarding the Sensitivity of a Gyroscope System during Tracking and Laser Illumination of a Ground Target

Let us assume that the initial conditions do not match set conditions:

$$\psi_{go} \neq \psi_{gzo},\ \vartheta_{go} \neq \vartheta_{gzo},\ \dot{\psi}_{go} \neq \dot{\psi}_{gzo},\ \dot{\vartheta}_{go} \neq \dot{\vartheta}_{gzo},$$

and are equal:

$$\psi_{go} = 5\ \text{deg},\ \vartheta_{go} = -5\ \text{deg},\ \dot{\psi}_{go} = 0,\ \dot{\vartheta}_{go} = 0.$$

Control moments were adopted in the following form:

$$M_b = -k_b e_\vartheta + k_c e_\psi - h_g \dot{e}_\vartheta,\ M_c = -k_c e_\vartheta - k_b e_\psi - h_g \dot{e}_\psi,$$

where:

$$e_\vartheta = \vartheta_g - \vartheta_{gz},\ \dot{e}_\vartheta = \dot{\vartheta}_g - \dot{\vartheta}_{gz},\ e_\psi = \psi_g - \psi_{gz},\ \dot{e}_\psi = \dot{\psi}_g - \dot{\psi}_{gz}.$$

It was also assumed that the task of gyroscope system control was displacement over a minimum time and maintaining the gyroscope axis in a position consistent with the target line of sight position, with an error below 0.5 degrees, i.e., 0.0087 rad. After satisfying this condition, it activates only the laser system to illuminate the target. The target should be steadily illuminated, with preset precision, regardless of the drone maneuvers, vibrations of its deck and other external disturbances, such as wind gusts [20,21] or projectile explosion. Such disturbances appeared within the simulation in question, in the period between 15 to 20 s.

Figures 3–6 contain simulation results for nonoptimal regulator coefficients:

$$k_b = 10,\ k_c = 100,\ h_g = 100.$$

The simulation results presented in Figures 3–6 clearly show that the controller parameters are chosen incorrectly. There are significant deviations of the actual values of the position angles of the GS axis ψ_g, ϑ_g from the pre-set values ψ_{gz}, ϑ_{gz} (Figures 3 and 4). Consequently, the control error reaches large values (Figure 6). The kinematic excitations occurring after 15 s of motion have a very negative effect on the motion of the gyroscope system, which can also be seen in Figure 5.

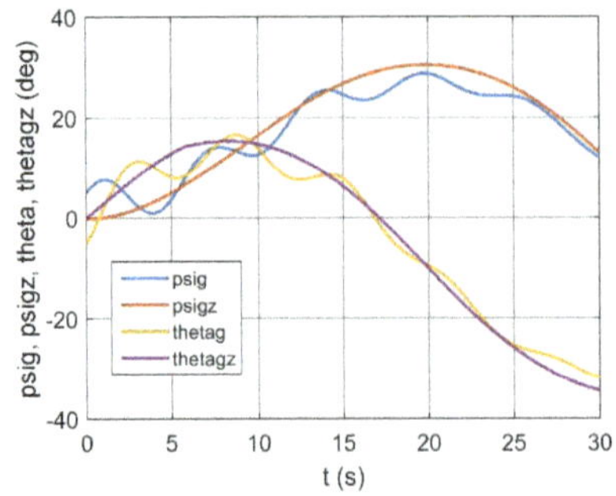

Figure 3. Real and set angle of gyroscope system (GS) deflection and inclination as a function of time.

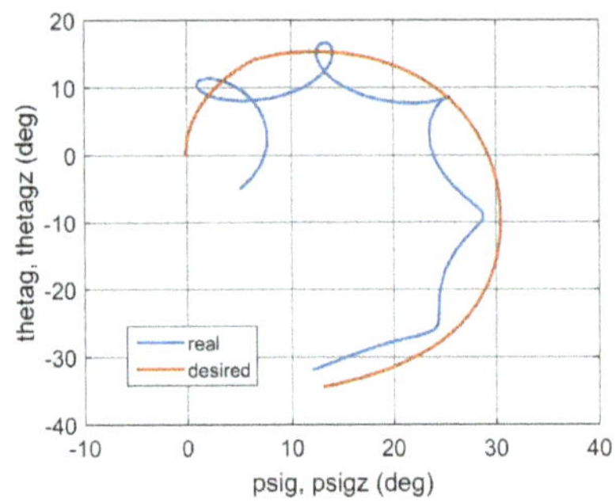

Figure 4. Real and desired motion trajectory.

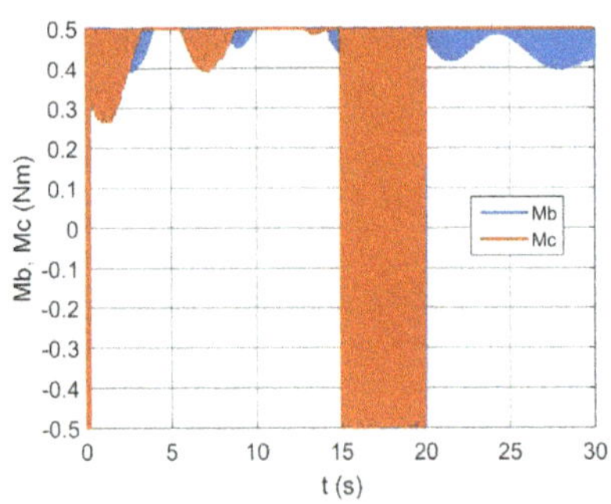

Figure 5. Control moments as a function of time.

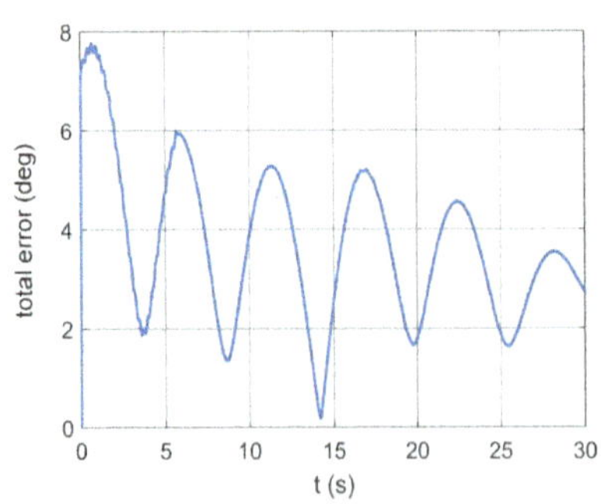

Figure 6. Total error as a function of time.

Figures 7–10 show simulation results for other nonoptimal values of regulator coefficients of

$$k_b = 10,\ k_c = 10,\ h_g = 10.$$

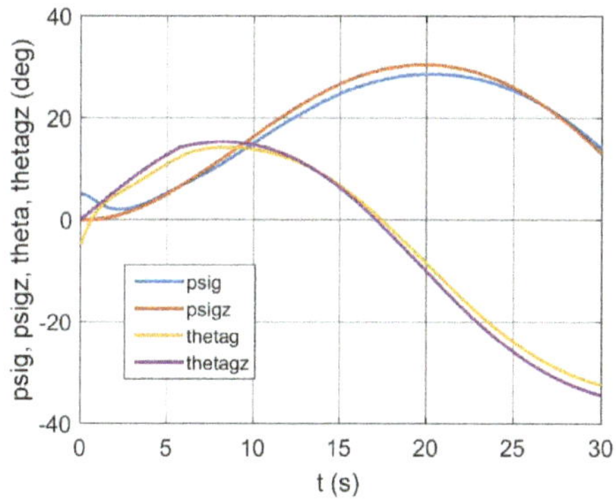

Figure 7. Real and set angle of GS deflection and inclination as a function of time.

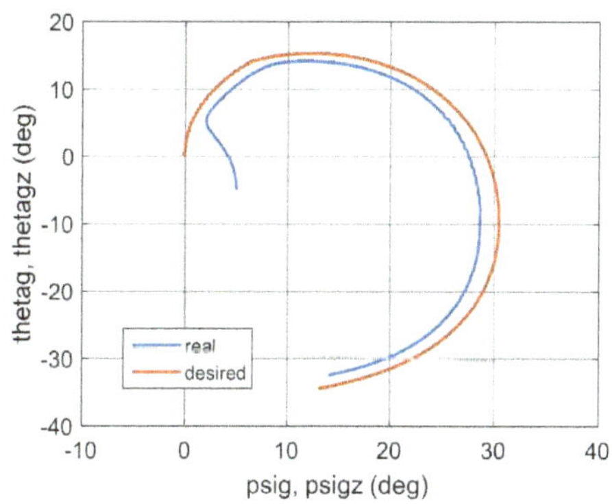

Figure 8. Real and desired motion trajectory.

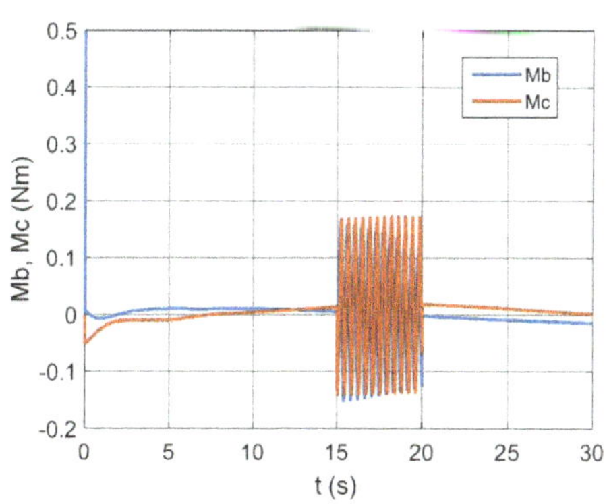

Figure 9. Control moments as a function of time.

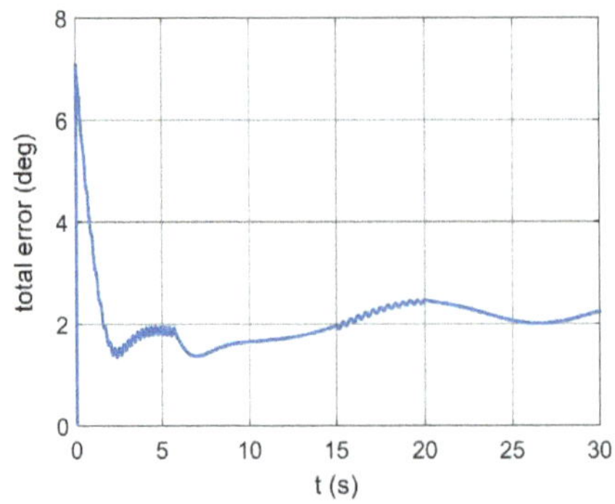

Figure 10. Total error as a function of time.

From the simulation results shown in Figures 7–10, it can be seen that the motion of the gyroscope system is more stable. The deviations from the set values are smaller than for the case shown in Figures 3–6. However, the control error is still too large, whereas Figures 11–14 contain test results for a regulator coefficient of $k_b = 10$, while k_c and h_g are determined based on relationships (13)–(14) (i.e., optimal).

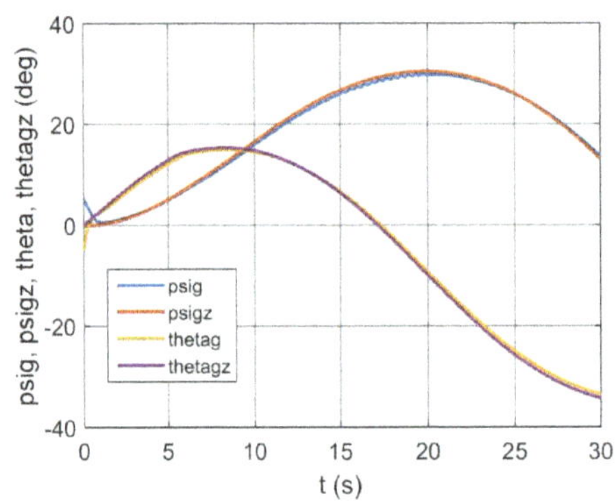

Figure 11. Real and set angle of GS deflection and inclination as a function of time.

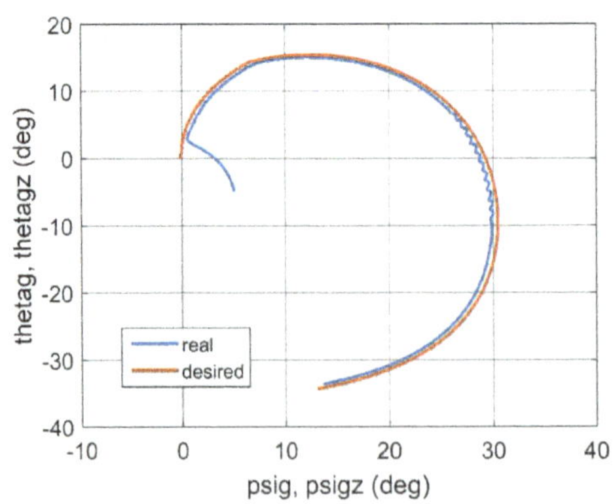

Figure 12. Real and desired motion trajectory.

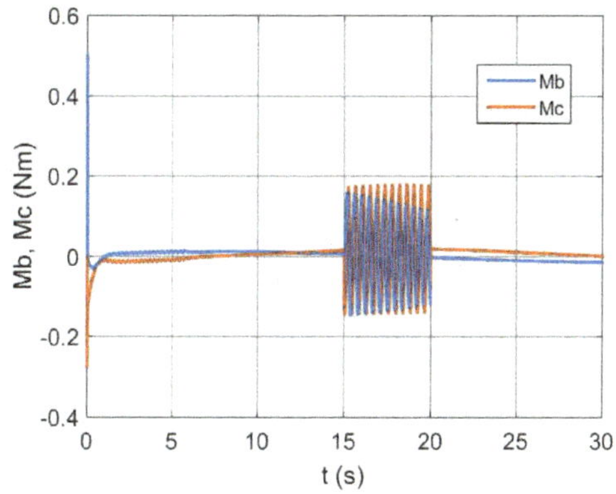

Figure 13. Control moments as a function of time.

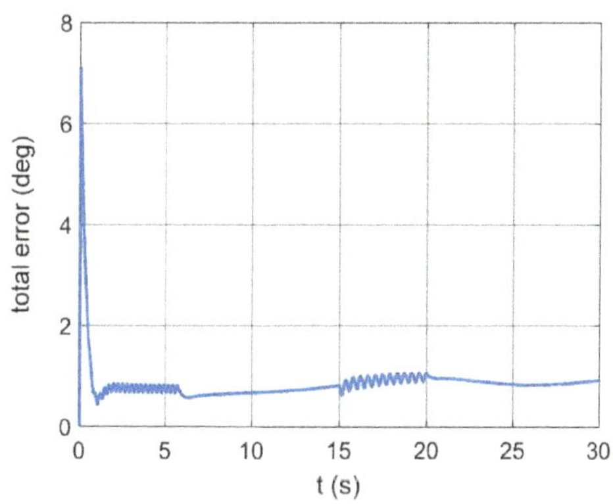

Figure 14. Total error as a function of time.

The results presented in Figures 11–14 show that the controller with the optimal parameters work properly. The trajectory of the actual motion almost coincides with the desired trajectory. The total control error for the steady-state motion oscillates around a value equal to 1 degree.

Figures 15–18 shows the test results for optimal regulator coefficients of $k_b = 348$ and k_c and h_g determined based on relationships (13)–(14).

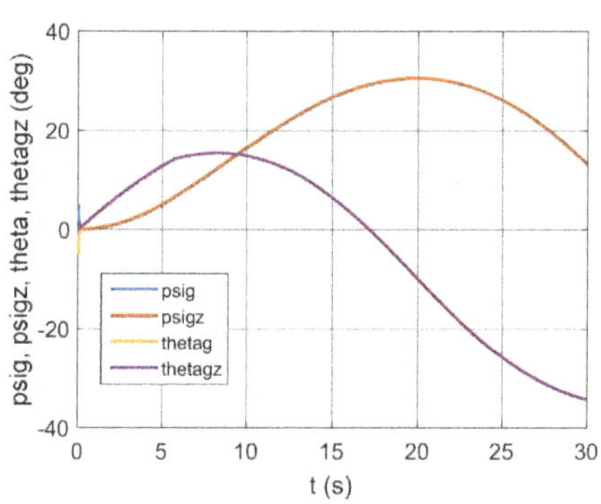

Figure 15. Real and set angle of GS deflection and inclination as a function of time.

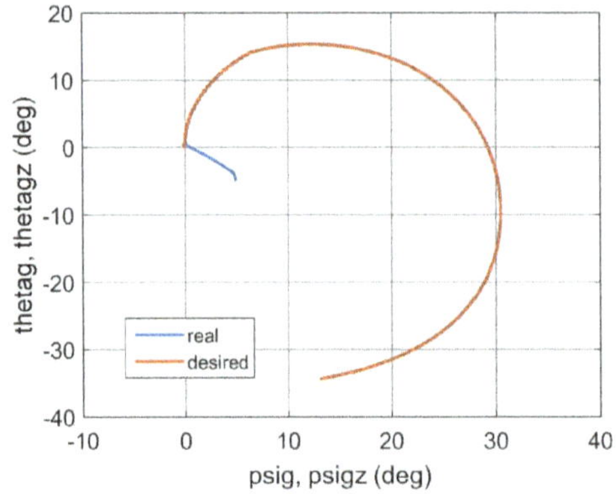

Figure 16. Real and desired motion trajectory.

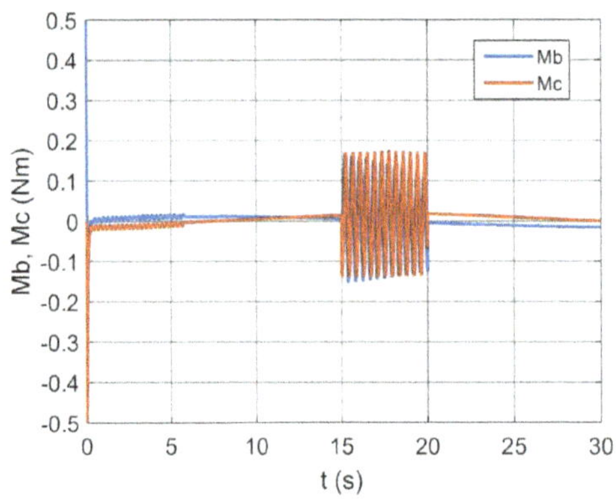

Figure 17. Control moments as a function of time.

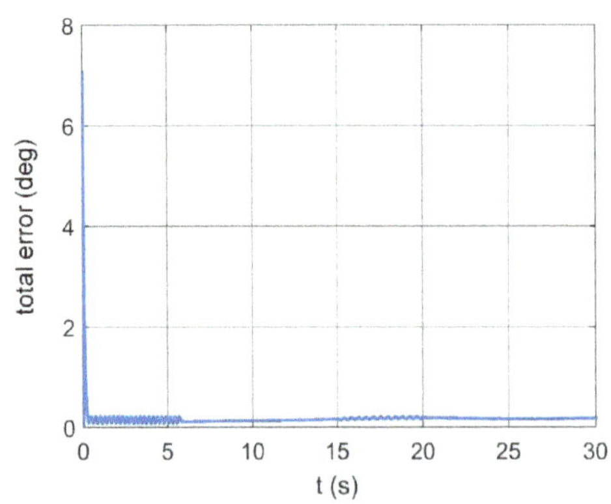

Figure 18. Total error as a function of time.

Such a selection of control coefficients causes the changes over time of the actual angles of GS axis position and the set angles to coincide almost perfectly, which can also be seen in Figure 16. The influence of kinematic excitations on GS motion is imperceptible. The total control error is close to zero.

Some selected test results shown in Figures 3–18 indicate that the most important regulator parameter is the k_b coefficient, which was ultimately selected so that the control moments did not exceed the permissible absolute values of 0.5 Nm on one hand, and the total error between the set and implemented gyroscope axis position was below 0.5 degrees (0.0087 rad) on the other. Other coefficients were determined based on the relationships (7)–(15). Numerous tests involving gyroscope system sensitivity indicated that with an optimally selected k_b coefficients, other coefficients, namely k_c and h_g, varying within 30% of the optimum values, did not cause significant errors in maintaining the gyroscope axis in accordance with the target line of sight. Errors exceeding permissible values, i.e., axis deviation from the set value higher than 0.0087 rad appeared after leaving the aforementioned change range of the coefficients. At the same time, the control moments reached unacceptable values.

3.2. Simulation Studies Involving the Control over an Optimum Gyroscope System for Homing onto a Ground Target from Onboard a Quadcopter

In order to test the operating effectiveness of a gyroscope system with optimally selected parameters when homing a quadcopter onto a moving ground target under conditions of external disturbance, the following controls, quality indicators and initial conditions of drone and target motion were adopted.

GS axis control moments:

$$M_b = -k_b e_\vartheta + k_c e_\psi - h_g \dot{e}_\vartheta,\ M_c = -k_c e_\vartheta - k_b e_\psi - h_g \dot{e}_\psi$$

where:

$$e_\vartheta = \vartheta_g - \varepsilon;\ \dot{e}_\vartheta = \dot{\vartheta}_g - \dot{\varepsilon};\ e_\psi = \psi_g - \sigma;\ \dot{e}_\psi = \dot{\psi}_g - \dot{\sigma}$$

ε, σ—deflection and inclination angles of the target line of sight, determined from homing kinematics equations.

In terms of selecting the optimum parameters for the gyroscope system in question, the following two quality indicators were adopted:

1. IAE (Integral Absolute Error) quality indicator:

$$\mathrm{IAE} = \int_0^\infty |e_c| dt$$

where:

$e_c = \sqrt{e_\vartheta^2 + e_\psi^2}$—total error.

2. ISSC (Integral Square State and Control) quality indicator:

$$\mathrm{ISSC} = \int_0^\infty \left(x^T x\right) dt + \int_0^\infty \left(u^T u\right) dt$$

where:

$$x = \begin{bmatrix} \vartheta_g & \psi_g & \dot{\vartheta}_g & \dot{\psi}_g \end{bmatrix},\ u = \begin{bmatrix} M_b & M_c \end{bmatrix}.$$

Initial positions and angular velocities of gyroscope system axis position:

$$\psi_g = \sigma + 0.2 \text{ rad},\ \vartheta_g = \varepsilon + 0.2 \text{ rad},\ \dot{\psi}_g = 0;\ \dot{\vartheta}_g = 0.$$

Initial ground target movement conditions:

$$x_c = 150 \text{ m}, y_c = 50 \text{ m}, z_c = 0.$$

$$V_c = 25 \text{ m/s},\ \gamma_c = 0,\ \chi_c = 0.$$

Initial drone flight parameters:

$$x_s = 0,\ y_s = 0,\ z_s = 500\ \text{m}.$$

$$V_s = 75\frac{\text{m}}{\text{s}},\ \gamma_s = \varepsilon,\ \chi_s = \sigma.$$

Coefficient k_b is selected heuristically, while the optimum PD regulator gain coefficients is calculated from relationships (13)–(14).

The analysis involved six variants with two GS control parameter values.

The test results shown in Figures 19–48 and determined quality indicator values in Table 1 indicate that effective control over a gyroscope system when homing a quadcopter onto a moving ground target under external disturbance conditions requires the application of optimum regulator parameters, with values from a range determined based on the tests discussed in Section 3.1.

Table 1. Quality indicators.

Variant	Regulator Parameters	ISSC	IAE
1	$k_b = 10,\ k_c = 10,\ h_g = 10$	9.2403×10^8	5.3726×10^3
2	$k_b = 10,\ k_c = 100,\ h_g = 100$	9.6861×10^8	7.2331×10^3
3	$k_b = 10,\ k_c,\ h_g$ – optimum	1.1070×10^9	4.3352×10^3
4	$k_b = 348,\ k_c,\ h_g$ – optimum	1.6691×10^9	1.7290×10^3
5	$k_b = 348,\ k_c,\ h_g = 0.7 *$ optimum	1.6845×10^9	1.8137×10^3
6	$k_b = 348,\ k_c, h_g = 1.3 *$ optimum	1.6263×10^9	1.9237×10^3

The tests also indicated that with an optimally selected coefficient k_b, the values of coefficients k_c and h_g can change over a broad range but cannot exceed 30% of the optimum values. Figures 36 and 46 clearly show that a gyroscope system in variant 6 was faster in "entering" the set trajectory but the IAE indicator reached a much higher value relative to variant 4, with comparable values of the ISSC indicator for both variants.

A similar situation applies to variant 5. The value of coefficients k_c and h_g were 30% lower than the optimum values, which significantly increased the gyroscope system homing error. This is particularly visible when comparing Figures 36 and 41.

In conclusion, it should be stated that the assumed homing accuracy is achieved for optimum gyroscope control system regulator values adopted in variant 4.

Variant 1.

$$k_b = 10; k_c = 10; h_g = 10.$$

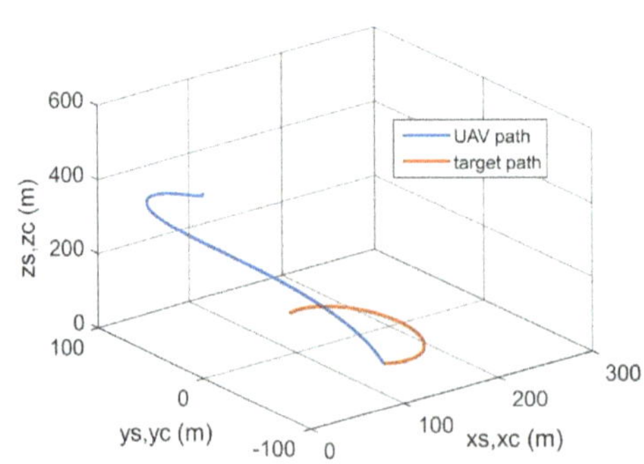

Figure 19. Trajectories of unmanned aerial vehicle (UAV) and target flight.

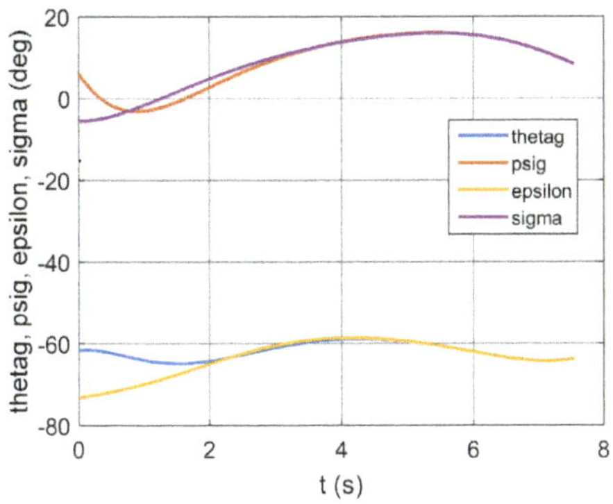

Figure 20. Changes of realizing and pre-set deflection and inclination angles of GS as a time function.

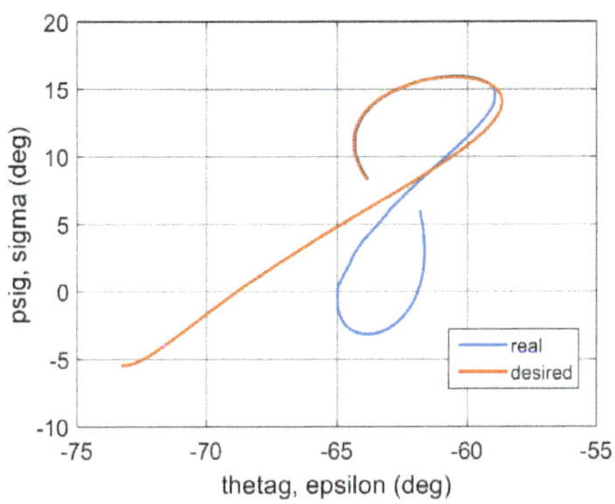

Figure 21. Trajectories of the real and desired motion.

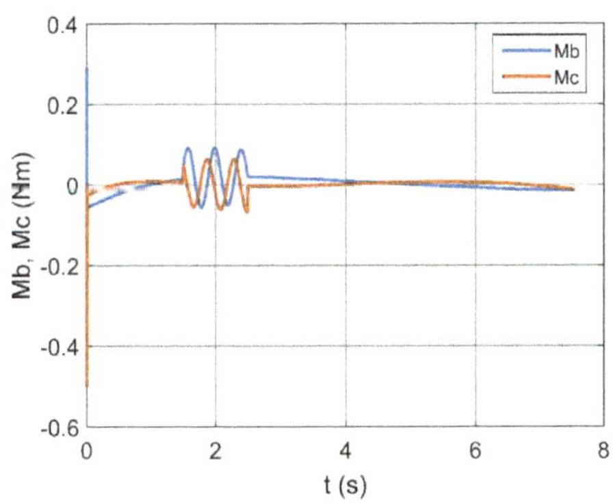

Figure 22. Control moments as a function of time.

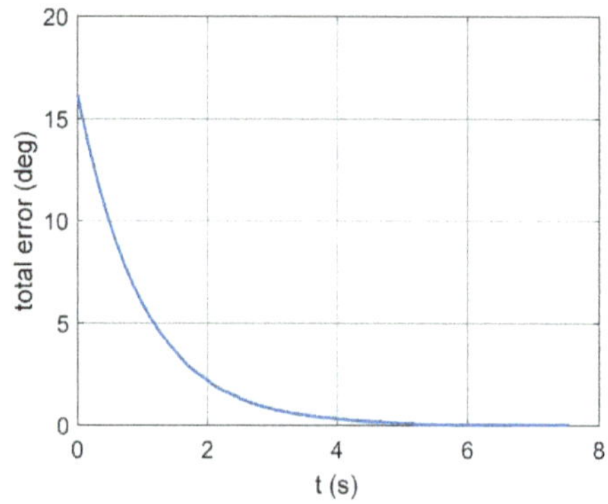

Figure 23. Total error as a function of time.

Variant 2.

$$k_b = 10; k_c = 100; h_g = 100.$$

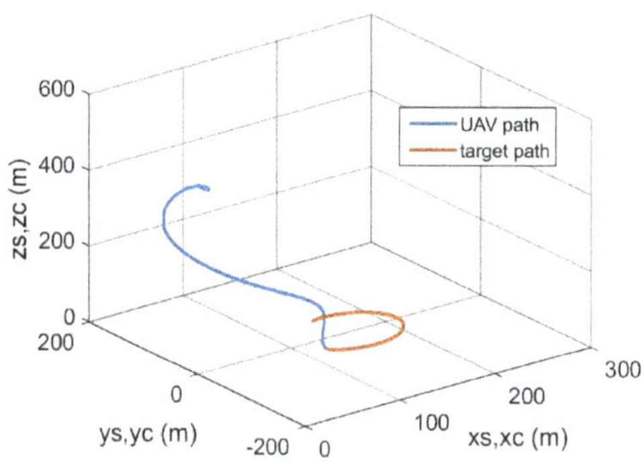

Figure 24. Trajectories of UAV and target flight.

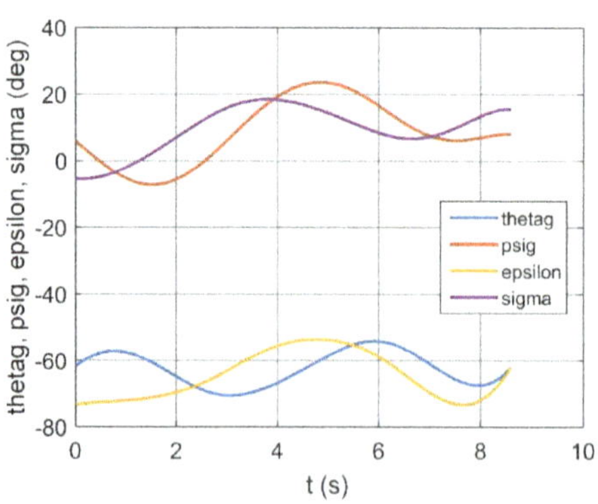

Figure 25. Changes of realizing and pre-set deflection and inclination angles of GS as a time function.

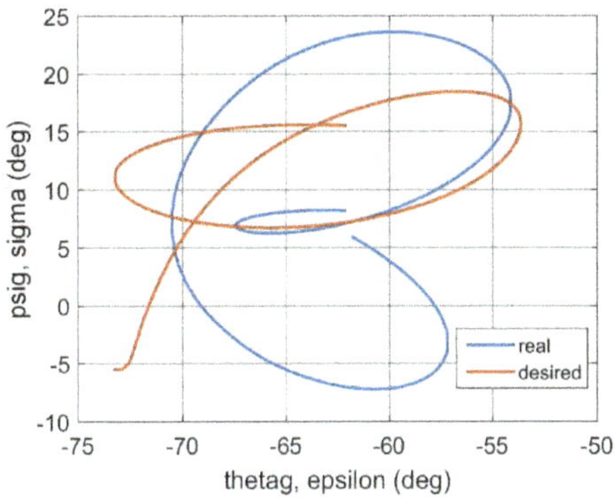

Figure 26. Trajectories of the real and desired motion.

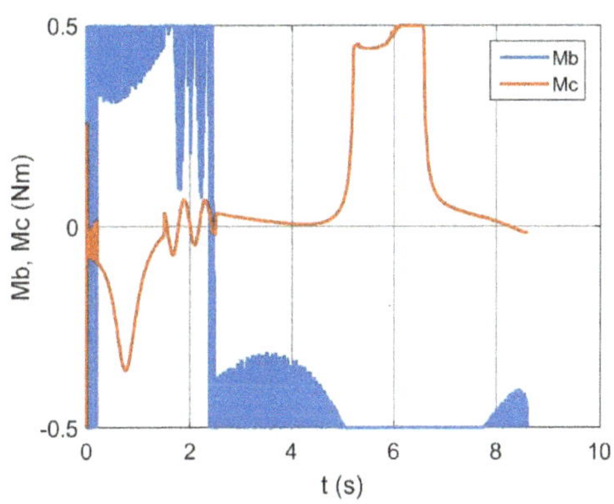

Figure 27. Control moments as a function of time.

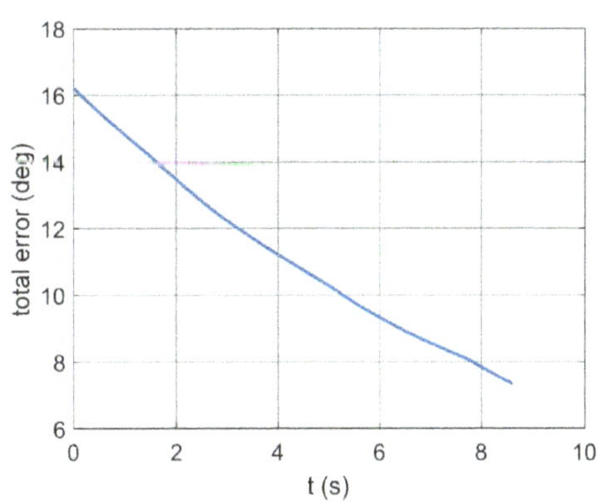

Figure 28. Total error as a function of time.

Variant 3.

$k_b = 10$, whereas k_c, h_g are adopted optimum values.

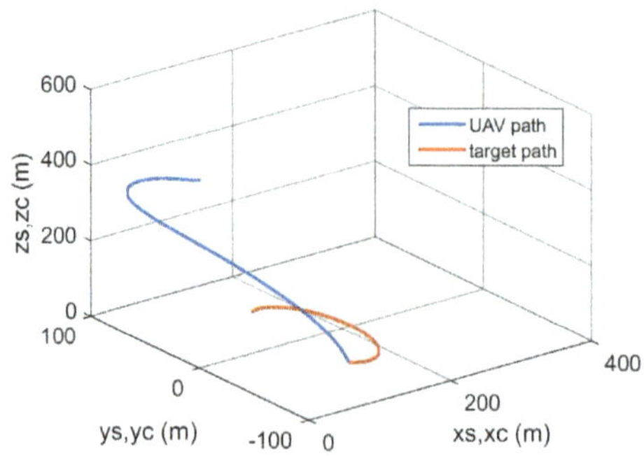

Figure 29. Trajectories of UAV and target flight.

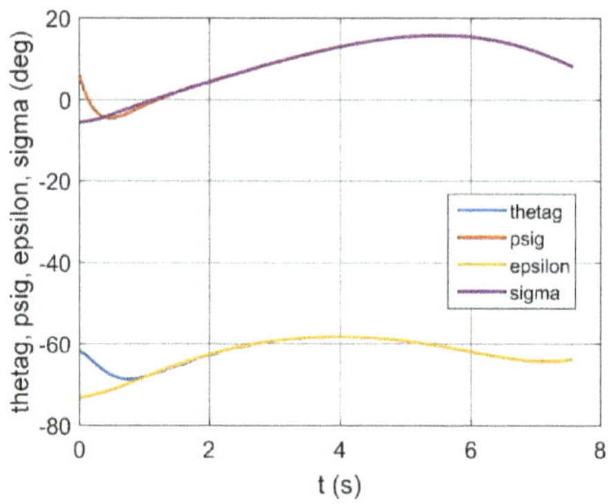

Figure 30. Changes of realizing and pre-set deflection and inclination angles of GS as a time function.

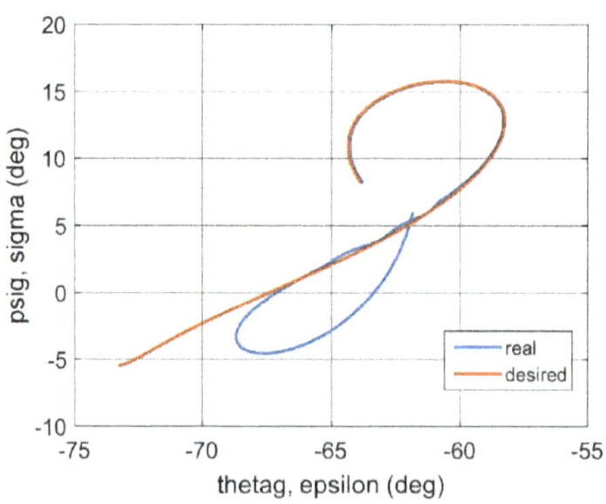

Figure 31. Trajectories of the real and desired motion.

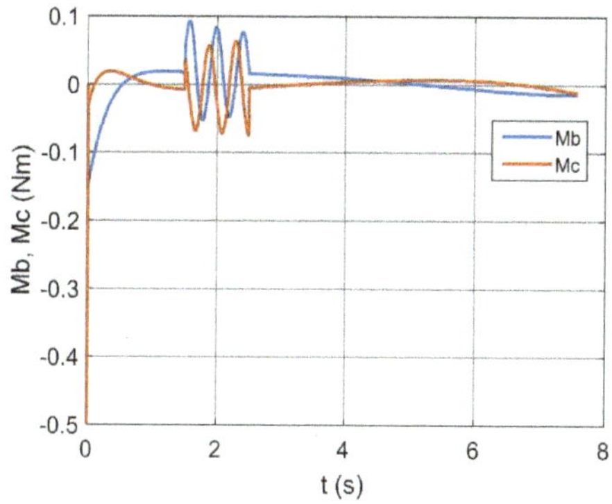

Figure 32. Control moments as a function of time.

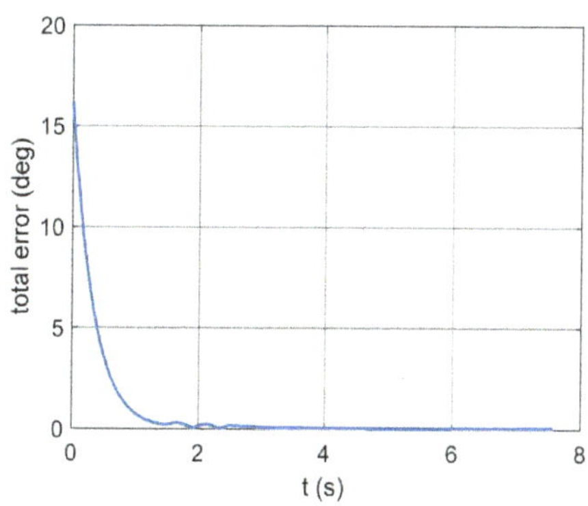

Figure 33. Total error as a function of time.

Variant 4.

$k_b = 348$, whereas k_c, h_g adopted optimum values.

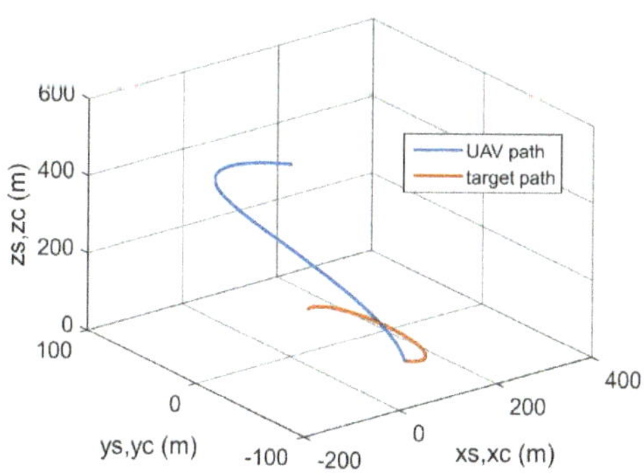

Figure 34. Trajectories of UAV and target flight.

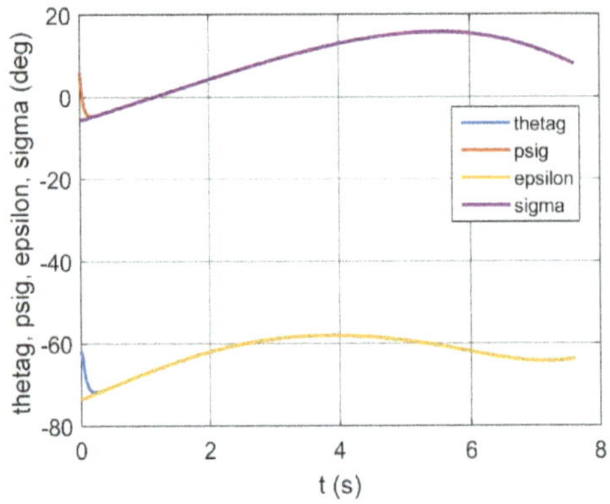

Figure 35. Changes of realizing and preset deflection and inclination angles of GS as a time function.

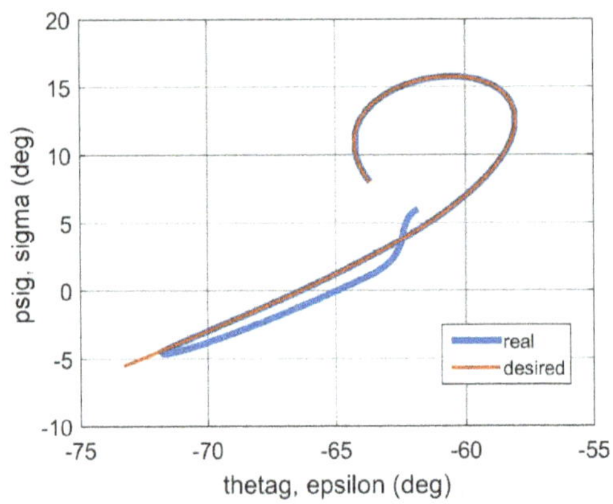

Figure 36. Trajectories of the real and desired motion.

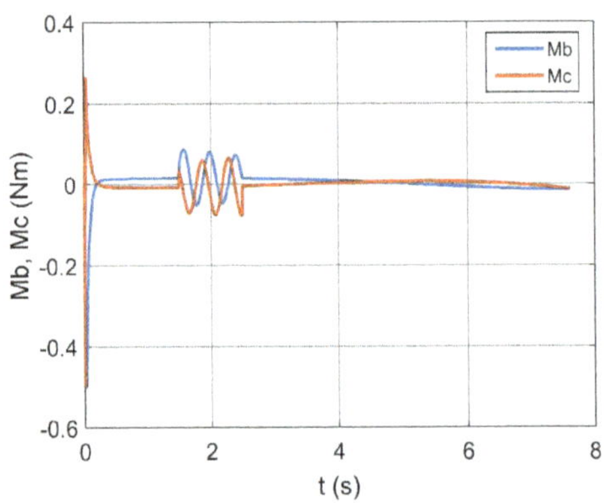

Figure 37. Control moments as a function of time.

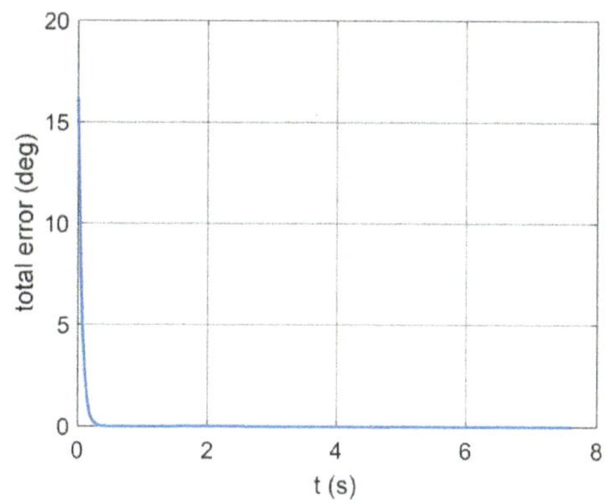

Figure 38. Total error as a function of time.

Variant 5.

$k_b = 348$, whereas k_c, h_g are adopted values 30% lower than optimum.

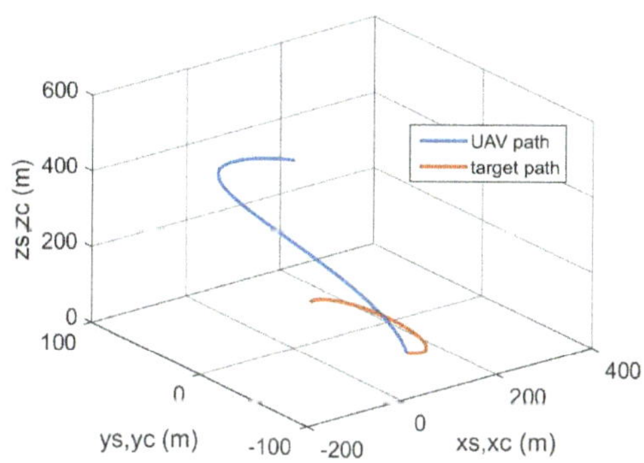

Figure 39. Trajectories of UAV and target flight.

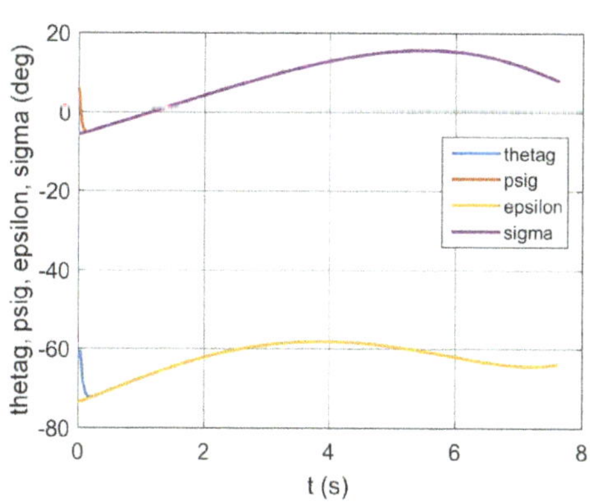

Figure 40. Changes of realizing and pre-set deflection and inclination angles of GS as a time function.

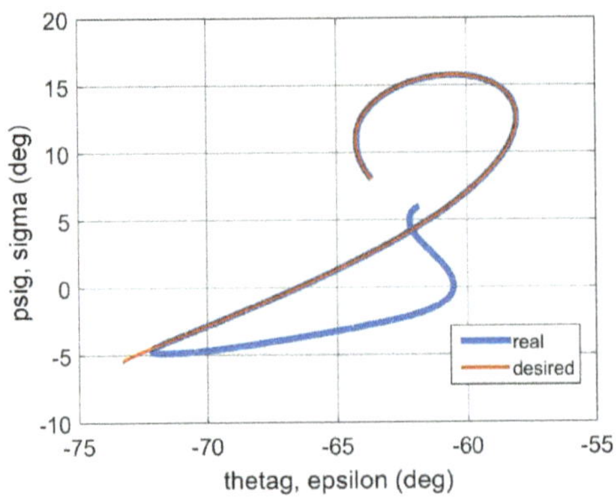

Figure 41. Trajectories of the real and desired motion.

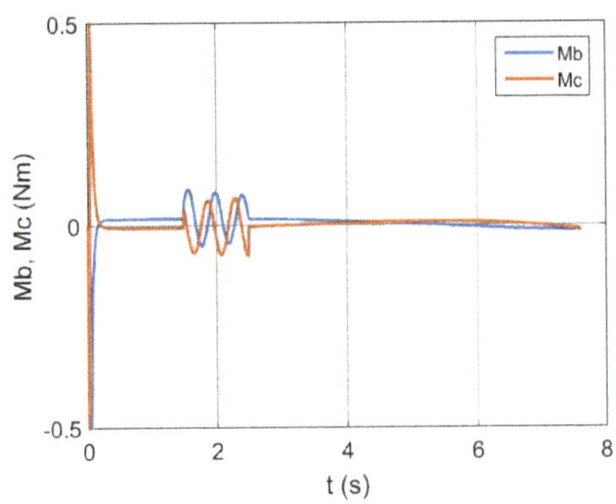

Figure 42. Control moments as a function of time.

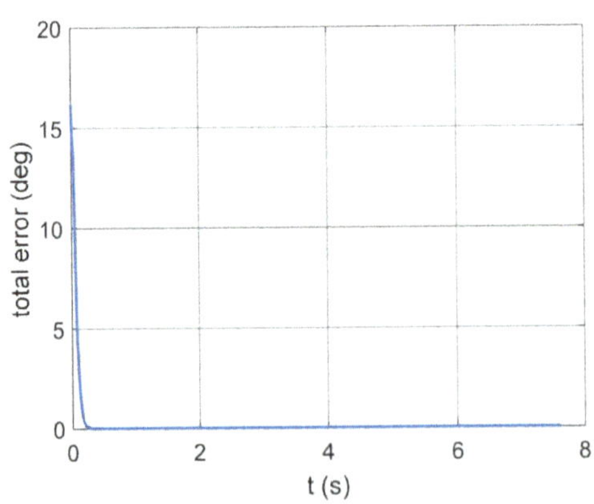

Figure 43. Total error as a function of time.

Variant 6.

$k_b = 348$, whereas k_c, h_g are adopted values 30% higher than optimum.

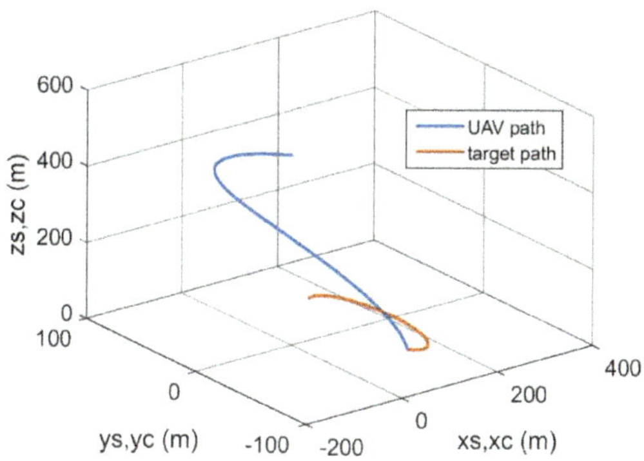

Figure 44. Trajectories of UAV and target flight.

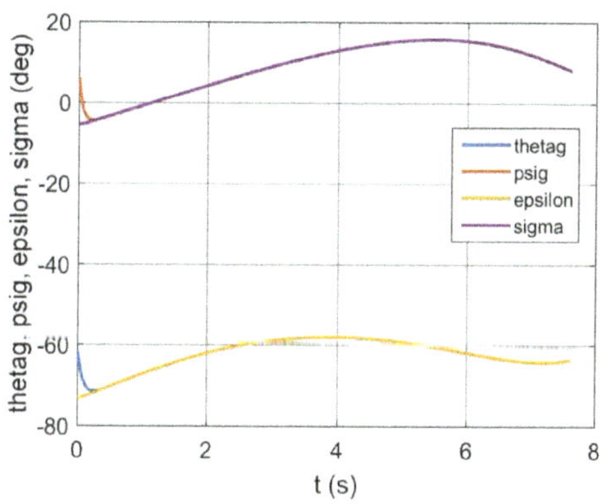

Figure 45. Changes of realizing and preset deflection and inclination angles of GS as a time function.

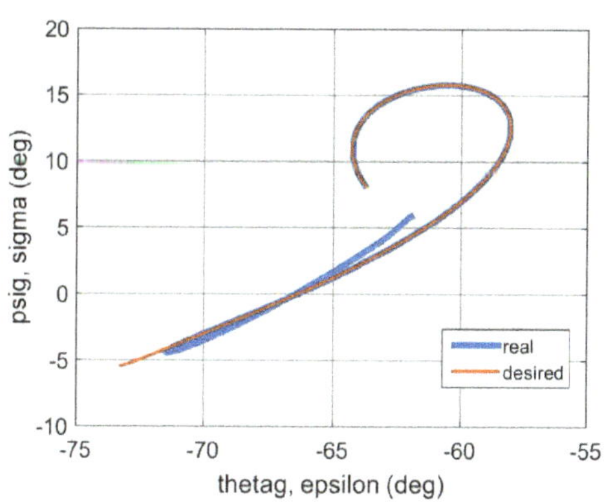

Figure 46. Trajectories of the real and desired motion.

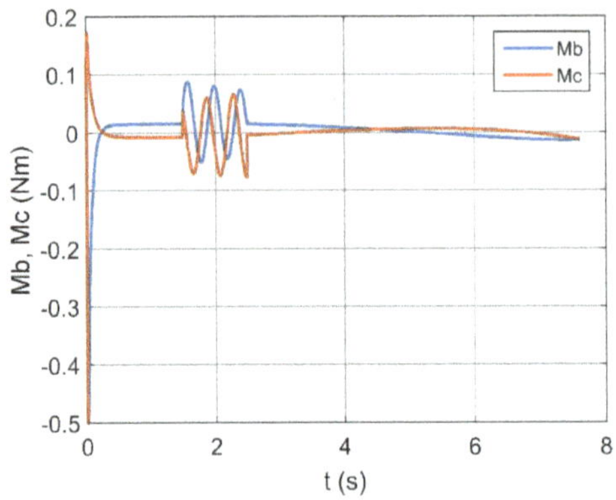

Figure 47. Control moments as a function of time.

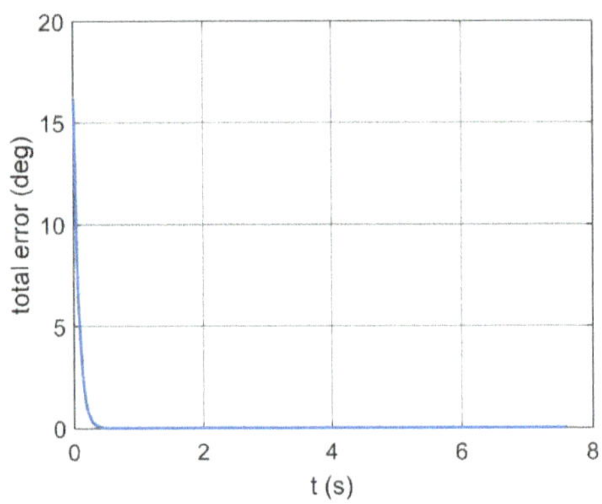

Figure 48. Total error as a function of time.

Table 1 shows the values of determined quality indicators.

The data presented in Table 1 show that the IAE indicator is the decisive criterion for the selection of regulator parameters. The ISSC does not give unambiguous answers. The lowest value of the IAE indicator was achieved for variant 4. The parameters of the regulator determined in this variant made it possible to obtain the highest accuracy of guiding the quadcopter to the moving target.

4. Conclusions

The optimum parameters for controlling the gyroscope axis position presented herein minimize the error between the set and desired movements to acceptable values and reduce the impact of kinematic excitation of the QUAV base and external disturbance acting on the drone. The authors studied sensitivity and determined the optimum parameter change ranges for stable, precise tracking, laser illumination and homing onto a moving ground target from onboard a drone.

Due to a minimum offset between the set and implemented gyroscope axis position, a specific optimum coefficient k_b was determined for the controlled gyroscope system in question. The GS is very sensitive to changes in this coefficient. Other coefficients, k_c and h_g, are functions of k_b and the GS is not really sensitive to changes in their values, since they can vary by up to 30% of optimum values without a significant impact on the precision of ground target tracking and laser illumination.

Summing up, it can be concluded that the application of a gyroscope system with optimally selected parameters ensures stable and precise QUAV homing onto a moving ground target under external disturbance.

The next stage of the research will be the application of the optimal parameters set out in this paper for the experimental verification of the operation of the gyro system guiding the quadcopter onto a moving ground target to prove ground conditions.

Author Contributions: Conceptualization, I.K. and Z.K.; methodology, I.K. and Z.K.; software, I.K.; validation, I.K. and Z.K.; formal analysis, I.K. and Z.K.; investigation, I.K.; resources, I.K. and Z.K.; data curation, I.K. and Z.K.; writing—original draft preparation, I.K.; writing—review and editing, I.K. and Z.K.; visualization, I.K.; supervision, Z.K. All authors have read and agreed to the published version of the manuscript.

Funding: This research received no external funding.

Institutional Review Board Statement: Not applicable.

Informed Consent Statement: Not applicable.

Data Availability Statement: Not applicable.

Conflicts of Interest: The authors declare no conflict of interest.

References

1. Gapinski, D.; Krzysztofik, I. The process of tracking an air target by the designed scanning and tracking seeker. In Proceedings of the 2014 15th International Carpathian Control Conference (ICCC), Velke Karlovice, Czech Republic, 28–30 May 2014; IEEE: New York, NY, USA; pp. 129–134. [CrossRef]
2. Stefański, K.; Koruba, Z. Analysis of the guiding of bombs on ground targets using a gyroscope system. *J. Theor. Appl. Mech.* **2012**, *50*, 967–973.
3. Chodnicki, M.; Bartnik, K.; Nowakowski, M.; Kowaleczko, G. Design and analysis of a feedback loop to regulate the basic parameters of the unmanned aircraft. *Aircr. Eng. Aerosp. Technol.* **2018**, *92*, 318–328. [CrossRef]
4. Kowaleczko, G.; Pietraszek, M.; Słonkiewicz, Ł. Analysis of the Impact of the Target Illumination Time on the Effectiveness of the Flight Trajectory Correction System. *J. Konbin* **2019**, *49*, 229–253. [CrossRef]
5. Hasseni, S.; Abdou, L. Decentralized PID Control by Using GA Optimization Applied to a Quadrotor. *J. Autom. Mob. Robot. Intell. Syst.* **2018**, *12*, 33–44. [CrossRef]
6. Praveen, V.; Pillai, A.S. Modeling and simulation of quadcopter using PID controller. *Int. J. Control Theory Appl.* **2016**, *9*, 7151–7158.
7. Zulu, A.; John, S. A Review of Control Algorithms for Autonomous Quadrotors. *Open J. Appl. Sci.* **2014**, *4*, 547–556. [CrossRef]
8. Xuan-Mung, N.; Hong, S.K. Robust Backstepping Trajectory Tracking Control of a Quadrotor with Input Saturation via Extended State Observer. *Appl. Sci.* **2019**, *9*, 5184. [CrossRef]
9. Xuan-Mung, N.; Hong, S.K.; Nguyen, N.P.; Ha, L.N.N.T.; Le, T.-L. Autonomous Quadcopter Precision Landing Onto a Heaving Platform: New Method and Experiment. *IEEE Access* **2020**, *8*, 167192–167202. [CrossRef]
10. Xuan-Mung, N.; Hong, S.-K. Improved Altitude Control Algorithm for Quadcopter Unmanned Aerial Vehicles. *Appl. Sci.* **2019**, *9*, 2122. [CrossRef]
11. Xuan-Mung, N.; Hong, S.-K. A Multicopter Ground Testbed for the Evaluation of Attitude and Position Controller. *Int. J. End. Technol.* **2018**, *7*, 65–73.
12. Shi, X.; Lu, L.; Jin, G.; Tan, L. Research on the attitude of small UAV based on MEMS devices. *AIP Conf. Proc.* **2017**, *1839*, 020094. [CrossRef]
13. Kumar, P.; Praveen, B.; Nisanth, U.; Hemanth, M. Development of Auto Stabilization Algorithm for UAV Using Gyro Sensor. *Int. J. Eng. Res. Technol.* **2017**, *5*, 1–3.
14. Awrejcewicz, J.; Koruba, Z. *Classical Mechanics. Applied Mechanics and Mechatronics*; Springer: New York, NY, USA, 2012; Volume 30.
15. Lewis, F.L.; Vrabie, D.L.; Syrmos, V.L. *Optimal Control*; John Wiley & Sons: Hoboken, NJ, USA, 2012.
16. Tewari, A. *Modern Control Design with Matlab and Simulink*; John Wiley & Sons: Chichester, UK, 2002.
17. Kaczorek, T.; Dzieliński, A.; Dąbrowski, W.; Łopatka, R. *Fundamentals of Control Theory*; WNT: Warsaw, Poland, 2013.
18. Krzysztofik, I.; Takosoglu, J.; Koruba, Z. Selected methods of control of the scanning and tracking gyroscope system mounted on a combat vehicle. *Annu. Rev. Control* **2017**, *44*, 173–182. [CrossRef]
19. Baranowski, L. Effect of the mathematical model and integration step on the accuracy of the results of computation of artillery projectile flight parameters. *Bull. Pol. Acad. Sci. Tech. Sci.* **2013**, *61*, 475–484. [CrossRef]

20. Grzyb, M.; Stefanski, K. Turbulence impact on the control of guided bomb unit. In Proceedings of the 23rd International Conference Engineering Mechanics 2017, Svratka, Czech Republic, 15–18 May 2017; Brno University of Technology: Brno, Czech Republic; pp. 358–361.
21. Kowaleczko, G.; Żyluk, A. Influence of atmospheric turbulence on bomb release. *J. Theor. Appl. Mech.* **2009**, *47*, 69–90.

energies

MDPI

Article

A New Uncertainty-Based Control Scheme of the Small Modular Dual Fluid Reactor and Its Optimization

Chunyu Liu [1], Run Luo [1,2,*] and Rafael Macián-Juan [1]

[1] Chair of Nuclear Technology, Department of Mechanical Engineering, Technical University of Munich (TUM), 85748 Garching, Germany; chunyu.liu@tum.de (C.L.); rafael.macian@ntech.mw.tum.de (R.M.-J.)

[2] School of Resource & Environment and Safety Engineering, University of South China, Hengyang 421001, China

* Correspondence: luorun@usc.edu.cn or run.luo@tum.de

Abstract: The small modular dual fluid reactor is a novel variant of the Generation IV molten salt reactor and liquid metal fast reactor. In the primary circuit, molten salt or liquid eutectic metal (U-Pu-Cr) is employed as fuel, and liquid lead works as the coolant in the secondary circuit. To design the control system of such an advanced reactor, the uncertainties of the employed computer model and the physicochemical properties of the materials must be considered. In this paper, a one-dimensional model of a core is established based on the equivalent parameters achieved via the coupled three-dimensional model, taking into account delayed neutron precursor drifting, and a power control system is developed. The performance of the designed controllers is assessed, taking into account the model and property uncertainties. The achieved results show that the designed control system is able to maintain the stability of the system and regulate the power as expected. Among the considered uncertain parameters, the reactivity coefficients of fuel temperature have the largest influence on the performance of the control system. The most optimized configuration of the control system is delivered based on the characteristics of uncertainty propagation by using the particle swarm optimization method.

Keywords: small module dual fluid reactor; delayed neutron precursor drifting; load regulation; uncertainty-based optimization; particle swarm optimization; uncertainty and sensitivity analysis

Citation: Liu, C.; Luo, R.; Macián-Juan, R. A New Uncertainty-Based Control Scheme of the Small Modular Dual Fluid Reactor and Its Optimization. *Energies* **2021**, *14*, 6708. https://doi.org/10.3390/en14206708

Academic Editor: Ryszard Dindorf

Received: 2 August 2021
Accepted: 10 October 2021
Published: 15 October 2021

1. Introduction

The dual fluid reactor, which adopts molten salt or liquid eutectic metal (U-Pu-Cr) as fuel and liquid lead as a coolant, has a considerable number of advantages over conventional reactors. It is a novel variant of the Generation IV molten salt reactor and liquid metal fast reactor. A relevant study can be traced back to 1966, when a two-fluid molten salt breeder reactor (MSBR) was designed at Oak Ridge National Laboratory [1]. The proposed MSBR was investigated for its load-following capability under various ramp rates of the power demand without a controller [2], and its enlarged version of 2 GW thermal output was proposed by Taube et al. in 1974 [3]. Later, a new design of the two-fluid molten salt reactor, named the dual fluid reactor (DFR), was proposed by researchers at the Institue for Solid-State Nuclear Physics (IFK) [4]. By using liquid lead, the dual fluid reactor has a high capability to transfer heat produced in the core and can be operated at a considerably high power density without any safety issues. The characteristics of the steady-state physics of the dual fluid reactor concept were investigated [5,6], and a comparative study of the two fuel options was conducted [7]. The distribution zone of the dual fluid reactor concept has been designed and analyzed from the perspective of thermal hydraulics [8,9]. According to previous studies, the dual fluid reactor concept has a unique feature. Based on this fact, special attention has to be paid to the design and optimization of its control system to ensure that the reactor is able to follow the load as desired and has little overshoot/fluctuation as possible. However, to date, research on the control system of the dual fluid reactor concept has not been conducted, and the strategy for optimization of the controller is unknown. Moreover, the uncertainties introduced during the modeling process must be considered to ensure that the optimized control system is always reliable in spite of the deviations between the numerical model and reality.

In this work, a one-dimensional model of the core of the small modular dual fluid reactor (SMDFR) is built based on the equivalent parameters achieved by the coupled three-dimentional model, taking into account delayed neutron precursor (DNP) drifting, and a power control system is developed. Then, the most representative uncertain parameters of this model are chosen based on the experience of a previous study [10], and their influences on the numerical model are quantified. Finally, the technique of particle swarm optimization (PSO) is implemented to optimize the controller of the core considering its uncertainties.

2. Description of SMDFR and Control System

2.1. Small Modular Dual Fluid Reactor

Unlike the original dual fluid reactor design, the nominal power of the small modular dual fluid reactor was reduced to: $P_0 = 0.1\ \text{GW} = 100\ \text{MW}$, which is much lower compared to the original designed thermal power of 3 GW. The schematic of the SMDFR is shown in Figure 1. There are three circuits in the schematic: the primary circuit of molten salt as fuel is depicted in deep red; the secondary circuit of liquid lead as the primary coolant is depicted in yellow; and the tertiary circuit of liquid lead as the secondary coolant is depicted in blue. The mass flow rate of the molten salt in the primary circuit is notably low and is intended solely for online fuel processing instead of heat removal. The heat produced by fission in the core is majorly taken by the liquid lead as the primary coolant in the secondary circuit and then transferred to the secondary coolant in the tertiary circuit for utilization, such as hydrogen production, electricity generation, and water desalination

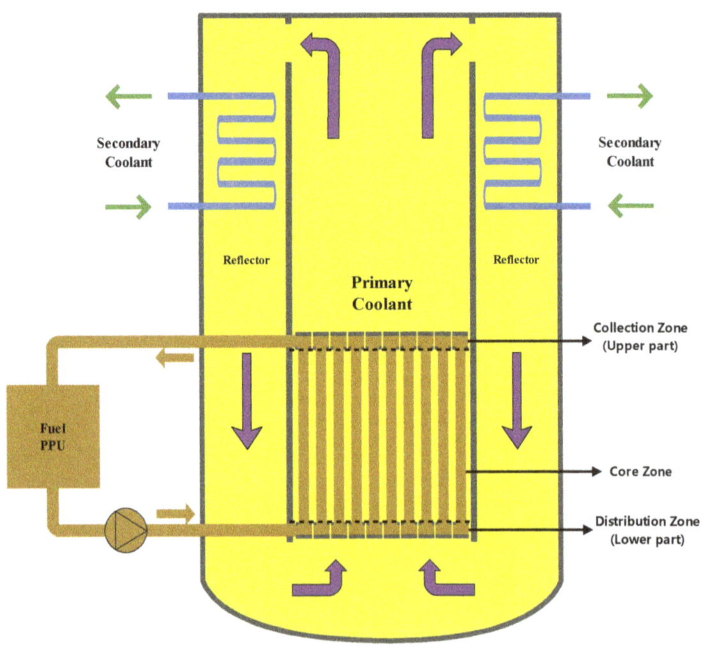

Figure 1. Schematic of the small modular dual fluid reactor.

In this work, the core of the SMDFR is chosen as the research object, which is made of three parts: the distribution zone, core zone, and collection zone. Starting from the bottom part of the core, the fuel (molten salt) goes into the distribution zone, flows through the core zone, and then leaves the collection zone. The coolant (liquid lead) flows in the same direction and takes heat generated by the fission reaction taking place in the molten salt. Since this work focuses on the one-dimensional model of the core, a detailed description of the geometry is beyond the scope of this paper, and the reader is referred to the literature [9].

The major design parameters are listed in Table 1. In this work, the molten salt, which is a mixture of uranium tetrachloride and plutonium tetrachloride, is selected as fuel, and silicon carbide is employed as the pipe wall.

Table 1. SMDFR design parameters.

Parameters	Values
Core zone D × H (m)	0.95 × 2.0
Distribution zone D × H (m)	0.95 × 0.2
Collection zone D × H (m)	0.95 × 0.2
Height of core (m)	2.4
Outer reflector diameter (m)	1.25
Tank D × H (m)	1.65 × 3.4
Number of fuel tubes	1027
Fuel pin pitch (m)	0.025
Outer/interior fuel tube diameter (m)	0.008/0.007
Outer/interior coolant tube diameter (m)	0.005/0.004
Mean linear power density (W/cm)	609
Fuel inlet/outlet temperature (K)	1300/1300
Coolant inlet/outlet temperature (K)	973/1100
Fuel inlet/in-core velocity (m/s)	3/0.5225
Coolant inlet/in-core velocity (m/s)	5/1.3488

2.2. Uncertainty-Based Particle Swarm Optimization of the Control System

A control system is designed for the control of power of the SMDFR. Its working principle is shown in Figure 2. The desired power is determined based on the load demand and then compared with the measured power. Their difference is sent to the PI controller to adjust the value of reactivity to be introduced. It can be observed that a limiter of ±15 pcm/s is applied to limit the introduction rate of reactivity, considering the criticality safety of the core and engineering feasibility [11]. The reactivity is assumed to be introduced by the control rod and applied uniformly to the whole core. As initial values, Kp is set to 200 and Ki is set to 50. Firstly, the feasibility of the designed control system is investigated, and then optimization of the examined control system is conducted.

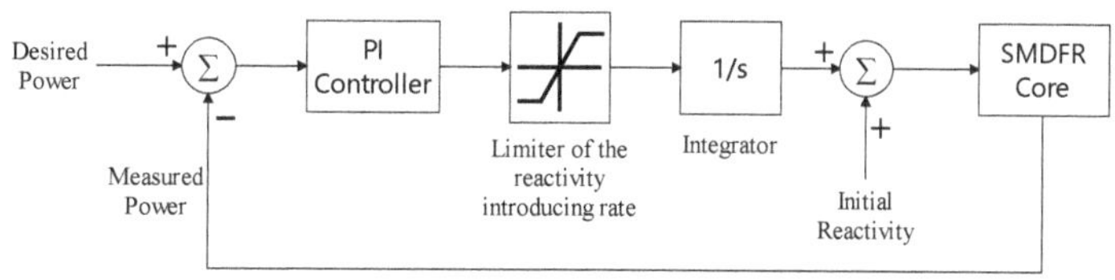

Figure 2. Working principle of the power control system.

To achieve the best performance, the particle swarm optimization technique is employed to settle the PI parameters for the control system. The uncertainties of the model and property have to be quantified to ensure that the optimization is always valid. The structure of the uncertainty-based PSO optimization of the PI controller is shown in Figure 3.

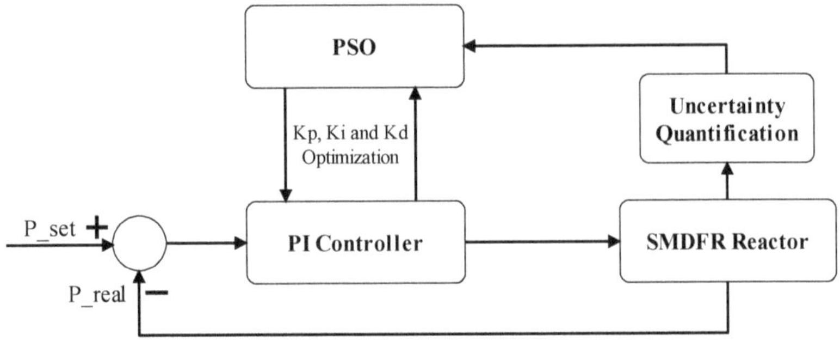

Figure 3. Structure of the uncertainty-based particle swarm optimization of the control system.

3. Methodology

In this work, a one-dimensional model of the SMDFR core is established based on the equivalent parameters achieved by the coupled three-dimensional model, taking into account the delayed neutron precursor drifting by modifying the point kinetic model. The reactivity feedback, resulting from the temperature and density change in the two fluids, is considered by a linearized correlation of the introduced reactivity and the changed temperatures. After finishing the core model, a PI controller is added to control the power via the introduction of positive or negative reactivity based on the difference between the set value and the measured value of core power. Finally, the uncertainty-based control system is designed and optimized using the PSO technique considering the uncertainty quantification.

3.1. Reactor Core Modeling

In order to obtain basic data for the establishment of the one-dimensional model of the SMDFR core, the parameters needed for the neutronic and thermodynamic equations are acquired from the coupled three-dimensional model. In addition, the point kinetic model has to be modified to accurately describe the process of delayed neutron precursor drifting. The thermodynamic model is built based on the assumption that the fuel, the piping wall, and the coolant in the core are divided into 12 parts and each part has lumped properties. The point kinetic model and the thermodynamic model are then linked by the reactivity feedback.

3.1.1. Data Used for Analysis

For the simulation, two types of data are required, the neutronics data (Table 2) and the thermodynamics data (Table 3). The neutronics data are obtained by using Serpent 2.1.31 with the ENDF/B-VII nuclear data library, applying the calculated temperature and density distributions of the fuel and coolant. The thermodynamics data are calculated using the fully resolved CFD model via COMSOL Multiphysics version 5.6 [12] together with its CFD module [13] and heat transfer module [14].

Table 2. Neutronics data for the point kinetic model.

Parameters	Value
Λ (s)	1.05×10^{-6}
β_i $(-)$	$(7.91 \times 10^{-5}$ 7.03×10^{-4} 5.04×10^{-4} 1.17×10^{-3} 4.57×10^{-4} $1.10 \times 10^{-4})$
β $(-)$	3.02×10^{-3}
λ $(1/s)$	$(1.27 \times 10^{-2}$ 3.00×10^{-2} 1.10×10^{-1} 3.19×10^{-1} 1.18 $7.02)$
ϕ_{tot} $(1/(m^2 \cdot s))$	3.61×10^{21}
$\frac{1}{v}$ (s/m)	4.16×10^{-9}
N_0 $(1/m^3)$	1.50×10^{13}
P^i (W)	$(1.58 \times 10^{-1}$ 6.42×10^{-2} 6.98×10^{-2} 7.10×10^{-2} 7.14×10^{-2} 7.12×10^{-2} 7.07×10^{-2} 7.00×10^{-2} 6.91×10^{-2} 6.76×10^{-2} 6.49×10^{-2} $1.52 \times 10^{-1})$
P_0 (W)	1.00×10^{8}
τ_c (s)	5.43
τ_e (s)	10.0
λ_c $(1/s)$	0.184
λ_e $(1/s)$	0.1
α_f $(1/K)$	-2.08×10^{-4}
α_c $(1/K)$	-8.28×10^{-6}
ρ_0 $(\$)$	1.11×10^{-3}
$C_{i,0}$ $(1/m^3)$	$(4.29 \times 10^{14}$ 4.15×10^{15} 3.83×10^{15} 1.16×10^{16} 5.70×10^{15} $1.53 \times 10^{15})$
$Ce_{i,0}$ $(1/m^3)$	$(7.01 \times 10^{14}$ 5.88×10^{15} 3.36×10^{15} 5.09×10^{15} 8.20×10^{14} $3.96 \times 10^{13})$

Table 3. Thermodynamics data.

Parameters	Distribution Zone	Core Zone (1 from 10 Nodes)	Collection Zone
Fuel pipes (#)	-	-	1027
Coolant pipes (#)	2166	2166	-
A_{fw} (m^2)	13.6	9.0	13.6
A_{wc} (m^2)	10.9	10.3	10.9
$\dot{m}_f$ (kg/s)	327.7	327.7	327.7
$\dot{m}_c$ (kg/s)	5550.5	5550.5	5550.5
M_f (kg)	307.7	116.4	307.7
M_w (kg)	39.3	31.1	39.3
M_c (kg)	219.9	770.5	219.9
$c_{p,f}$ $(J/(kg \cdot K))$	400	400	400
$c_{p,w}$ $(J/(kg \cdot K))$	690	690	690
$c_{p,c}$ $(J/(kg \cdot K))$	140.2	140.2	140.2
Heat taken by coolant (J)	18,744,000	5,970,000	16,956,000
h_{fw} $(W/(K \cdot m^2))$	6546.0	4305.1	7629.5
h_{wc} $(W/(K \cdot m^2))$	27,405.4	13,833.4	32,177.3
T_f^{in} (K)	1300.0	-	-
T_c^{in} (K)	973.0	-	-

3.1.2. Modified Point Kinetic Model

In order to accurately capture the neutronic behavior of the SMDFR, the point kinetic model [15] with 6 delayed neutron groups (i = 1, ..., 6) is adopted and modified. Although the fuel outside of the core is kept subcritical, the decay process of delayed neutron precursors has to be considered to describe its influence on neutron distribution in the core. Traditionally, the effect of the delayed neutrons due to the flowing fuel is described by two additional terms [16]: the precursor loss when the fuel leaves the core, and the precursor gain when the fuel re-enters the core, which are the 3rd and the 4th terms on the right side of the delayed neutron precursors equations, respectively, as shown in Equation (1). However, the time delay term introduces some complexities to the modeling process, and some special treatments have to be applied to solve this kind of equation. To eliminate this time delay term, the concentrations of the six groups of delayed neutron precursors outside of the core, C_ei, are defined, and then the balance of delayed neutron precursors can be described as shown in Figure 4. The concentrations of the six group delayed neutron precursors inside the core can be given by Equation (2). Since six depend variables are introduced, an additional six equations (Equation (3)) describing the evolution of the DNP concentrations outside of the core, together with the equation of the neutron density (Equation (4)), have to be added to close the set of equations for the point kinetic model.

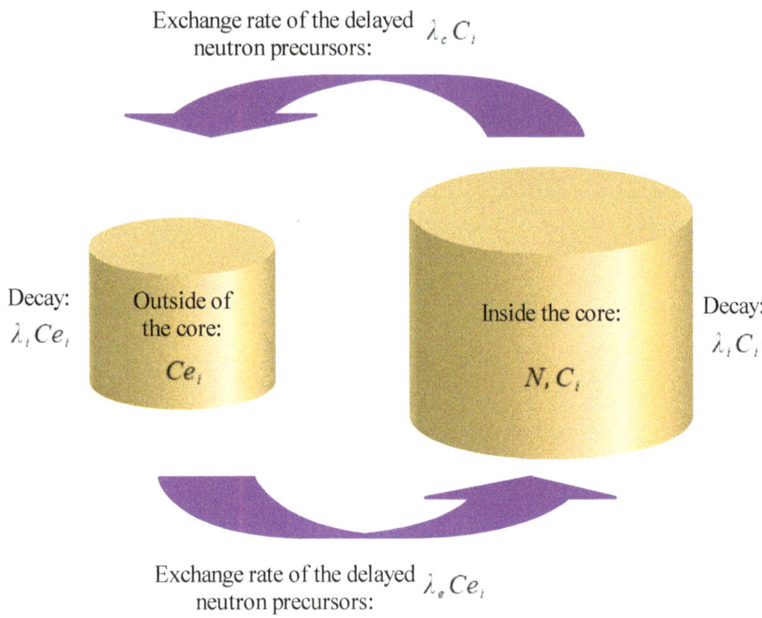

Figure 4. Schematic of the balance of delayed neutron precursors.

$$\frac{dC_i(t)}{dt} = \frac{\beta_i}{\Lambda} \cdot N(t) - \lambda_i C_i(t) - \lambda_c C_i(t) + \lambda_c C_i(t-\tau_e) \cdot e^{-\lambda_i \tau_e} \tag{1}$$

$$\frac{dC_i(t)}{dt} = \frac{\beta_i}{\Lambda} \cdot N(t) - \lambda_i C_i(t) - \lambda_c C_i(t) + \lambda_e Ce_i(t) \tag{2}$$

$$\frac{dC_e i(t)}{dt} = -\lambda_i Ce_i(t) - \lambda_e Ce_i(t) + \lambda_c C_i(t) \tag{3}$$

$$\frac{dN(t)}{dt} = \frac{(\rho(t) - \beta)}{\Lambda} \cdot N(t) + \sum_{i=1}^{6} \lambda_i C_i(t) \tag{4}$$

The initial value of ρ, C_i and Ce_i can be calculated by solving the governing equations Equations (2)–(4), applying a stationary state condition by making the time derivative terms equal to zero, as shown in Equations (5)–(7).

$$\rho_0 = \beta - \sum_{i=1}^{6} \frac{\beta_i \cdot (\lambda_i + \lambda_e)}{\lambda_i + \lambda_e + \lambda_c} \tag{5}$$

$$C_{i,0} = \frac{\beta_i}{\Lambda} \cdot N_0 \cdot \frac{\lambda_i + \lambda_e}{(\lambda_i + \lambda_e + \lambda_c) \cdot \lambda_i} \tag{6}$$

$$Ce_{i,0} = C_{i,0} \cdot \frac{\lambda_c}{\lambda_i + \lambda_e} \tag{7}$$

3.1.3. Thermodynamic Model

Considering the geometry structure of the core, 12 nodes are defined: Node 1 for the distribution zone, Node 2–11 for the core zone, and Node 12 for the collection zone, as shown in Figure 5. Each node has a height of 0.2 m. The energy balance of Node 1 is described by Equations (8)–(10). For the remaining nodes (Node 2–12, i = 2, ..., 12), the heat transfer process is governed by Equations (11)–(13). As shown in Figure 5, the heat generated by the nuclear fission is transferred from the fuel to the piping wall and then to the primary coolant. Finally, the fission energy is taken by the primary coolant and transferred to the secondary coolant for utilization. The power is assumed to be proportional to the neutron density, as shown in Equation (14).

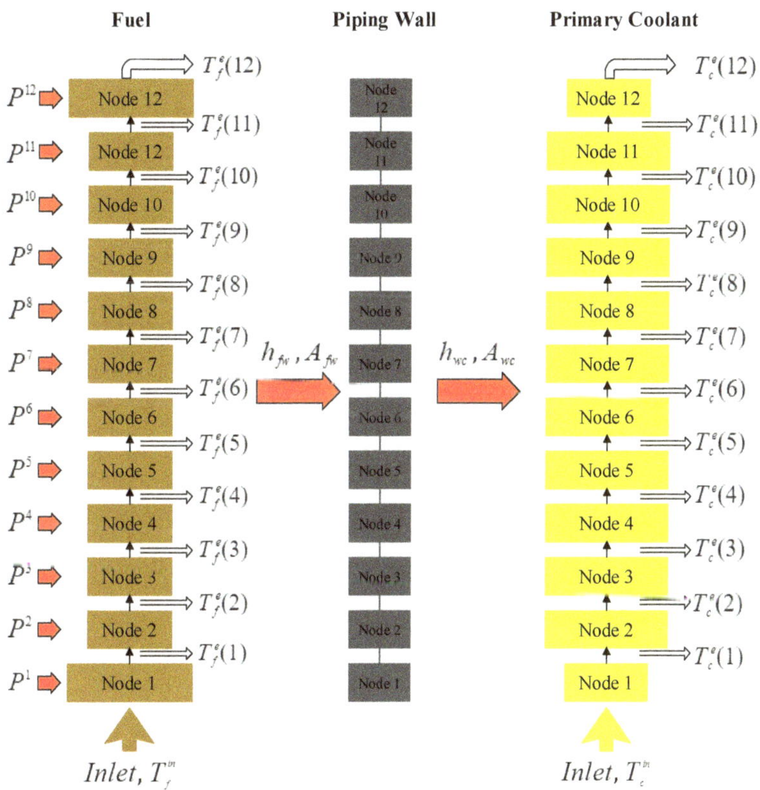

Figure 5. Schematic of nodalization and energy balance.

$$M_f^1 \cdot c_{p,f}^1 \cdot \frac{dT_f^1(t)}{dt} = P^1(t) + \dot{m}_f \cdot c_{p,f}^1 \cdot (T_f^{in} - T_f^e(1)) - h_{fw}^1 \cdot A_{fw}^1 \cdot (T_f^1 - T_w^1) \tag{8}$$

$$M_w^1 \cdot c_{p,w}^1 \cdot \frac{dT_w^1(t)}{dt} = h_{fw}^1 \cdot A_{fw}^1 \cdot (T_f^1 - T_w^1) - h_{wc}^1 \cdot A_{wc}^1 \cdot (T_w^1 - T_c^1) \tag{9}$$

$$M_c^1 \cdot c_{p,c}^1 \cdot \frac{dT_c^1(t)}{dt} = \dot{m}_c \cdot c_{p,c}^1 \cdot (T_c^{in} - T_c^e(1)) + h_{wc}^1 \cdot A_{wc}^1 \cdot (T_w^1 - T_c^1) \tag{10}$$

$$M_f^i \cdot c_{p,f}^i \cdot \frac{dT_f^i(t)}{dt} = P^i(t) + \dot{m}_f \cdot c_{p,f}^i \cdot (T_f^e(i-1) - T_f^e(i)) - h_{fw}^i \cdot A_{fw}^i \cdot (T_f^i - T_w^i) \tag{11}$$

$$M_w^i \cdot c_{p,w}^i \cdot \frac{dT_w^i(t)}{dt} = h_{fw}^i \cdot A_{fw}^i \cdot (T_f^i - T_w^i) - h_{wc}^i \cdot A_{wc}^i \cdot (T_w^i - T_c^i) \tag{12}$$

$$M_c^i \cdot c_{p,c}^i \cdot \frac{dT_c^i(t)}{dt} = \dot{m}_c \cdot c_{p,c}^i \cdot (T_c^e(i-1) - T_c^e(i)) + h_{wc}^i \cdot A_{wc}^i \cdot (T_w^i - T_c^i) \tag{13}$$

$$P(t) = P_0 \cdot \frac{N(t)}{N_0} \tag{14}$$

3.1.4. Reactivity Feedback

The change in temperatures of the fuel or coolant introduces additional reactivity, which is called reactivity feedback, and it has an important impact on the operation of the reactor. The reactivity $\rho(t)$ consists of the initial reactivity, the reactivity introduced by the variation in temperatures of the fuel and coolant, and the externally inserted reactivity from control rods or other sources, as shown by Equation (15):

$$\rho(t) = \rho_0 + \alpha_f(\overline{T}_f(t) - \overline{T}_{f,0}) + \alpha_c(\overline{T}_c(t) - \overline{T}_{c,0}) + \rho_{insert}, \tag{15}$$

where α_f is the temperature feedback coefficient of fuel; α_c is the temperature feedback coefficient of the coolant; $\overline{T}_{f,0}$ and $\overline{T}_{c,0}$ are the initial mean temperatures of the fuel and coolant, respectively; and $\overline{T}_f(t)$ and $\overline{T}_c(t)$ denote the real time mean temperatures.

3.2. PI Controller

In this work, the PI controller is adopted for the control system. As a variation of proportional integral derivative (PID) control, only the proportional and integral terms are used in proportional integral (PI) control. The difference between the set value and the real (measured) value, $e(t)$, is given to the PI controller as the feedback error, and then the value of the controller output $u(t)$ is calculated in the time domain from the feedback error by Equation (17) and then fed into the system as the manipulated variable input. It is clear that the two parameters K_p and K_i have a significant influence on the system response and, thus, should be optimized for the best control performance.

$$e(t) = P_{set} - P_{real} \tag{16}$$

$$u(t) = K_p e(t) + K_i \int_0^t e(t)dt \tag{17}$$

3.3. Uncertainty Quantification

In order to quantify the system uncertainties, the input uncertainties of the system must be identified, and their ranges and probability densities function must be determined. Several methods are available for the quantification of uncertainty, but the nondestructive technique, tolerance limits [17], is selected in this work considering its reduced consumption of computational resources. The Monte-Carlo method, which is much more computationally expensive, is used to verify the performance of the delivered control system.

3.3.1. Input Uncertainties of the Numerical Model

Regardless of the accuracy that the numerical model can achieve, since approximations and assumptions are essential in the modeling and calculation process, a degree of uncertainty is always expected and has to be well quantified to deliver reliable results with adequate assurance. However, since not all of the uncertain parameters can be considered due to the limited computational resources and the limited knowledge, only the most representative uncertain parameters are taken into account based on the experience of a previous study [10]. Their influences on the numerical model have to be quantified and considered during the design and optimization of the control system.

The selected uncertain parameters, which are considered as the input of the uncertainty quantification procedure, were assumed to be uniformly distributed in their ranges and are given in Table 4. The ranges are determined in the following manner:

- Thermal power, mass flow rates, and inlet temperatures are specified with reference to the reported experimental uncertainty in the literature [18].
- The heat capacity of liquid lead (coolant) is specified according to the uncertainty bounds found in [19].
- The heat capacity of molten salt (fuel) is specified referring to the value of liquid lead. Since no data are available for the molten salt, and there is a heat transfer process between the two fluids, a consistent uncertainty range is applied to both of them in order to achieve a conservative assumption for the heat transfer process. In the future, its value will be adapted when experimental data of the molten salt are generated/available.
- Temperature feedback coefficients and heat transfer coefficients are taken with a $\pm10\%$ uncertainty range from the default value. Since no data are available, this assumption is considered to be sufficiently conservative. When more data are obtained, their ranges will be updated accordingly.

Table 4. Uncertain parameters.

#	Parameters	Probability Density Function	Min.	Max.
1	Thermal power (factor)	Uniform	0.98	1.02
2	Temperature feedback coefficient of fuel (factor)	Uniform	0.9	1.1
3	Temperature feedback coefficient of coolant (factor)	Uniform	0.9	1.1
4	Mass flow rate of fuel (factor)	Uniform	0.95	1.05
5	Mass flow rate of coolant (factor)	Uniform	0.95	1.05
6	Heat capacity of fuel (factor)	Uniform	0.95	1.05
7	Heat capacity of coolant (factor)	Uniform	0.95	1.05
8	Heat transfer coefficient between fuel and wall (factor)	Uniform	0.9	1.1
9	Heat transfer coefficient between wall and coolant (factor)	Uniform	0.9	1.1
10	Fuel inlet temperature (additive)	Uniform	−2.2	2.2
11	Coolant inlet temperature (additive)	Uniform	−2.2	2.2

3.3.2. Tolerance Limits

The idea of applying tolerance limits for uncertainty quantification was proposed by Glaeser [17]. By employing this technique, the required sample size, which is the required runs of the computer model, is reduced, while the resulting statistical interval binds with confidence (usually 95%). According to the definition made by Krishnamoorthy and Mathew [20], two types of tolerance limits can be achieved: one-sided tolerance limits and two-sided tolerance limits.

The upper/lower tolerance limit can be explained as follows: the $1-\alpha$ upper/lower confidence limit can be granted for at least the $(100 \times p)^{th}$ percentile of the population. A $100 \times p$ percentage of the population would be bounded by the upper/lower tolerance limit with a confidence of $1-\alpha$.

The two-sided tolerance limits can be explained as follows: The $1-\alpha$ two-sided confidence limits can be granted for at least the $(100 \times p)^{th}$ percentile of the population. A $100 \times p$ percentage of the population would be bounded by the two-sided tolerance limits with a confidence of $1-\alpha$.

The confidence level α can be calculated by [21]

$$\alpha \leq \sum_{k=0}^{r+s-1} \binom{N}{k} (1-p)^k p^{N-k} \tag{18}$$

When a one-sided upper/lower tolerance limit is considered, r/s has to be set to 0. When two-sided tolerance limits are expected, normally, r is set to be equal to s to ensure a consistent grade of upper and lower limits. Given r, s, p (population proportion of interest) and α, the required sample size (N) is given by Equation (18).

In this work, 59 runs were performed to obtain the one-sided upper tolerance limit based on the 1D computer model. Making the confidence level equal to 0.95, the population proportion of interest p would be 0.95 according to Equation (18).

3.3.3. Sensitivity Quantification

The effect of the sensitivity of input uncertainty on the system output can be investigated by the following four types of methods: graphical methods, screening analysis, regression-based analysis, and variance-based analysis. The sensitivity is quantitatively analyzed by the latter two methods in this paper, as the former two methods are qualitative.

Regression-based techniques are often used to quantitatively analyze sensitivity when the system has a linear response between the input and output. The computational cost is lower than that of the variance-based technique. In this work, the standardized regression coefficient (SRC), which is given by Person's ordinary correlation coefficients (Equation (19)) based on standardized variables (Equation (20)), is selected as the index to quantify the importance of input variables. The squared value of SRC is considered to be the fraction of the output variance linearly explained by the input [22]. The coefficient of multiple determination [23] (Equation (21)) is used to examine the linearity of this relationship. Only when it is larger than 0.6 (cut-off value) can a high enough fraction of the output variability be explained by the input variability under a linear relation. The closer its value to 1.0, the more significant the linear relationship.

$$r(X_j, Y) = \frac{\sum_{i=1}^{N}(y_i - \bar{Y})(x_{i,j} - \bar{X}_j)}{\left[\sum_{i=1}^{N}(x_{i,j} - \bar{X}_j)^2 \sum_{i=1}^{N}(y_i - \bar{Y})^2\right]^{\frac{1}{2}}} \tag{19}$$

$$\widetilde{Z} = \frac{Z - E(Z)}{\sigma(Z)} \tag{20}$$

$$R^2 = (r(Y, X_1), \ldots, r(Y, X_k))[Corr(X)]^{-1}(r(Y, X_1), \ldots, r(Y, X_k))^T \tag{21}$$

The variance-based technique (Sobol indices) was proposed by Ilya M. Sobol [24,25]. The output variance is decomposed, and the contribution of each input variable to output variance is identified by a certain percentage. This technique can quantify the global sensitivity regardless of whether or not the system is linear, while the interaction between input variables properly occurs. The major drawback of this method is the high computational cost: $(k+2) \cdot N$, where k denotes the amount of input variables and N denotes the sample size, with a value typically larger than 1000. By defining two commonly used Monte-Carlo estimators, S1 and ST, the first-order sensitivity index and the total effect index [25,26], the sensitivity can be quantified taking into account the linear and non-linear relationships. A detailed discussion of the Sobol method is beyond the scope of this paper; more details on its algorithm and estimators can be found in the literature [10].

3.4. *PSO Optimization with Uncertainties*

Particle swarm optimization is a technique used to optimize a problem [27]. By moving the candidate particles in the search space according to the defined formula, the particles are guided toward the best positions based on a given criterion. Finally, the best position will be found as the optimal solution. Since this paper focuses on the uncertainty-based design and optimization of the control system, a more detailed discussion of the algorithm of particle swarm optimization and its development is beyond the scope of this work. Further details can be found in the literature [27–29]. In this work, the Matlab toolbox, Constrained Particle Swarm Optimization version 1.31.4 [30], is used for the PSO algorithm.

The process of the PSO algorithm with uncertainties of the control system is shown in Figure 6. Firstly, the input parameters are sampled based on the given distributions shown in Table 4, and then the initial conditions of the computer models are set. After setting the controller parameters, the system uncertainty is quantified, and then the performance of the given control system is assessed. By continuously adjusting controller parameters using the PSO algorithm, the optimized control system considering the existence of uncertainty will be achieved. Notably, not only can this working flow be applied to the design of the control system of the nuclear system, but it can also be used for the control system of any other energy systems subjected to uncertainties.

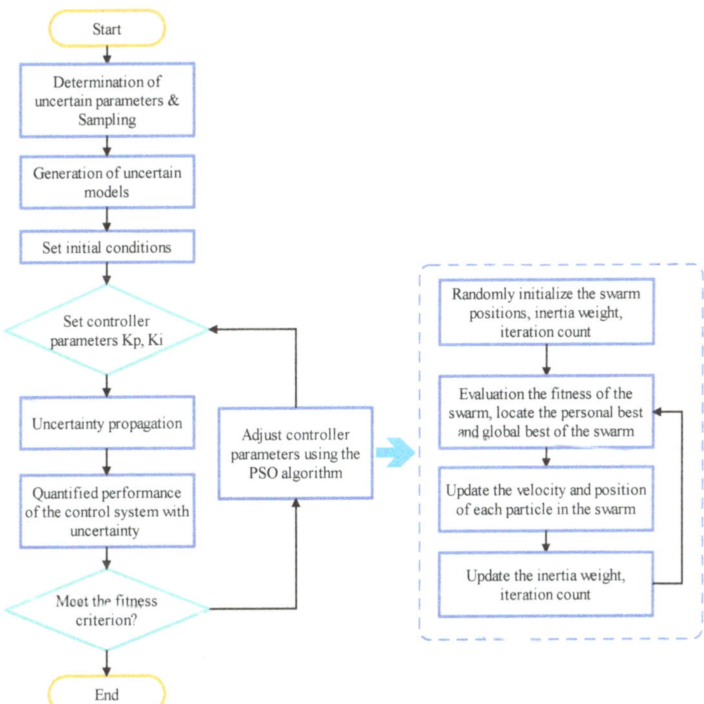

Figure 6. Flowchart of the design of the uncertainty-based control system.

4. Results and Discussion

The Results and Discussion Section is divided into four parts: verification, feasibility of the control system, optimization, and performance assessment.

4.1. Verification

Since the SMDFR is a new concept, and no experimental data are available at this stage, the 1D Matlab model has to be verified against the high-fidelity 3D COMSOL model. A comparison is shown in Figure 7, where asterisks represent the temperature of fuel, the downward-pointing triangle represents the temperature of piping wall, and the cross represents the temperature of the coolant. The results from COMSOL model are depicted in red and those from the Matlab model in black. Taking the COMSOL model as a reference, it can be seen that the shapes of temperature distributions of fuel, the piping wall, and the coolant are generally well captured by the Matlab model. Besides Node 12, the temperature differences in all of the other nodes are within ±20°C. Matlab model was demonstrated to have sufficient capacity to capture the system behaviors and correctly predict the temperature distributions, which means that this 1D Matlab model can be used for the design and investigation of the control system.

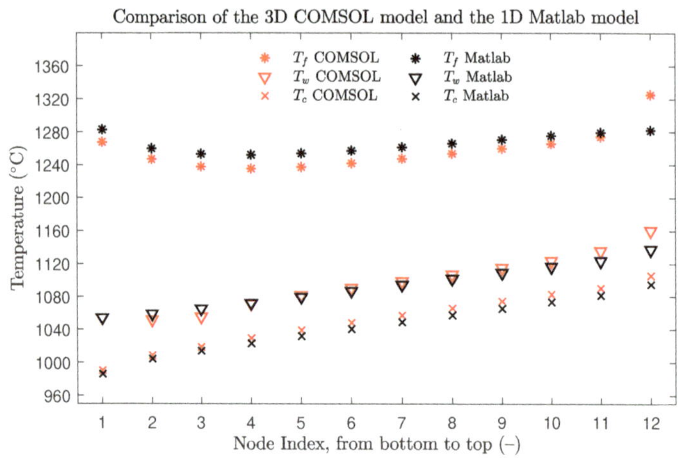

Figure 7. Verification of the 1D Matlab model against the 3D COMSOL model.

4.2. Feasibility of the Control System

In order to investigate the feasibility of the control system, three transients, with and without the control system, are calculated: ±100 pcm insertion of reactivity, ±20 °C variation in the inlet temperature, and ±10% variation in the coolant mass flow rate.

The results of the system response for ±100 pcm insertion of reactivity are shown in Figure 8. The power evolution of the system without a control system is depicted in red and that with a control system in black. Without a control system, as shown in Figure 8a, 100 pcm reactivity is inserted into the core at t = 10 s, causing the power to increase by around 6%. Since more power is generated to heat the fuel and coolant, their temperatures are increased. Due to the negative reactivity feedback of the fuel and coolant, a negative reactivity is introduced, and, thus, the power returns to 102.5% of its nominal power and stays constant at this level. However, the power level changes due to this perturbation, which is not preferred during the normal operation when an unchanged power level is expected despite perturbation. With the control system, the power goes back to its nominal value after several oscillations, and the peak value of the power is 1% lower than that without a control system. For the transient with –100 pcm reactivity insertion (Figure 8b), the system with a control system is also capable of retaining the nominal power after perturbation. For both cases, the time needed for the system to go back to its original state is within 150 s.

The results of the system response for ±20 °C variation in the inlet temperature are shown in Figure 9. Without a control system, as shown in Figure 9a, the coolant inlet temperature increases by 20 °C at t = 10 s, causing the power to decrease by around 8% due to the negative reactivity feedback of the coolant. The decreased power results in a decrease in fuel temperature. Thus, the negative reactivity introduced by the coolant is countered by the positive reactivity introduced by the decrease in fuel temperature. Finally, the power stays constant at the level of 92.5% of its nominal power. With a control system, the power returns to its nominal value after several oscillations, and the peak value of the power is 1% lower than that without a control system. For the transient with the decrease of

20 °C (Figure 9b), the system with a control system is also capable of retaining the nominal power after perturbation. For both cases, the time needed for the system to return to its original state is 150 s.

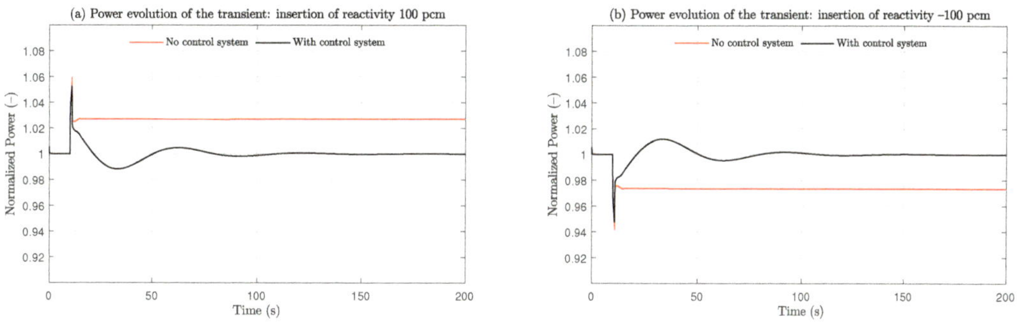

Figure 8. System response of the insertion of reactivity: (**a**) insertion of reactivity at 100 pcm; (**b**) Insertion of reactivity at −100 pcm.

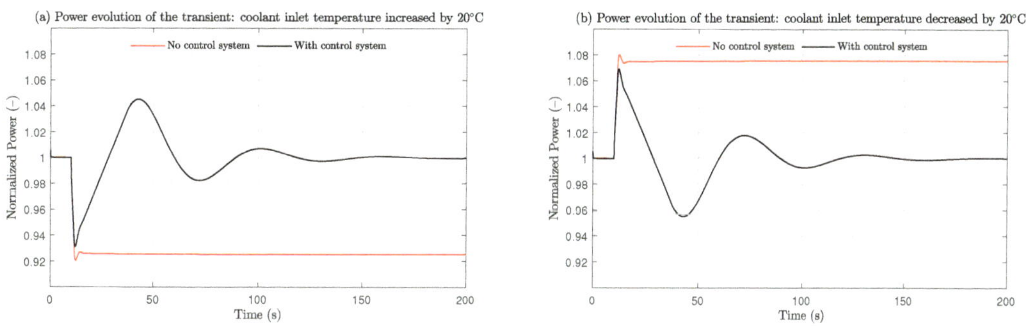

Figure 9. System response of variation in inlet temperature: (**a**) inlet temperature increased by 20 °C; (**b**) inlet temperature decreased by 20 °C.

The results of the system response for ±10% variation in the coolant mass flow rate are shown in Figure 10. Without a control system, as shown in Figure 10a, the coolant mass flow rate increases by 10% at t = 10 s, causing the power to increase by around 4%. As soon as the coolant mass flow rate increases, the heat transfer between the fuel and coolant enhances, and their temperatures thus decrease, resulting in an introduction of positive reactivity due to the negative reactivity feedback. Once the power increases, the temperatures of fuel and coolant increase, and, then, the introduced positive reactivity is countered by the negative reactivity introduced due to the temperature increase in the fuel and coolant. Finally, the power stays constant at the level of 100.4% of its nominal power. With a control system, the power returns to its nominal value after several oscillations, and the peak value of the power is 1% lower than that without a control system. For the transient with the decrease in the coolant mass flow rate (Figure 10b), the system with a control system is also capable of retaining the nominal power after perturbation. For both cases, the time needed for the system to return to its original state is 150 s.

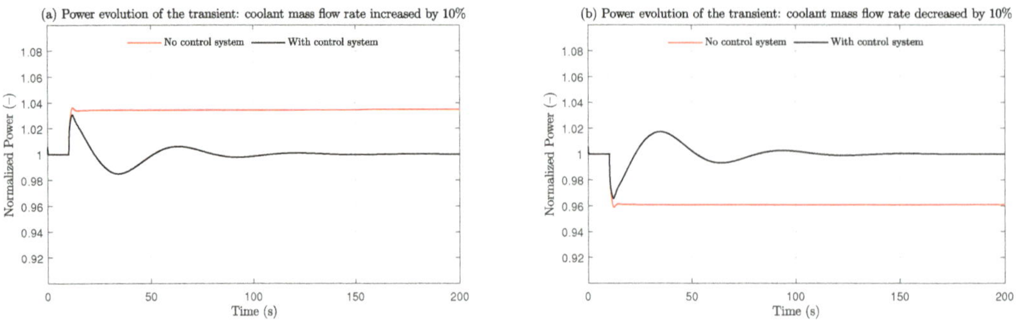

Figure 10. System response of the variation in the coolant mass flow rate: (**a**) coolant mass flow rate increased by 10%; (**b**) coolant mass flow rate decreased by 10%.

4.3. Optimization

Although the control system is proved to be capable of handling various transients, its parameters have to be optimized to achieve the best performance. For the optimization of the control system, two scenarios are selected as benchmark cases: step load change: 100% FP to 90% FP; linear load change: 100% FP to 50% FP to 100% FP, as shown in Figure 11.

The integral time-weighted absolute error (ITAE), as defined by Equation (22), is selected as the criterion of performance of the control system for the case of step load change, since the errors that exist after a long time have to be weighted much more heavily than those at the start of the transient. For linear load change, the integral absolute error (IAE), as defined by Equation (23), is chosen as the criterion of performance, since the error should not be time weighted in this case any more. For both cases, the uncertain parameters in Table 4 are used to calculate the one-sided upper tolerance limit of the ITAE, which is chosen as the output of the fitness function of the PSO process. The achieved optimized solution has the following meaning: under the given uncertainty parameter ranges, at least 95% of the possible system states would have a better performance than that of the final optimized value achieved from the PSO process, with a probability of 95%. By applying the optimized parameters, the control system has a probability of 95% to deliver an optimized performance for more than 95% of the system states.

$$ITAE = \int_0^{t_{max}} t|e(t)|dt, \tag{22}$$

$$IAE = \int_0^{t_{max}} |e(t)|dt, \tag{23}$$

where $e(t)$ is the error between the measured value of the controlled variable and its desired value.

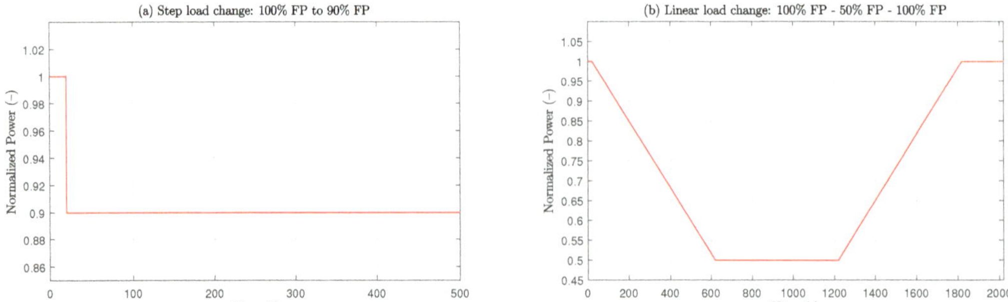

Figure 11. Two benchmark cases: (**a**) step load change: 100% FP to 90% FP; (**b**) linear load change: 100% FP to 50% FP to 100% FP.

4.3.1. Step Load Change

The major parameters for the PSO process are shown in Table 5. After 232 generations, the average cumulative change in the value of the fitness function over 50 generations is less than 1×10^{-6}, and the final best point is: Kp = 7723.2, Ki = 872.29, with a fitness value of 70.78. The evolution of fitness value during the PSO process is shown in Figure 12, where the maximum and minimum values of the particles in each generation are depicted by the error bar. For the first 60 generations, large variations can be observed, as the particles were going through the entire domain to identify the best point (Figure 13). Then, they started to focus on a smaller region, where the best point can be found. After 130 generations, all particles were close to the final best point, and the optimization process finished after 232 generations.

Table 5. Major parameters for the PSO process of step load change.

Parameters	Value
Fitness function	$\int_0^{t_{max}} t\lvert e(t)\rvert dt$
Number of variables	2
Constraints	none
Kp	10~10,000
Ki	10~1000
Population size	20
Generations	300
Social attraction	1.25

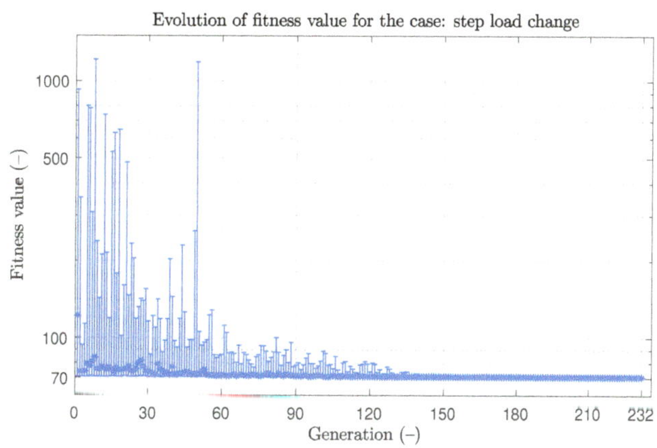

Figure 12. Evolution of fitness value for the case of step load change.

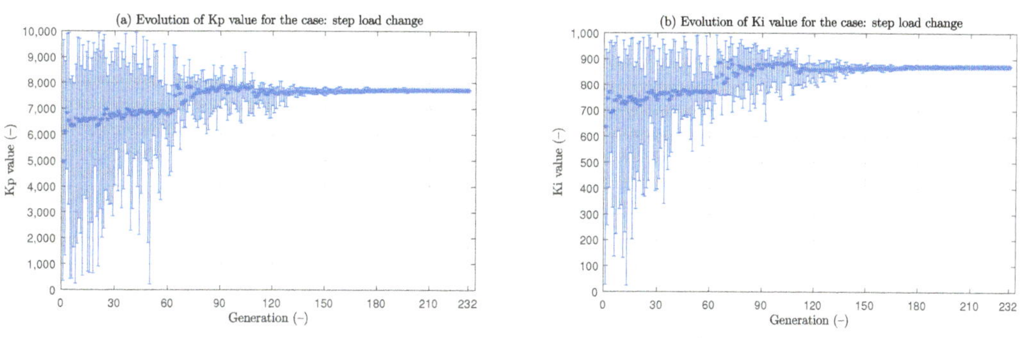

Figure 13. Evolution of parameters for the case of step load change: (**a**) Kp; (**b**) Ki.

4.3.2. Linear Load Change

The major parameters for the PSO process are kept the same as those for the step load change, as shown in Table 5. After 66 generations, the average cumulative change in the value of the fitness function over 50 generations is less than 1×10^{-6}, and the final best point is: Kp = 10,000.0, Ki = 1000.0, with a fitness value of 0.014. The evolution of fitness value during the PSO process is shown in Figure 14, where the maximum and minimum values of the generation are depicted by the error bar. Unlike the step load change, the particles found the location of the best point after only 20 generations (Figure 15), and the optimization process was terminated after 66 generations. The amount of generation needed to identify the best point is much lower, since it is not so challenging for the control system to follow the load for the linear load change. Without the preset limits, Kp and Ki would obtain other values since they already reached their limits. However, the integral absolute error considering the limits lies below 0.015, which means that a well-optimized control system is obtained and no further optimization is necessary.

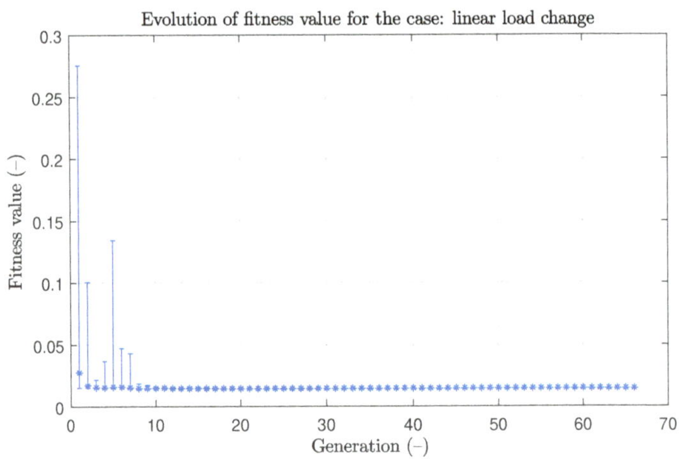

Figure 14. Evolution of fitness value for the case of linear load change.

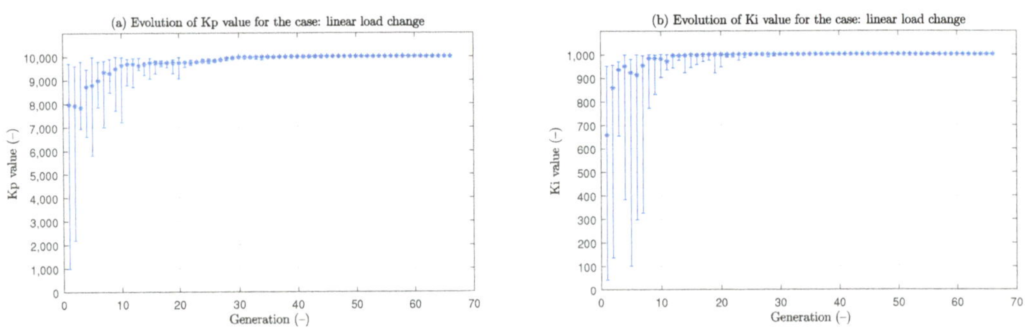

Figure 15. Evolution of parameters for the case of linear load change: (**a**) Kp; (**b**) Ki.

4.4. Performance Assessment

The performance of the optimized control system must be assessed considering its uncertainty and compared with its performance before optimization. Unlike the tolerance limits technique used in the optimization process, a Monte-Carlo method is adopted to assess the performance before and after optimization: 1000 uncertain runs are performed, and their confidence intervals (95% confidence) for quantiles 95% and 5% are chosen as the upper and lower boundaries, respectively.

For step load change, the control system-introduced reactivity during the transient is shown in Figure 16, and the system responses before and after optimization are shown in Figure 17. The desired value is depicted in red; the median value of the real power (system response) is depicted in black; the blue and green lines depict the upper (95%) and lower (5%) uncertain boundaries of the system response, respectively.

The evolution of the introduced reactivity after optimization can be found in Figure 16. In the first stage of the transient (from t = 10 s to t = 60 s), the introduced reactivity reached its lower limit (−15 pcm/s), as the PI controller tried to eliminate the error introduced by the step signal as much as possible and thus made the output value equal to its lower limit to decrease the power. In the second stage (from t = 60 s to t = 90 s), since the reactor power was decreased by the introduced negative reactivity in the first stage, the fuel and coolant temperatures were lower than their original values. Due to the negative reactivity feedback coefficient of both fluids (Table 2), an additional positive reactivity was introduced and thus made the power higher than the set value. In order to counteract this effect, the PI controller switched its output from the lower limit to the upper limit (15 pcm/s) to reduce the reactor power. Finally, the power reached its set value at a new steady-state, and then the output signal of the PI controller returned to zero.

It is clear that the system response cannot be presented by a single fixed curve in time but by a region that is bounded by the upper and lower uncertain boundaries, which means that the system performance cannot be assessed by a simple curve but can be achieved based on the results of statistics with a certain confidence. However, after optimization, the performance of the control system is largely improved, as shown in Figure 17b, despite the existence of uncertainty, as it has a lower overshoot, less oscillation, and less time consumed to reach the new steady state.

The integral time-weighted absolute errors before and after optimization are shown in Figure 18. Before optimization, the upper boundary, the largest value that the system could deliver taking into account the existence of uncertainty with a confidence of 95%, is around 252.6 s^2, and its value is reduce to 70.5 s^2 after optimization, which is consistent with the results presented in Section 4.3. Considering the uncertainties of the computer model and the physicochemical properties, the integral time-weighted absolute error of the optimized system has a probability of 95% to stay below 71 s^2 with a confidence of 95%.

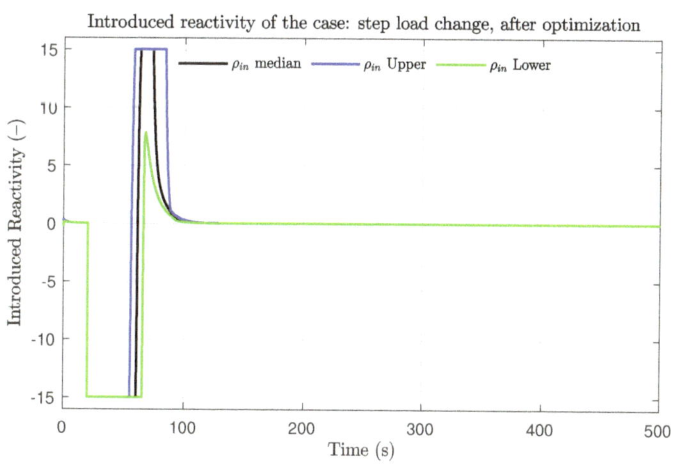

Figure 16. Introduced reactivity of the case of step load change.

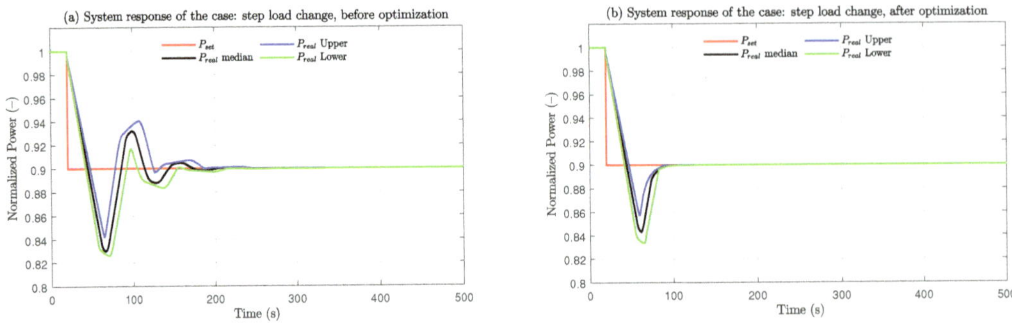

Figure 17. System responses of the case of step load change: (**a**) before optimization; (**b**) after optimization.

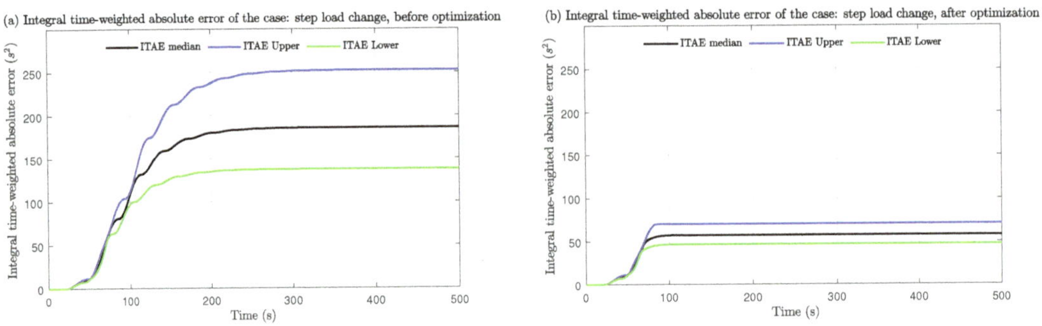

Figure 18. Integral time-weighted absolute error of the case of step load change: (**a**) before optimization; (**b**) after optimization.

The system responses and IAE of the second case: linear load changes are shown in Figures 19 and 20. It can be seen that the system response matches the desired power very well, and there is almost no discrepancy between the curves, which means that the control system can adjust the power according to the linear load change without any significant deviations or oscillations. However, before optimization, minor oscillations are observed right after the power ramps, which are eliminated after optimization.

The upper boundary of IAE is reduced from 0.7 s to 0.014 s through optimization. Considering the uncertainties, the integral absolute error of the optimized system has a probability of 95% to stay below 0.014 s with a confidence of 95%.

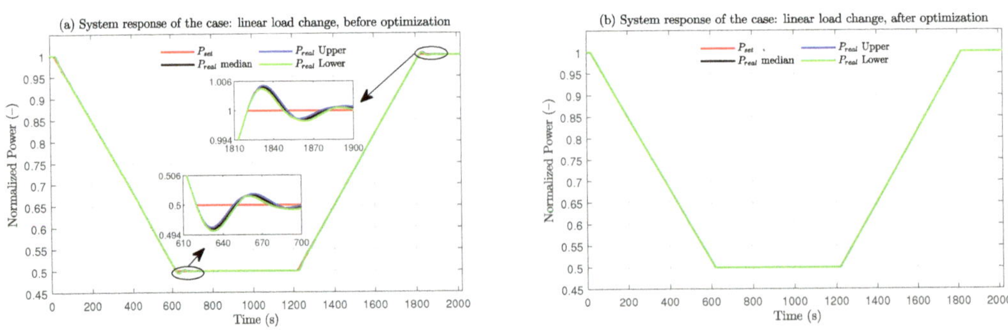

Figure 19. System responses of the case of linear load change: (**a**) before optimization; (**b**) after optimization.

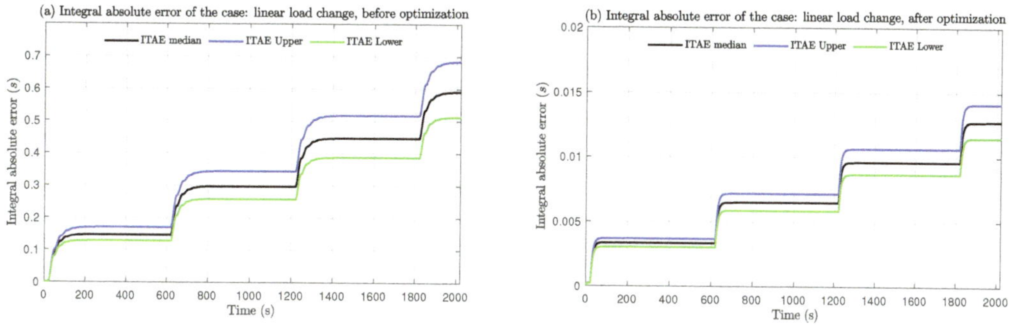

Figure 20. Integral absolute error of the case of linear load change: (**a**) before optimization; (**b**) after optimization.

4.5. Sensitivity Analysis

As shown in Figure 21, parameter 2, the temperature feedback coefficient of fuel, has a major influence on the performance of the control system. The reason is that this feedback on reactivity plays an important role in adjusting neutron flux, and, thus, the power changes. Moreover, since its value is several times larger than that of other reactor concepts due to the large expansion rate of molten salt, its impact on power is significant, and determining its value should be prioritized to deliver a reliable model with less uncertainty. Moreover, parameter 8, the heat transfer coefficient between fuel and wall, has a non-negligible influence on output power. Careful calibration of the heat transfer coefficient has to be conducted in order to reduce the output uncertainty introduced by the heat transfer process. Since the coefficient of multiple determination R^2 is fairly close to 1.0, the system response can be assumed to be totally linear. This means that the results delivered by the regression-based technique have the same high quality as those from the variance-based technique. However, this does not imply that the system model is linear. This linear relationship can be only applied between the input variables and the selected system response: performance of the control system. From Figure 21b it can be observed that for parameter 2 and 8, the total effect indices, ST, are slightly higher than the first-order sensitivity indices, S1, which means these two parameters have certain co-effects on the system response together with other parameters, and these effects are captured by the total-effect indices.

Another option to reduce uncertainty is the application of a high-fidelity model, such as the CFD model or Monte-Carlo model. However, this will considerably increase the computational cost and, thus, is not suitable for the design and optimization of the control system under the given computational resources. A compromise between model accuracy and its computational cost is the employment of a fast model together with the corresponding uncertainty quantification, which could perfectly envelop the introduced error by uncertainty boundaries.

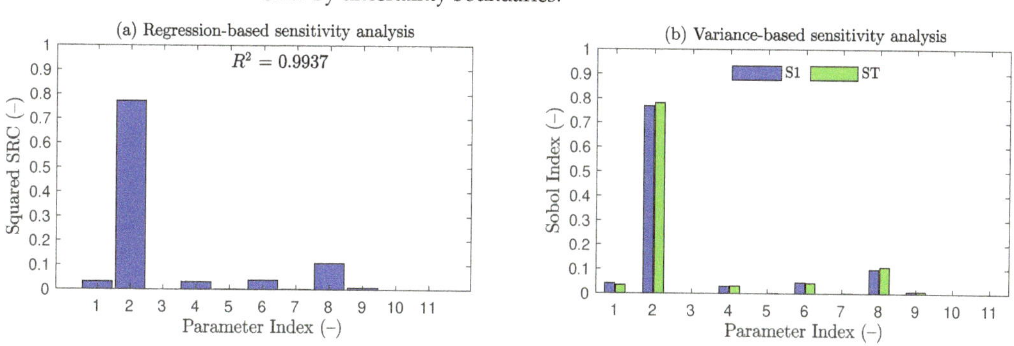

Figure 21. Sensitivity analysis: (**a**) regression-based: SRC; (**b**) variance-based S1/ST.

5. Conclusions

A one-dimensional model of the SMDFR core was established based on the equivalent parameters achieved by the coupled three-dimensional model, taking into account delayed neutron precursor drifting, and its accuracy was verified against the high-fidelity 3D COMSOL model. It was demonstrated that the built one-dimensional model in Matlab had sufficient capacity to capture and predict the system responses. A control system was designed, and its feasibility was verified under various scenarios.

The methodology of uncertainty based optimization of the control system was proposed and conducted for two cases, namely, step load change and linear load change, during which uncertainty was quantified by the tolerance limits technique, and its influence was quantitatively considered during the optimization process. The performances before and after optimization were assessed and compared with the uncertainty quantified by the Monte-Carlo method. Through the usage of different uncertainty propagation methods, namely, the tolerance limits and Monte-Carlo methods, the validity of the uncertainty quantification process in this work was cross-validated. It must be noted that the developed methodology of uncertainty-based optimization is not only suitable for the design of control systems of nuclear systems, but it is also applicable to the design of control systems of other energy systems subjected to uncertainties.

The achieved results showed that the designed control system was able to maintain the stability of the system and regulate the power as expected. By applying the methodology of uncertainty-based optimization, the delivered control system was optimal and demonstrated to have the best performance, demonstrating lower overshoot, less oscillation, and less time needed to reach the new steady state after optimization.

By performing quantitative sensitivity analysis, two key parameters were identified. A reference for the dimension reduction was delivered, and more efforts in this area must to be made in future modeling to effectively reduce the overall uncertainty of the system.

Author Contributions: Conceptualization, C.L., R.L. and R.M.-J.; methodology, C.L. and R.L.; software, C.L.; validation, C.L.; formal analysis, C.L.; investigation, C.L. and R.L.; resources, C.L. and R.M.-J.; data curation, C.L.; writing—original draft preparation, C.L.; writing—review and editing, C.L., R.L. and R.M.-J.; visualization, C.L. and R.L.; supervision, R.M.-J.; project administration, R.M.-J.; funding acquisition, R.L. All authors have read and agreed to the published version of the manuscript.

Funding: This research was funded by the China Scholarship Council (Grant No. 201908430062) and Scientific Research Foundation of the Education Department of Hunan Province, China (Grant No. 18B265).

Data Availability Statement: Not applicable.

Acknowledgments: We would like to thank the research group at IFK for providing basic data and fruitful discussions on the design characteristics of the DFR concept.

Conflicts of Interest: The authors declare no conflicts of interest.

Abbreviations

The following abbreviations are used in this manuscript:

SMDFR	Small modular dual fluid reactor
PSO	Particle swarm optimization
DNP	Delayed neutron precursors
FP	Full power
IAE	Integral absolute error
ITAE	Integral time-weighted absolute error
CFD	Computational fluid dynamics
SRC	Standardized regression coefficient
S1	First-order sensitivity index
ST	Total effect index

All of the quantities in this work are expressed according to the International System of Units (SI) and to the nomenclature listed below:

Λ	Mean neutron generation time
β_i	Fraction of the i-th DNP group
β	Overall fraction of the DNP groups
λ	Decay constant
ϕ_{tot}	Total neutron flux
$\frac{1}{v}$	Inverse of the neutron velocity
N_0	Initial neutron density
P_0	Nominal power
τ_c	Transit time inside the core
τ_e	Transit time outside of the core
λ_c	Drifting-induced decay constant inside the core
λ_e	Drifting-induced decay constant outside of the core
α_f	Reactivity feedback coefficient of fuel
α_c	Reactivity feedback coefficient of coolant
ρ_0	Initial reactivity
$C_{i,0}$	Initial concentration of the i-th DNP group inside the core
$Ce_{i,0}$	Initial concentration of the i-th DNP group outside of the core
A_{fw}	Heat transfer area between fuel and pipe wall
A_{wc}	Heat transfer area between coolant and pipe wall
$\dot{m}_f$	Mass flow rate of fuel
$\dot{m}_c$	Mass flow rate of coolant
M_f	Mass of fuel
M_w	Mass of pipe wall
M_c	Mass of coolant
$c_{p,f}$	Heat capacity of fuel
$c_{p,w}$	Heat capacity of pipe wall
$c_{p,c}$	Heat capacity of coolant
h_{fw}	Heat transfer coefficient between fuel and pipe wall
h_{wc}	Heat transfer coefficient between coolant and pipe wall
T_f	Fuel temperature
T_w	Pipe wall temperature
T_c	Coolant temperature
T_f^{in}	Fuel inlet temperature
T_c^{in}	Coolant inlet temperature
$e(t)$	Difference between the set value and the real (measured) value
K_p	Proportional gain
K_i	Integral gain

References

1. Robertson, R.; Briggs, R.; Smith, O.; Bettis, E. *Two-Fluid Molten-Salt Breeder Reactor Design Study (Status as of 1 January 1968)*; Technical Report; Oak Ridge National Lab: Oak Ridge, TN, USA, 1970.
2. Singh, V.; Lish, M.R.; Wheeler, A.M.; Chvála, O.; Upadhyaya, B.R. Dynamic modeling and performance analysis of a two-fluid molten-salt breeder reactor system. *Nucl. Technol.* **2018**, *202*, 15–38. [CrossRef]
3. Taube, M.; Ligou, J. Molten plutonium chlorides fast breeder reactor cooled by molten uranium chloride. *Ann. Nucl. Sci. Eng.* **1974**, *1*, 277–281. [CrossRef]
4. Huke, A.; Ruprecht, G.; Weißbach, D.; Gottlieb, S.; Hussein, A.; Czerski, K. The Dual Fluid Reactor—A novel concept for a fast nuclear reactor of high efficiency. *Ann. Nucl. Energy* **2015**, *80*, 225–235. [CrossRef]
5. Wang, X.; Macian-Juan, R. Steady-state reactor physics of the dual fluid reactor concept. *Int. J. Energy Res.* **2018**, *42*, 4313–4334. [CrossRef]
6. Wang, X.; Zhang, Q.; He, X.; Du, Z.; Seidl, M.; Macian-Juan, R.; Czerski, K.; Dabrowski, M. Neutron physical feasibility of small modular design of Dual Fluid Reactor. In *Proceedings of the International Conference on Nuclear Engineering (ICONE) 2019.27*; The Japan Society of Mechanical Engineers: Tsukuba, Ibaraki, Japan, 2019; p. 1229.
7. Sierchuła, J.; Weissbach, D.; Huke, A.; Ruprecht, G.; Czerski, K.; Da̧browski, M.P. Determination of the liquid eutectic metal fuel dual fluid reactor (DFRm) design—Steady state calculations. *Int. J. Energy Res.* **2019**, *43*, 3692–3701. [CrossRef]

8. Wang, X.; Liu, C.; Macian-Juan, R. Preliminary hydraulic analysis of the distribution zone in the Dual Fluid Reactor concept. *Prog. Nucl. Energy* **2019**, *110*, 364–373. [CrossRef]
9. Liu, C.; Li, X.; Luo, R.; Macian-Juan, R. Thermal Hydraulics Analysis of the Distribution Zone in Small Modular Dual Fluid Reactor. *Metals* **2020**, *10*, 1065. [CrossRef]
10. Liu, C.; Papukchiev, A.; Macián-Juan, R. Uncertainty and sensitivity analysis of coupled multiscale simulations in the context of the SESAME EU-Project. *Int. J. Adv. Nucl. React. Des. Technol.* **2020**, *2*, 117–130. doi: 10.1016/j.jandt.2020.11.002. [CrossRef]
11. Luo, R.; Liu, C.; Macián-Juan, R. Investigation of Control Characteristics for a Molten Salt Reactor Plant under Normal and Accident Conditions. *Energies* **2021**, *14*, 5279. [CrossRef]
12. Introduction to COMSOL Multiphysics. COMSOL; 2020. Available online: https://cdn.comsol.com/doc/5.5/IntroductionToCOMSOLMultiphysics.pdf (accessed on 10 October 2021)
13. CFD Module User's Guide. COMSOL; 2020. Available online: https://doc.comsol.com/5.4/doc/com.comsol.help.cfd/CFDModuleUsersGuide.pdf (accessed on 10 October 2021)
14. Heat Transfer Module User's Guide. COMSOL; 2020. Available online: https://doc.comsol.com/5.4/doc/com.comsol.help.heat/HeatTransferModuleUsersGuide.pdf (accessed on 10 October 2021)
15. Keepin, G. *Physics of Nuclear Kinetics*; Addison-Wesley Series in Nuclear Science and Engineering; Addison-Wesley Publishing Company: Boston, MA, USA, 1965.
16. Wang, X.; Liu, C.; Macián-Juan, R. Dynamics and stability analysis of DFT using U-Pu and Tru fuel salts. In Proceedings of the 2018 International Congress on Advances in Nuclear Power Plants, ICAPP 2018, Charlotte, NC, USA, 8–11 April 2018..
17. Glaeser, H. GRS method for uncertainty and sensitivity evaluation of code results and applications. *Sci. Technol. Nucl. Install.* **2008**, *2008*. [CrossRef]
18. Grishchenko, D.; Papukchiev, A.; Liu, C.; Geffray, C.; Polidori, M.; Kööp, K.; Jeltsov, M.; Kudinov, P. TALL-3D open and blind benchmark on natural circulation instability. *Nucl. Eng. Des.* **2020**, *358*, 110386. [CrossRef]
19. Fazio, C.; Sobolev, V.; Aerts, A.; Gavrilov, S.; Lambrinou, K.; Schuurmans, P.; Gessi, A.; Agostini, P.; Ciampichetti, A.; Martinelli, L.; et al. *Handbook on Lead-Bismuth Eutectic Alloy and Lead Properties, Materials Compatibility, Thermal-Hydraulics and Technologies-2015 Edition*; Organisation for Economic Co-Operation and Development: Paris, France, 2015.
20. Krishnamoorthy, K.; Mathew, T. *Statistical Tolerance Regions: Theory, Applications, and Computation*; John Wiley & Sons: Hoboken, NJ, USA, 2009; Volume 744.
21. Noether, G.E. *Elements of Nonparametric Statistics*; Wiley: Hoboken, NJ, USA, 1967.
22. Helton, J.C.; Johnson, J.D.; Sallaberry, C.J.; Storlie, C.B. Survey of sampling-based methods for uncertainty and sensitivity analysis. *Reliab. Eng. Syst. Saf.* **2006**, *91*, 1175–1209. [CrossRef]
23. Glaeser, H.; Krzykacz-Hausmann, B.; Luther, W.; Schwarz, S.; Skorek, T. Methodenentwicklung und exemplarische Anwendungen zur Bestimmung der Aussagesicherheit von Rechenprogrammergebnissen. *GRS Rep.* **2008**, *1*, 86–90.
24. Saltelli, A.; Ratto, M.; Andres, T.; Campolongo, F.; Cariboni, J.; Gatelli, D.; Saisana, M.; Tarantola, S. *Global Sensitivity Analysis: The Primer*; John Wiley & Sons: Hoboken, NJ, USA, 2008.
25. Sobol, I.M. Global sensitivity indices for nonlinear mathematical models and their Monte Carlo estimates. *Math. Comput. Simul.* **2001**, *55*, 271–280. [CrossRef]
26. Saltelli, A.; Annoni, P.; Azzini, I.; Campolongo, F.; Ratto, M.; Tarantola, S. Variance based sensitivity analysis of model output. Design and estimator for the total sensitivity index. *Comput. Phys. Commun.* **2010**, *181*, 259–270. [CrossRef]
27. Bonyadi, M.R.; Michalewicz, Z. Particle Swarm Optimization for Single Objective Continuous Space Problems: A Review. *Evol. Comput.* **2017**, *25*, 1–54. [CrossRef]
28. Kennedy, J.; Eberhart, R. Particle swarm optimization. In Proceedings of the ICNN'95—International Conference on Neural Networks, Perth, WA, Australia, 27 November–1 December 1995; Volume 4, pp. 1942–1948. [CrossRef]
29. Poli, R.; Kennedy, J.; Blackwell, T. Particle swarm optimization. *Swarm Intell.* **2007**, *1*, 33–57. [CrossRef]
30. Chen, S. Constrained Particle Swarm Optimization. *MATLAB File Exch.* 2009–2018. Available online: https://www.mathworks.com/matlabcentral/fileexchange/25986 (accessed on 10 October 2021).

energies

MDPI

Article

Investigation of Control Characteristics for a Molten Salt Reactor Plant under Normal and Accident Conditions

Run Luo [1,2], Chunyu Liu [2,*] and Rafael Macián-Juan [2]

1 School of Resource & Environment and Safety Engineering, University of South China, Hengyang 421001, China; luorun@usc.edu.cn or run.luo@tum.de
2 Chair of Nuclear Technology, Department of Mechanical Engineering, Technical University of Munich, 85748 Garching, Germany; rafael.macian@ntech.mw.tum.de
* Correspondence: chunyu.liu@tum.de

Abstract: A molten salt reactor (MSR) has unique safety and economic advantages due to the liquid fluoride salt adopted as the reactor fuel and heat carrier fluid. The operation scheme and control strategy of the MSR plant are significantly different from those of traditional solid-fuel reactors because of the delayed neutron precursors drift with the liquid-fuel flow. In this paper, a simulation platform of the MSR plant is developed to study the control characteristics under normal and accident conditions. A nonlinear dynamic model of the whole system is built in the platform consisting of a liquid-fuel reactor with a graphite moderator, an intermediate heat exchanger and a steam generator. A new control strategy is presented based on a feed-forward and feedback combined scheme, a power control system and a steam temperature control system are designed to regulate load changes of the plant. Three different types of operation conditions are simulated with the control systems, including transients of normal load-follow operation, a reactivity insertion accident and a loss of flow accident. The simulation results show that the developed control system not only has a fast load-follow capability during normal operation, but also has a good control performance under accident conditions.

Keywords: molten salt reactor plant; nonlinear dynamic model; control system design; transient characteristics analysis; normal and accident conditions

Citation: Luo, R.; Liu, C.; Macián-Juan, R. Investigation of Control Characteristics for a Molten Salt Reactor Plant under Normal and Accident Conditions. *Energies* **2021**, *14*, 5279. https://doi.org/10.3390/en14175279

Academic Editors: Ryszard Dindorf, Jakub Takosoglu, Piotr Wos and Jong-Il Yun

Received: 2 August 2021
Accepted: 20 August 2021
Published: 25 August 2021

Publisher's Note: MDPI stays neutral with regard to jurisdictional claims in published maps and institutional affiliations.

1. Introduction

The initial concept design of the molten salt reactor (MSR) was developed by the Oak Ridge National Laboratory (ORNL) for the purpose of driving a nuclear powered aircraft, which adopted liquid uranium fluoride as both reactor fuel and coolant [1]. A test loop of the aircraft reactor was developed to explore the preliminary feasibility of molten fluoride fuel at ORNL [2]. By the 1960s, the eighth MW Molten Salt Reactor Experiment (MSRE) was built and underwent critical operation. Based on the experience with the MSRE's technology, more conceptual designs of MSRs were presented, with single-fluid and two-fluid circuits, thermal-spectrum and fast-spectrum reactors [3,4]. Due to high melting point and high viscosity of the fuel salt, its long-term operation may cause irradiation and corrosion of structural materials. Another difficulty is that the liquid fuel flowing from the reactor core needs to be purified quickly, which poses a challenge to the chemical treatment of nuclear fuel [5]. However, the on-line fuel supply and reprocessing systems of the MSR bring a low excess reactivity to the reactor, in which the liquid fuel is separated from the fission products and the highly radioactive waste can be removed from the primary loop. In particular, the overheated nuclear fuel can be automatically discharged into the safety storage tanks under severe accident conditions [5]. The boiling point of the fuel salt is high and the operating pressure of the reactor vessel is close to atmospheric pressure. Furthermore, the fluoride fuel has a higher heat capacity than the traditional reactor coolant, which means that the molten liquid fuel could enhance core power density and make the reactor system more compact [6]. Because of its great safety and economic advantages compared with solid-fuel reactors, the MSR was chosen as one of the six reactor types by the international forum of Generation IV nuclear energy systems [7].

In recent years, there has been a rapid development of the design, simulation and operation of the MSR all over the world. A high-temperature coated particle fuel with a salt-coolant reactor (FHRs) was presented by US researchers, both preliminary conceptual design of the reactor core, and an experimental test facility of liquid fluoride-salt were completed [8,9]. In Europe, the EURATOM framework programs were carried out to develop two fast-spectrum MSR concepts for fuel breeding and nuclear waste burning, including a Molten Salt Actinide Recycler and Transmuter (MOSART) and a Molten Salt Fast Reactor (MSFR) [10,11]. The SPHINX project in the Czech Republic was performed for the technologies of thorium–uranium fuel cycle, in which the reactor physics and fuel chemistry, on-line reprocessing technology, and structural material were studied by theoretical and experimental methods [12]. Huke et al. [13] proposed a Dual Fluid Reactor (DFR) at the Institute for Solid-State Nuclear Physics (IFK), which adopts two-fluid molten salt fuel and liquid lead coolant in the core. The reactor neutronics/thermal-hydraulics of the DFR was simulated and analyzed in [14,15], and Liu et al. [16] presented a Small Modular Dual Fluid Reactor (SMDFR) based on the DFR concept. A combined MSR system was designed by Furukawa et al. [17] in Japan, in which the power is generated by molten-salt reactors (FUJI) and the nuclide ^{233}U is produced by a spallation reaction in Accelerator Molten-Salt Breeders (AMSB). A Thorium Molten Salt Reactor (TMSR) project was implemented by the Chinese Academy of Sciences for the utilization of thorium resources and hydrogen production [18]. For the developed MSR concepts worldwide, the safety operation and stable control of the reactors play a key role in their technology roadmap and engineering implementation.

The dynamic characteristics and control strategy of the MSR are significantly different from those of traditional solid-fuel reactors because of delayed neutron drift caused by the liquid-fuel flow in the primary loop. Therefore, it is necessary to develop a real-time and accurate dynamic model for the control system design and simulation of the MSR. Simplified dynamic models and control systems for MSR concepts have been developed by several researchers. Sides Jr. [19] carried out a preliminary investigation of the dynamics and control for a single-fluid Molten-Salt Breeder Reactor (MSBR) based on a lumped-parameter system model. Tripodo et al. [20] developed a control-oriented power plant simulator for the molten salt fast reactor, in which a one-dimensional flow model of a liquid fuel with a neutron point-kinetics model was built for the reactor taking into account the drift of the delayed neutron precursors along the fuel circuit and the consequent reactivity insertion. A point reactor kinetics model and heat transfer nodal model of the MSBR core were built for the controller design of the fuel temperature [21]. A non-linear dynamic model of the MSR core was developed by Zarei [22] and a power control scheme with the classic PID controller has also been proposed. Zeng et al. [23] presented a fuzzy-PID composite control approach based on a linear model of the MSBR core, and the Internal Model Control (IMC) method was used for the improvement of the PID controller performance for the reactor power control [24]. However, these simplified lumped parameter models, especially the linearized models, could not describe the real-time spatial distribution of the reactor parameters during transient operation, thereby affecting the control performance and regulation precision of the designed controller. These studies mainly focused on the power control of the MSR core without a simulation platform of the whole-plant system; thus, the control system could not reflect the coupling effects between the various subsystems in the actual operation of the plant. In addition, the designs of these controllers do not take into account the input saturation and dead zone of the control mechanism, which are necessary for its engineering realization.

The objective of this study is to develop a simulation platform of the entire system to study the control characteristics of the MSR plant under normal and accident conditions. The 2250 MW MSR concept presented by Robertson [25] was chosen as the reference configuration for the system modeling and controller design. A non-linear distributed parameter model of the whole system was built in the platform consisting of a liquid-fuel reactor with a graphite moderator, an intermediate heat exchanger and a steam generator. A new control strategy is presented based on a feed-forward and feedback combined scheme, in which the reactor power control system and steam temperature control system, considering the input saturation and dead zone, are designed to adjust load changes of the plant. In order to investigate the operation and control characteristics of the MSR plant, three different transients have been simulated with the new simulation platform, which include normal load-follow operation, reactivity insertion accident and loss of flow accident.

2. MSR Plant Description

A conceptual design of 2250 MW MSR plant was presented by ORNL, in which a liquid salt of Uranium and Thorium (U–Th) fluoride was adopted as the reactor fuel and heat carrier fluid [25]. This thermal-spectrum reactor concept was chosen as the reference configuration to study the transient operation and control characteristics of the MSR plant in this work. The flow diagram of the whole system is shown in Figure 1.

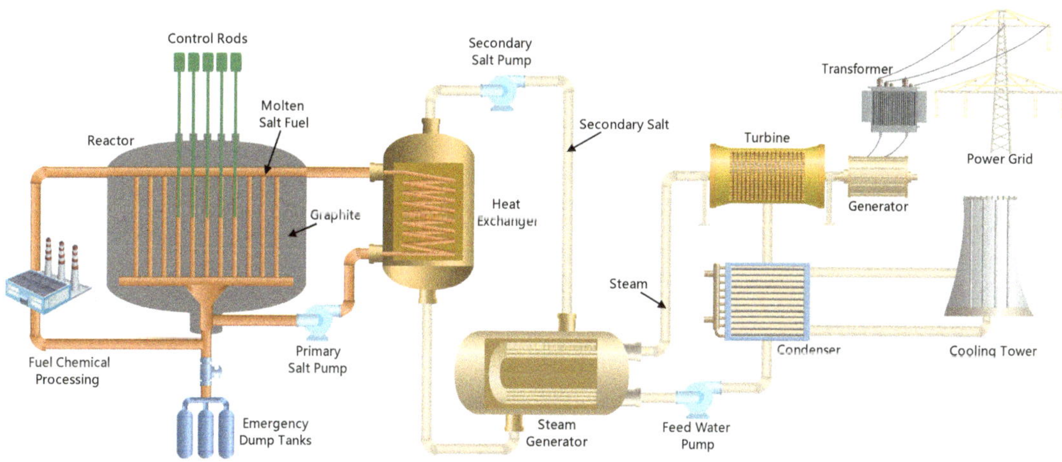

Figure 1. Whole system diagram of the MSR plant.

The MSR plant was composed of a thermal-spectrum molten salt reactor with graphite moderator, 4 tubular heat exchangers (IHX) and 16 steam generators (SG). The reactor core was divided into two zones: a central zone and an outer zone. The mass flow rate fractions of the primary fuel salt in the central zone and outer zone were 81.4% and 18.6%, respectively, which were determined by the heat generation rate in each zone to keep the temperature rise equal. The on-line fuel processing system resulted in a low excess reactivity in the reactor, in which the fission products and the minor actinides could be separated from the liquid fuel salt in the primary loop. In particular, under severe accident conditions, the overheated nuclear fuel would be automatically discharged into the emergency dump tanks so that the reactor could be shut down safely. The control rod could move up and down in the reactor core through the drive mechanism to introduce reactivity in order to adjust the core power. The primary fuel salt flowed upward from the bottom of the core through the graphite-moderated region, and then entered the tube side of heat exchangers. The design point temperature of the fuel salt entering the core was 838.7 K and that at the core outlet was 977.6 K. The fission energy generated by the reactor core was transferred from the primary fuel salt of the tube side to the counter-current secondary salt of the shell side in the heat exchangers. The cold-leg temperature of the secondary salt at the design point was 727.6 K, and the hot-leg temperature was 894.3 K. The temperature of secondary salt could be regulated by changing the pump speed under various operating power points. This salt flowed in each heat exchanger to form a closed secondary loop, and each loop supplied heat to four steam generators. Thus, there was a total of 16 steam generators in the plant. The high-temperature secondary salt flowed into the shell side of horizontal supercritical steam generators to heat the feedwater, in which the pressure of water/steam was close to 25 MPa in the tube side of steam generators. The feedwater entered the U-tube of steam generators and became superheated steam after being heated by the secondary salt, and finally flowed into the turbine to generate electricity. The feedwater pump was used to adjust the flow rate to keep the outlet temperature of the steam at the design value of 810.9 K under steady-state conditions and maintain fluctuations as small as possible during transient operations, thereby providing a stable steam quality to the turbine. The pressure in the SG-tube side and the inlet temperature of the feedwater were also held constant during transient operations. The main design parameters used in this work are summarized in Table 1.

Table 1. Main design parameters of the MSR plant [25].

Parameters	Values
Total thermal power (MW)	2250
Core height (m)	3.96
Central zone/Outer zone diameters (m)	4.39/5.15
Total delayed neutron fraction (pcm)	264
Core transit time of fuel salt τ_c (s)	3.57
External loop transit time of fuel salt τ_L (s)	6.05
Reactivity coefficient of fuel (1/K)	-2.4×10^{-5}
Reactivity coefficient of graphite (1/K)	1.9×10^{-5}
Primary fuel salt	LiF-BeF_2-ThF_4-UF_4
Core inlet temperature (K)	838.7
Core outlet temperature (K)	977.6
Fuel salt flow rate (kg/s)	11945
Secondary salt	$NaBF_4$-NaF
Secondary salt cold-leg temperature (K)	727.6
Secondary salt hot-leg temperature (K)	894.3
Secondary salt flow rate (kg/s)	8971
Feed water inlet temperature (K)	644.3
Steam outlet temperature (K)	810.9
Steam flow rate (kg/s)	1487
Steam pressure (MPa)	24.82

3. Development of the MSR Plant Simulation Platform

In this section, the nonlinear distributed parameter model of the whole system, the reactor power control system and the steam temperature control system for the MSR plant developed in the simulation platform are described.

3.1. Whole System Nonlinear Model

In order to avoid the simplifications and approximations introduced in the process of model linearization, this paper directly solves the nonlinear and time-varying system model of the MSR plant, which can more accurately reflect the dynamic behavior and the coupling effects between the various subsystems in the actual operation of the plant. The nonlinear distributed parameter model consists of a molten salt reactor model, a heat exchanger model and a steam generator model. A schematic diagram of the model node division is shown in Figure 2. In order to obtain the temperature distribution of the liquid fuel and the graphite, the core is divided into N nodes in the axial direction and two zones in the radial direction. The heat transfer coefficient of the primary fuel salt and secondary salt in the nonlinear model change with the mass flow rate and the salt temperatures. The molten salt fuel, tube wall and secondary salt in the heat exchanger are also divided into N nodes in the axial direction, and the node division of the steam generator is determined in a similar manner. In the process of numerical calculation, the spatial term of the differential equations of the whole system is first discretized, and then, the time-dependent term of the differential equations is calculated by a backward differentiation formulas (BDFs) method [26].

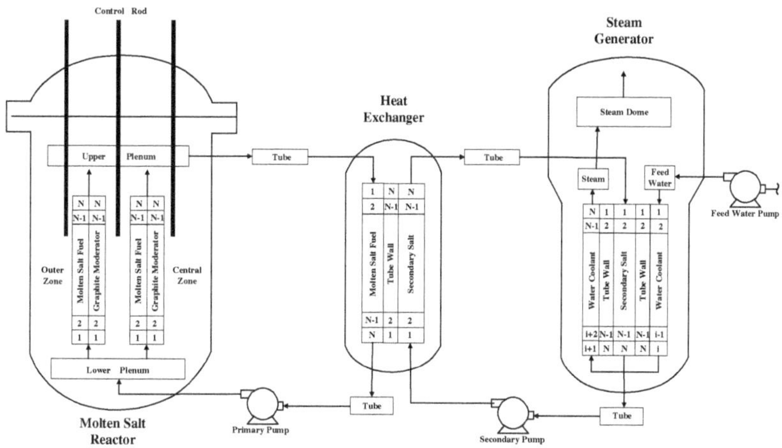

Figure 2. System nodalization of the nonlinear distributed parameter model.

3.1.1. Molten Salt Reactor

The neutron kinetics model of the reactor is built by using the point-reactor kinetics equations with the drift of delayed neutron precursors and temperature reactivity feedbacks of the liquid fuel and the graphite. The equations are as follows:

$$\frac{dN(t)}{dt} = \frac{\rho(t) - \beta}{\Lambda} N(t) + \sum_{j=1}^{2} \lambda_j C_j(t) \tag{1}$$

$$\frac{dC_j(t)}{dt} = \frac{\beta_j}{\Lambda} N(t) - \lambda_j C_j(t) - \frac{C_j(t)}{\tau_c} + \frac{C_j(t-\tau_L)}{\tau_c} e^{-\lambda_j \tau_L}; \quad j = 1,2 \tag{2}$$

where N is the neutron density, t is the time, β is the total delayed neutron fraction, β_j is the delayed neutron fraction of group j, Λ is the neutron generation time, C_j and λ_j are the delayed neutron precursor concentration and decay constant of group j, respectively, τ_c denotes the transit time of liquid fuel through the core, τ_L denotes the transit time of liquid fuel through the external loop. The values of the transit time (τ_c and τ_L) are presented in Table 1 [27]. Taking into account temperature reactivity feedbacks of the fuel and the graphite and the reactivity variation introduced by the control rod, the reactivity model $\rho(t)$ could be written as:

$$\rho(t) = \rho_0 + \alpha_f(\overline{T}_f(t) - \overline{T}_f(0)) + \alpha_g(\overline{T}_g(t) - \overline{T}_g(0)) + \rho_{rod} \tag{3}$$

where α_f and α_g are temperature reactivity coefficients of the molten salt fuel and the graphite, respectively, $\overline{T}_f$ and $\overline{T}_g$ are the average temperatures of the fuel and the graphite in the core, respectively, ρ_{rod} denotes the reactivity variation of the control rod movement. Calculating Equations (1) and (2) in steady state, the initial reactivity ρ_0 could be derived by the following formula:

$$\rho_0 = \beta - \sum_{j=1}^{2} \frac{\lambda_j \beta_j}{\lambda_j + (1/\tau_c)(1 - e^{-\lambda_j \tau_L})} \tag{4}$$

The magnitude of the total thermal power is proportional to the neutron density in Equation (1), and the axial distribution of the power in the core is approximately a cosine distribution with a 2.4 m extrapolation distance [28]. The fraction of energy generated in the liquid fuel and in the graphite moderator are 91.9% and 8.1%, respectively.

The energy released by the fission reactions in the core is removed by the flowing molten-salt fuel. In order to obtain the spatial distribution of temperatures in the fuel and in the graphite, the core is divided into N nodes in the axial direction and two zones in the radial direction, as shown in Figure 2. According to the conservation principle of mass and energy, the heat transfer equation for the molten salt fuel in the central zone is as follows:

$$M_{f,i}C_{f,i}\frac{dT_{f,i}}{dt} = w_{cz}C_{f,i}(T_{f,i-1} - T_{f,i}) + h_{fg}A_{fg,i}(T_{g,i} - T_{f,i}) + k_{f,i}P_t \tag{5}$$

where $M_{f,i}$, $C_{f,i}$ and $T_{f,i}$ are the mass, specific heat and temperature of the fuel in the axial node i, respectively, w_{cz} denotes the mass flow rate in the central zone, h_{fg} is the heat transfer coefficient between the fuel and the graphite, $A_{fg,i}$ is the heat transfer area of node i, $k_{f,i}$ is the fraction of fission heat generated in fuel node i, P_t is the total thermal power of the reactor.

For the graphite, the heat transfer equation is described as:

$$M_{g,i}C_{g,i}\frac{dT_{g,i}}{dt} = h_{fg}A_{fg,i}(T_{f,i} - T_{g,i}) + k_{g,i}P_t \tag{6}$$

where $M_{g,i}$, $C_{g,i}$ and $T_{g,i}$ are the mass, specific heat and temperature of the graphite in the axial node i, respectively, $k_{g,i}$ is the fraction of heat generation in graphite node i.

The heat transfer equations of the fuel and the graphite in the outer zone is derived in the same way. The molten salt fuel from the two zones of the core is mixed in the reactor upper plenum, the outlet temperature $T_{f,out}$ is given by:

$$T_{f,out} = f_r T_{f,cz} + (1 - f_r)T_{f,oz} \tag{7}$$

where f_r is the fraction of mass flow rate of the fuel in the central zone, $T_{f,cz}$ and $T_{f,oz}$ are the fuel temperatures at the node N of the central zone and the outer zone, respectively.

3.1.2. Heat Exchanger

The heat exchanger adopts a shell-and-tube structure in which the primary fuel salt flows downwards in the tube side and the secondary salt flows upwards in the shell side. The node divisions of primary fuel salt, tube wall and secondary salt in the heat exchanger are shown in Figure 2. The heat transfer model of the primary fuel salt is as follows:

$$M_{p,i}C_{p,i}\frac{dT_{p,i}}{dt} = w_p C_{p,i}(T_{p,i-1} - T_{p,i}) + h_{pt}A_{pt,i}(T_{t,i} - T_{p,i}) \tag{8}$$

where $M_{p,i}$, $C_{p,i}$ and $T_{p,i}$ are the mass, specific heat and temperature of the primary fuel salt in the axial node i respectively, w_p denotes the mass flow rate in the tube side, h_{pt} is the heat transfer coefficient between the primary fuel salt and the tube wall, $A_{pt,i}$ is the heat transfer area of node i.

For the tube wall:

$$M_{t,i}C_{t,i}\frac{dT_{t,i}}{dt} = h_{pt}A_{pt,i}(T_{p,i} - T_{t,i}) + h_{st}A_{st,i}(T_{s,i} - T_{t,i}) \tag{9}$$

where $M_{t,i}$, $C_{t,i}$ and $T_{t,i}$ are the mass, specific heat and temperature of the tube wall in the axial node i, respectively.

For the secondary salt:

$$M_{s,i}C_{s,i}\frac{dT_{s,i}}{dt} = w_s C_{s,i}(T_{s,i-1} - T_{s,i}) + h_{st}A_{st,i}(T_{t,i} - T_{s,i}) \tag{10}$$

where $M_{s,i}$, $C_{s,i}$ and $T_{s,i}$ are the mass, specific heat and temperature of the secondary salt in the axial node i, respectively, w_s denotes the mass flow rate in the shell side, h_{st} is the heat transfer coefficient between the secondary salt and the tube wall, $A_{st,i}$ is the heat transfer area of node i.

3.1.3. Steam Generator

The steam generator is a U-tube counter-current exchanger, with the high-temperature secondary salt flowing into the shell side and the feedwater entering the U-tube side, which becomes overheated steam after being heated by the secondary salt. The heat transfer model of the secondary salt is written as:

$$M_{ss,i}C_{ss,i}\frac{dT_{ss,i}}{dt} = w_{ss}C_{ss,i}(T_{ss,i-1} - T_{ss,i}) + h_{su}A_{su,i}(T_{u,i} - T_{su,i}) \tag{11}$$

where $M_{ss,i}$, $C_{ss,i}$ and $T_{ss,i}$ are the mass, specific heat and temperature of the secondary salt in the axial node i of steam generators, respectively, w_{ss} denotes the mass flow rate in the shell side, h_{su} and $A_{su,i}$ denote the heat transfer coefficient and the heat transfer area between the secondary salt and the U-tube wall.

For the U-tube wall:

$$M_{u,i}C_{u,i}\frac{dT_{u,i}}{dt} = h_{su}A_{su,i}(T_{su,i} - T_{u,i}) + h_{wu}A_{wu,i}(T_{w,i} - T_{u,i}) \tag{12}$$

where $M_{u,i}$, $C_{u,i}$ and $T_{u,i}$ are the mass, specific heat and temperature of the U-tube wall in the axial node i, respectively.

The water and steam are in a supercritical state due to their high temperature and pressure in the U-tube of the steam generator. Thus, the water/steam mixture is a single-phase fluid, and the heat transfer model of the water/steam could be described as:

$$M_{w,i}C_{w,i}\frac{dT_{w,i}}{dt} = w_wC_{w,i}(T_{w,i-1} - T_{w,i}) + h_{wu}A_{wu,i}(T_{u,i} - T_{w,i}) \tag{13}$$

where $M_{w,i}$, $C_{w,i}$ and $T_{w,i}$ are the mass, specific heat and temperature of the water/steam in the axial node i, respectively, w_w denotes the mass flow rate in the tube side, h_{wu} is the heat transfer coefficient between the water/steam and the tube wall, $A_{wu,i}$ is the heat transfer area of node i.

The thermophysical properties of materials in Equations (5)–(13) were inserted into the thermal-hydraulic model based on the experimental data and empirical correlation in the design report [25], which are presented in Table 2. The relation of heat transfer coefficients to mass flow rate is shown in Figure 3 obtained from the empirical formula in the literature [28].

Table 2. Thermophysical Properties of Fuel Salt, Graphite, Secondary salt, Tube Wall and Water/Steam.

Parameters	Values
Fuel salt density (kg·m^{-3})	$3752 - 0.668(T - 273)$
Fuel salt specific heat (J·kg^{-1}·K^{-1})	1357
Graphite density (kg·m^{-3})	1843
Graphite specific heat (J·kg^{-1}·K^{-1})	1760
Fuel-graphite heat transfer coefficient (W·m^{-2}· K^{-1})	6047
Secondary salt density (kg·m^{-3})	$2252 - 0.711(T - 273)$
Secondary salt specific heat (J·kg^{-1}·K^{-1})	1507
Tube wall density (kg·m^{-3})	8671
Tube wall specific heat (J·kg^{-1}·K^{-1})	569
IHX fuel salt-wall heat transfer coefficient (W·m^{-2}· K^{-1})	13,786
IHX secondary salt-wall heat transfer coefficient (W·m^{-2} · K^{-1})	9533
Supercritical water/steam density (kg·m^{-3})	$2.144 \times 10^{-2}T^2 - 33T + 12{,}748$

Table 2. *Cont.*

Parameters	Values
Supercritical water/steam specific heat ($J \cdot kg^{-1} \cdot K^{-1}$)	$1.797T^2 - 2802T + 1.095 \times 10^6$
SG secondary salt-wall heat transfer coefficient ($W \cdot m^{-2} \cdot K^{-1}$)	4860
SG water inlet-wall heat transfer coefficient ($W \cdot m^{-2} \cdot K^{-1}$)	10,055
SG steam outlet-wall heat transfer coefficient ($W \cdot m^{-2} \cdot K^{-1}$)	8265

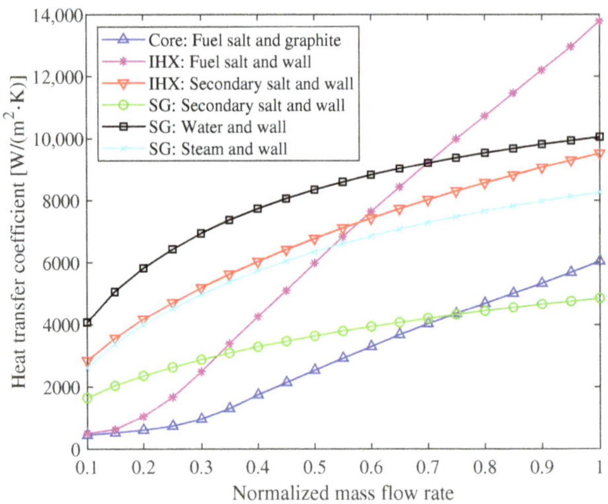

Figure 3. Variation of heat transfer coefficients with mass flow rate in the reactor core, heat exchanger (IHX) and steam generator (SG).

3.1.4. Transport Delays between Various Subsystems

Taking into account the transit time of the fluid flow in the pipes connecting the various subsystems, transport delay equations of the primary fuel salt between the reactor core and the heat exchanger are used in the dynamic simulation platform of the MSR plant:

$$T_{p,in}(t) = T_{f,out}(t - \tau_{d1}) \tag{14}$$

$$T_{f,in}(t) = T_{p,out}(t - \tau_{d2}) \tag{15}$$

where $T_{f,in}$ and $T_{f,out}$ are the inlet and outlet temperatures of the primary fuel salt in the core, respectively, $T_{p,in}$ and $T_{p,out}$ are the inlet and outlet temperatures of the primary fuel salt in the heat exchanger, respectively, the values of delay times τ_{d1} and τ_{d2} are 2 s and 2.3 s, respectively.

For the secondary-salt loop, the transport delays in the hot and cold legs can be written as:

$$T_{ss,in}(t) = T_{s,out}(t - \tau_{d3}) \tag{16}$$

$$T_{s,in}(t) = T_{ss,out}(t - \tau_{d4}) \tag{17}$$

where T_s and T_{ss} are the secondary-salt temperatures in the heat exchanger and the steam generator, respectively, the values of delay times τ_{d3} and τ_{d4} are 14.5 s and 11.9 s, respectively [28].

3.2. *Plant Control System*

Before designing the control system, the operation parameters should be obtained by a series of steady state calculations at different plant loads. Figure 4 shows the results for the temperature as a function of the plant load. To protect the turbine blades, the steam outlet temperature remains constant at 810.9 K. Thermal cycle efficiency and control (such as turbine model, condenser model, etc.) are not discussed in this work; thus, a simplified assumption is adopted that the feedwater temperature maintains a constant value of 644.3 K during different operation transients. The variation of the core outlet temperature is a linear function of the load demand. The reactor inlet temperature remains constant when the load is below 50% full power (FP) and linearly changes from 810.9 K to 838.7 K for loads above 50% FP. The cold-leg and hot-leg temperatures of the secondary salt are determined by the steady-state heat balance points, in which the minimum operation temperature is higher than its melting point.

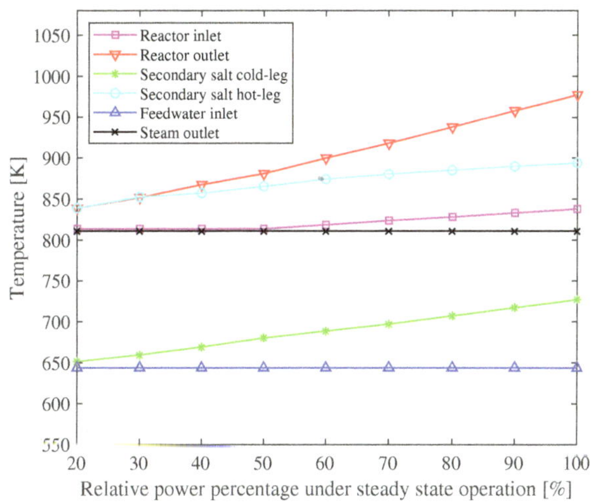

Figure 4. The operation temperature as a function of load.

A new control strategy was proposed to realize the desired operation conditions, which mainly includes two control modules for load tracking: the power control system and the steam temperature control system. The objective of the power control system in this work is to make the total thermal power follow the variation of the load demand as quickly as possible. The steam temperature control system is designed to keep the outlet temperature of the steam at the design value of 810.9 K under steady-state conditions and maintain the fluctuations as small as possible during transient operations, thereby providing a stable steam quality to the turbine. A schematic diagram of the plant control system is shown in Figure 5. The measured values of neutron flux and reactor outlet temperature are compared with those of set-values for the plant load demand. The deviations of neutron flux and outlet temperature (ΔN_r and ΔT_{fo}) are fed back to the control rod servo for suitable reactivity adjustments. In order to avoid neutronic fluctuations due to the drift of the delayed neutron precursors, the drift times (τ_c and τ_L) are kept constant during transient operations, which means the mass flow of the primary pump is always maintained at the 100% FP value. At the same time, a feed-forward control scheme is adopted to match the secondary salt temperature to its set-point by regulating the secondary pump, in which the feed-forward tuning parameter of the pump is programmed as a function of the load demand. The steam temperature control system is designed to adjust the mass flow rate of the feedwater pump when the measured steam outlet temperature deviates from its set-point.

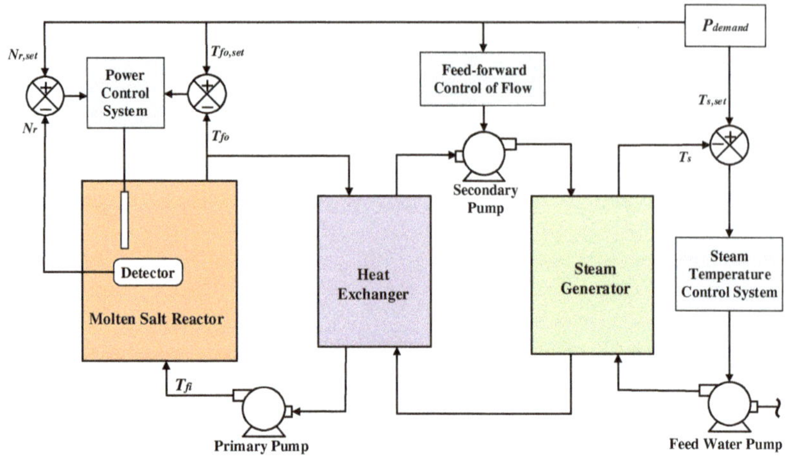

Figure 5. Control strategy of the MSR plant.

3.2.1. Reactor Power Control System

The reactor power control system is a three input and single output nonlinear controller, the block diagram of the control principle is shown in Figure 6.

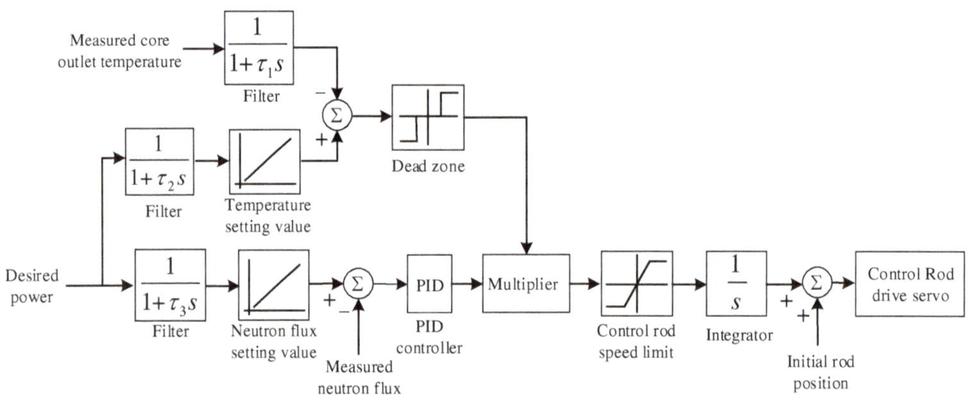

Figure 6. The block diagram of power control system.

According to the demand of the electric grid load, the desired power could be determined and converted to the set-value of the neutron flux. The deviation signal between this set-value and the measured value is sent to the PID controller. After the error signal is processed by the PID controller, the adjustment signal is transmitted to the rod speed controller for calculating the rod movement speed and direction. When we are designing a PID controller for a given system, we follow the steps shown below to obtain a desired response. The first step is to obtain an open-loop response and determine what needs to be improved. The second step is to add a proportional control parameter (K_P) to improve the rise time. The third step is to add a derivative control parameter (K_I) to reduce the overshoot, and then add an integral control (K_D) to reduce the steady-state error. Finally, adjust each of the gains K_P, K_I and K_D until we obtain a desired overall response. The designed values of PID parameters are 0.025, 0.0079 and 0.0028, respectively. Considering the criticality safety of the reactor and the engineering feasibility of the control rod mechanism, the change rate of the controlled reactivity is limited to ± 10 pcm/s. In addition, the power control system includes a dead zone block of outlet temperature to avoid frequent actions of the control mechanism caused by signal noise and by extremely small deviations. When the error of the measured outlet temperature and its set-value is below the threshold in the dead zone, a lock signal will be sent to the control rod mechanism to prevent unnecessary actions of the rod drive mechanism in order to extend its service life.

The feed-forward controller is designed to match the secondary salt temperature to its set-point, in which the tuning parameter of the secondary pump is programmed as a function of the load demand. The block diagram of the controller is shown in Figure 7. The mass flow rate of the secondary salt is a nonlinear fitting function $\psi(P)$ based on solutions of the steady state equations for the entire system at different power levels, as shown in Figure 8.

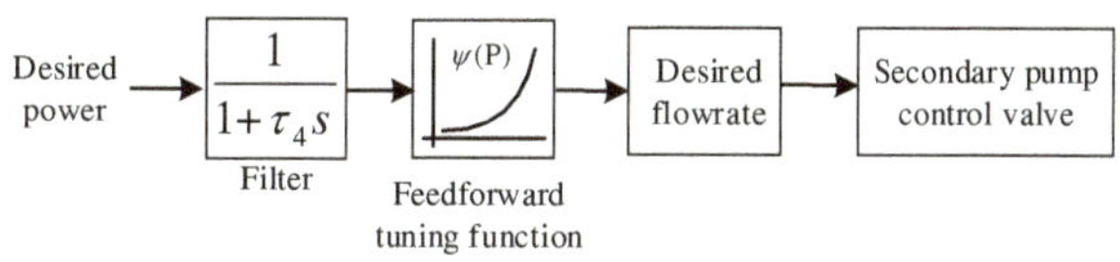

Figure 7. The block diagram of the secondary salt temperature control.

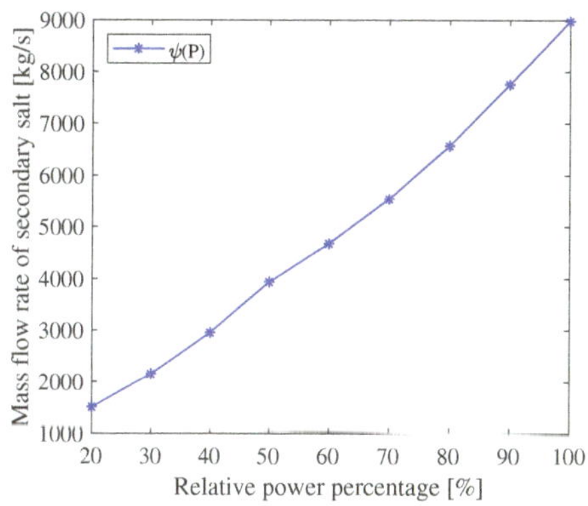

Figure 8. Feedforward tuning function of the mass flow rate of the secondary salt corresponding to the desired power.

3.2.2. Steam Temperature Control System

In the process of the plant load variation, the steam temperature deviates from its set value due to the energy imbalance between the various subsystems, thereby generating a temperature error signal. After the deviation signal is processed by the proportional controller, it is transmitted to a limiter of the feedwater flow rate to avoid the temperature oscillation caused by the sharp changes of the pump valve position. The value of K_p is 0.006 and the change rate of the mass flow rate is limited to $\pm 0.5\%/\text{s}$. Meanwhile, the feedback signal of steam temperature will produce a hysteresis effect due to the system thermal inertia. Thus, a lead-lag controller is designed to compensate for the thermal inertia which could improve the response speed of the steam temperature control. The parameter values of the lead-lag controller τ_5 and τ_6 are 20 and 1, respectively. The final output signal is sent to the flow-control valve for the adjustment of the feedwater flow rate to keep the steam outlet temperature constant, as shown in Figure 9.

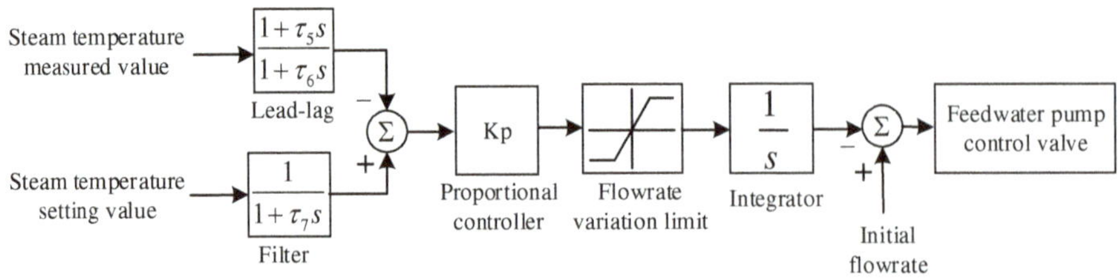

Figure 9. The block diagram of steam temperature control system.

3.3. Implementation of the Plant Nonlinear Model and Control System

The Matlab/Simulink program has been widely applied to the field of system transient simulation and controller design due to its powerful capabilities for model calculation and numerous toolboxes for simulation analysis. The nonlinear distributed parameter model, the reactor power control system and the steam temperature control system for the MSR plant were developed in the Matlab/Simulink environment. The developed simulation platform is presented in Figure 10, which mainly consists of five parts: the molten salt reactor system, the heat exchanger system, the steam generator system, the power control system, and the steam temperature control system. Based on the developed platform, the nonlinear dynamic model of the whole system could be applied under different transient conditions. With the designed reactor power and steam temperature controllers, the feed-forward and feedback combined control strategy can realize the load tracking of the MSR plant under normal and accident conditions.

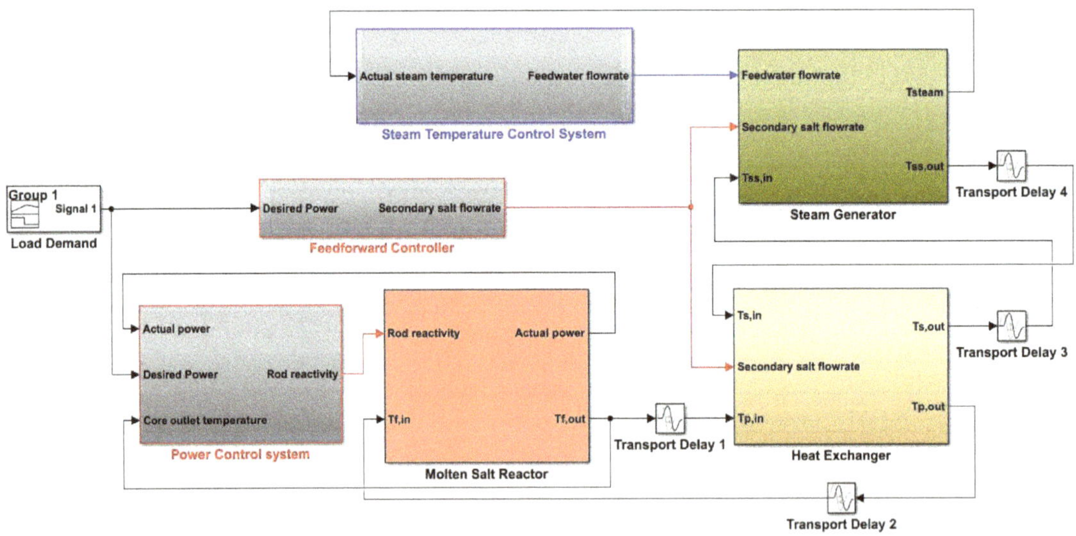

Figure 10. Simulation platform for the MSR plant nonliear model and control system in the Matlab/Simulink environment.

4. Results and Discussions

Based on the established dynamic model of the whole system, the reactor power control system and the steam temperature control system were designed in the simulation platform. Three different transients were simulated to investigate the operation and control characteristics of the MSR plant, namely normal load-follow operation, a reactivity insertion accident and a loss-of-flow accident. The following four criteria are often used to quantitatively assess the performance of control systems under different operating conditions [29]:

(i) Integral time absolute error (ITAE) is defined as:

$$ITAE = \int_0^{t_{max}} t|e(t)|dt \quad (18)$$

where $e(t)$ is the error between the measured value of the controlled variable and its desired value;

(ii) Mean square error (MSE) is defined as:

$$MSE = \frac{1}{t_{max}} \int_0^{t_{max}} e^2(t)dt; \quad (19)$$

(iii) Maximum percentage deviation (MPD) is defined as:

$$MPD = \max\left|\frac{V_m(t) - V_d(t)}{V_d(t)}\right| \times 100\% \quad (20)$$

where V_m and V_d denote the measured value and the desired value of the controlled variable, respectively;

(iv) Settling time (ST) is the total time required for the response curve to reach and stay within a small range of the final value.

4.1. Load-Follow Operation

A linear load change transient is investigated to test the load-follow capability of the control system, in which the load demand decreases from 100% FP to 40% FP at a speed of 10% FP/min after the steady-state operation for 100 s, and then, the load demand returns to 100% FP at the same rate after equilibrium has been established at 40% FP. The transient responses of the system parameters are presented in Figure 11.

The load demand signal is transmitted into the power control system, as shown in Figure 5. The error signals between the desired values and the measured values of the neutron flux and the fuel outlet temperature (ΔN_r and ΔT_{fo}) drive the control rod movement into the core to adjust the reactor power. As shown in Figure 11a,g, the reactivity insertion of the control rod makes the actual power quickly follow the desired power. The actual operating value of the reactor power is almost consistent with the load demand, and the maximum relative deviation does not exceed 1%, as presented in Table 3. Meanwhile, the mass flow rate of the secondary salt serves as a feed-forward tuning parameter which is programmed as a function of the demand load, as shown in Figure 11h. The reactor outlet temperature decreases with the reduction of the reactor power, and the reactor inlet temperature gradually decreases after a slight increase because less heat is transferred into the secondary salt with the reduction of its flow rate in the heat exchanger, as shown in Figure 11c,d.

Table 3. Quantitative evaluation results of the control performance.

Parameters	Load Follow Transient	Reactivity Insertion Transients		Loss of Flow Transients	
		Rod Ejection Accident	Rod Drop Accident	Primary Fuel Pump Trip	Secondary Salt Pump Trip
$IATE_P$	6.13	133.87	101.94	39.32	2.07
$IATE_T$	178.01	1.52	0.78	37.32	93.69
MSE_P	7.29×10^{-8}	1.16×10^{-3}	7.03×10^{-4}	2.44×10^{-4}	3.15×10^{-7}
MSE_T	4.07×10^{-8}	2.12×10^{-8}	4.93×10^{-9}	1.81×10^{-5}	8.89×10^{-5}
MPD_P(%)	0.78	29.07	20.51	12.49	1.13
MPD_T(%)	0.04	0.09	0.04	1.93	3.86
ST_P(s)	–	25	21	16	19
ST_T(s)	–	63	62	105	112

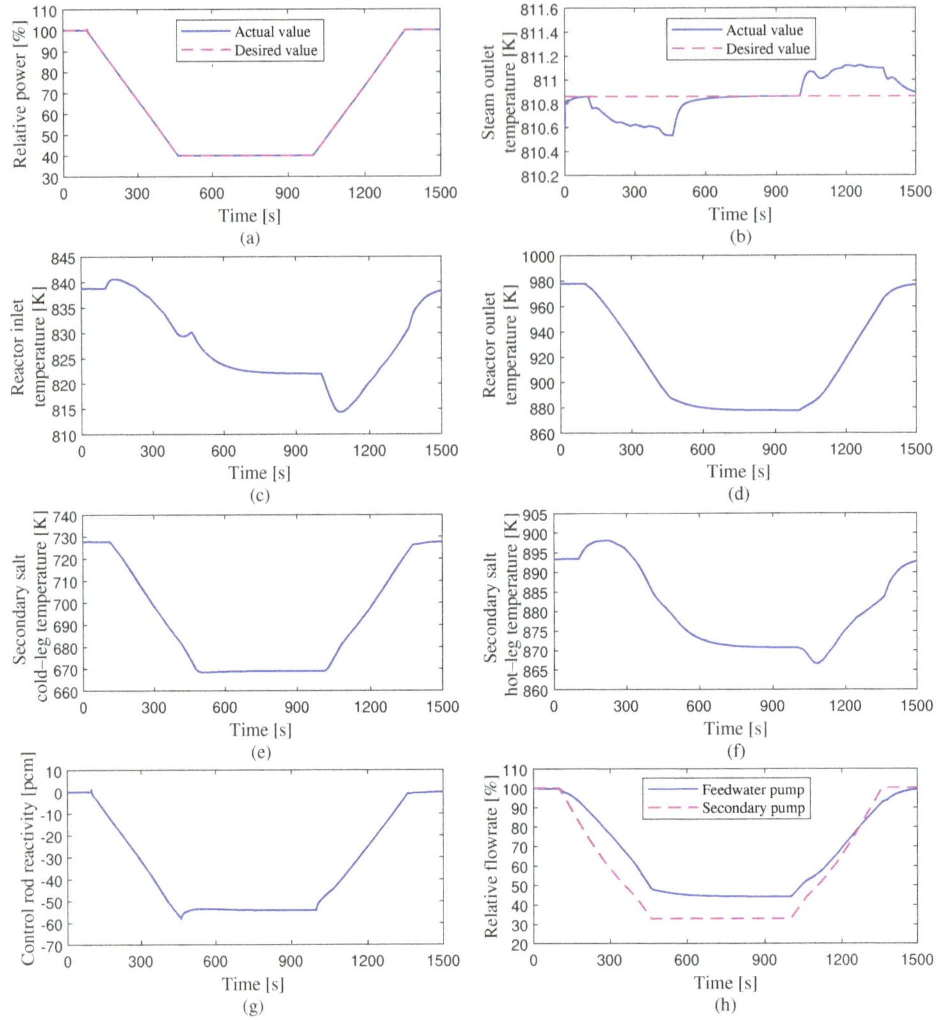

Figure 11. Transient responses of the system parameters during a load–follow change from 100% FP to 40% FP and, then, return at a speed of 10% FP/min. (**a**) Relative power; (**b**) Steam outlet temperature; (**c**) Reactor inlet temperature; (**d**) Reactor outlet temperature; (**e**) Secondary salt cold–leg temperature; (**f**) Secondary salt hot–leg temperature; (**g**) Control rod reactivity; (**h**) Relative flowrate.

The load demand signal is also sent to the steam temperature control system to adjust the feedwater flow rate, as shown in Figure 11b,h. With the rapid decrease of reactor power, the heat transferred to the steam generator reduces rapidly, so that the steam temperature drops. Then, the actual value of the steam temperature is compared with its desired value, and the error signal transmitted into the feedwater pump to reduce the mass flow rate in the steam generator. It can be seen from Figure 11b,h that the steam temperature decreases slightly with the drop of reactor power and then increases gradually with the reduction of the feedwater flow rate before reaching a steady-state point around its desired value, and the range of steam temperature variation does not exceed 1 K during the transient. The cold-leg temperature of the secondary salt decreases with the reduction of the reactor power, and the hot-leg temperature of the secondary salt gradually decreases after a small rise with the rapid reduction of the secondary salt flow rate, as shown in Figure 11e,f.

Based on this control scheme, the 40% load condition is followed quickly for this transient and the system parameters stabilize quickly at a new steady-state point. The reactor power and the steam temperature are well controlled during the process of the load increase from 40% FP to 100% FP, as shown in Table 3, where the subscript $_P$ of the performance parameters represents the reactor power and $_T$ represents the steam outlet temperature. The simulation results show that the designed control system for the MSR plant has a good load-follow capability, which can quickly adjust the reactor power and the steam temperature with satisfactory accuracy.

4.2. Reactivity Insertion Accident

During the operation of a nuclear power plant, due to the failure of mechanical devices or the actions of the operators, partially inserted shutdown rods can accidentally eject from the reactor or fall into the reactor, and then, positive or negative reactivity is introduced, causing the reactor power to deviate from normal operating conditions. In this work, accident transients caused by both rod ejection (60 pcm positive reactivity) and rod drop (60 pcm negative reactivity) are investigated to test the control system performance.

4.2.1. Rod Ejection Accident

Figure 12 shows the comparative results of the rod ejection transient with and without the plant control system. A step of positive reactivity of 60 pcm is suddenly introduced into the reactor at 50 s, and the load demand maintained at 100% FP level. The transient responses of the system parameters with the control system are compared with those of the case that the control system is inactive under the rod ejection accident.

For the controlled case, the reactor power rapidly increases by 29% when the step positive reactivity of 60 pcm is suddenly introduced into the reactor at 50 s, as shown in Figure 12a. The sudden rise of the reactor power leads to the reactor outlet temperature to increase rapidly from its initial steady-state value. The other temperature parameters also ascend with the increase of the reactor power. Because the load demand signal is kept at 100% FP level, the steam outlet temperature set-point is maintained at 810.9 K. Thus, the load deviation signal is sent to the power control system, and the control rod begins to insert negative reactivity into the reactor to reduce the power until it returns to the 100% FP level. The mass flow rate of the secondary pump is constant since the load demand remains 100% FP. At the same time, the increasing reactor temperature causes an increase of the steam temperature, which is delayed due to the transit time of the heat between different subsystems. The error signal of the steam temperature is transmitted to the control system to adjust the feedwater flow rate. It can be seen from Figure 12b that the steam outlet temperature quickly reaches the steady-state point around its desired value. The performance parameters of the control system during this transient are summarized in Table 3. The simulation results show that the developed control system could effectively deal with the rod ejection accident, thereby maintaining the stable operation of the MSR plant.

For the uncontrolled case, it can be seen from Figure 12 that the power increases by about 51% after 60 pcm positive reactivity is introduced into the reactor, and then, the power decreases gradually due to the negative reactivity introduced by the fuel Doppler feedback effect and the graphite temperature feedback effect, and finally reaches a new equilibrium value of 135% FP. Meanwhile, the temperatures of the primary fuel, the secondary salt and the steam also ascend with the power increase and then gradually stabilize to a new equilibrium point. The results show that the reactor power and steam temperature could not be restored to their desired values without the control system.

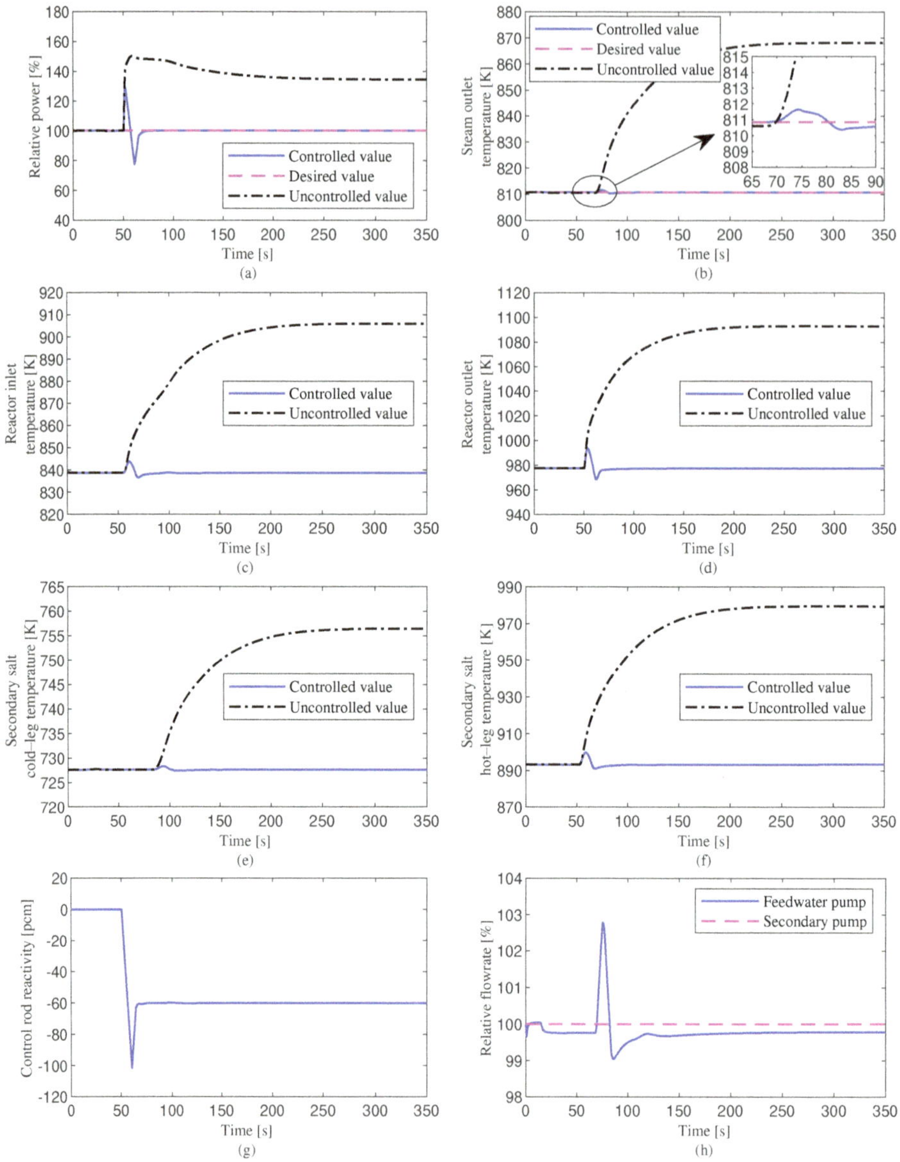

Figure 12. Transient responses of the system parameters under rod ejection accident. (**a**) Relative power; (**b**) Steam outlet temperature; (**c**) Reactor inlet temperature; (**d**) Reactor outlet temperature; (**e**) Secondary salt cold–leg temperature; (**f**) Secondary salt hot–leg temperature; (**g**) Control rod reactivity; (**h**) Relative flowrate.

4.2.2. Rod Drop Accident

The simulation results of the rod drop transient with and without the control system are shown in Figure 13. A step negative reactivity of 60 pcm is suddenly introduced into the reactor at 50 s, and the load demand maintained at 100% FP level. The transient responses of the system parameters with the control system are compared with those of the case that the control system is inactive under rod drop accident.

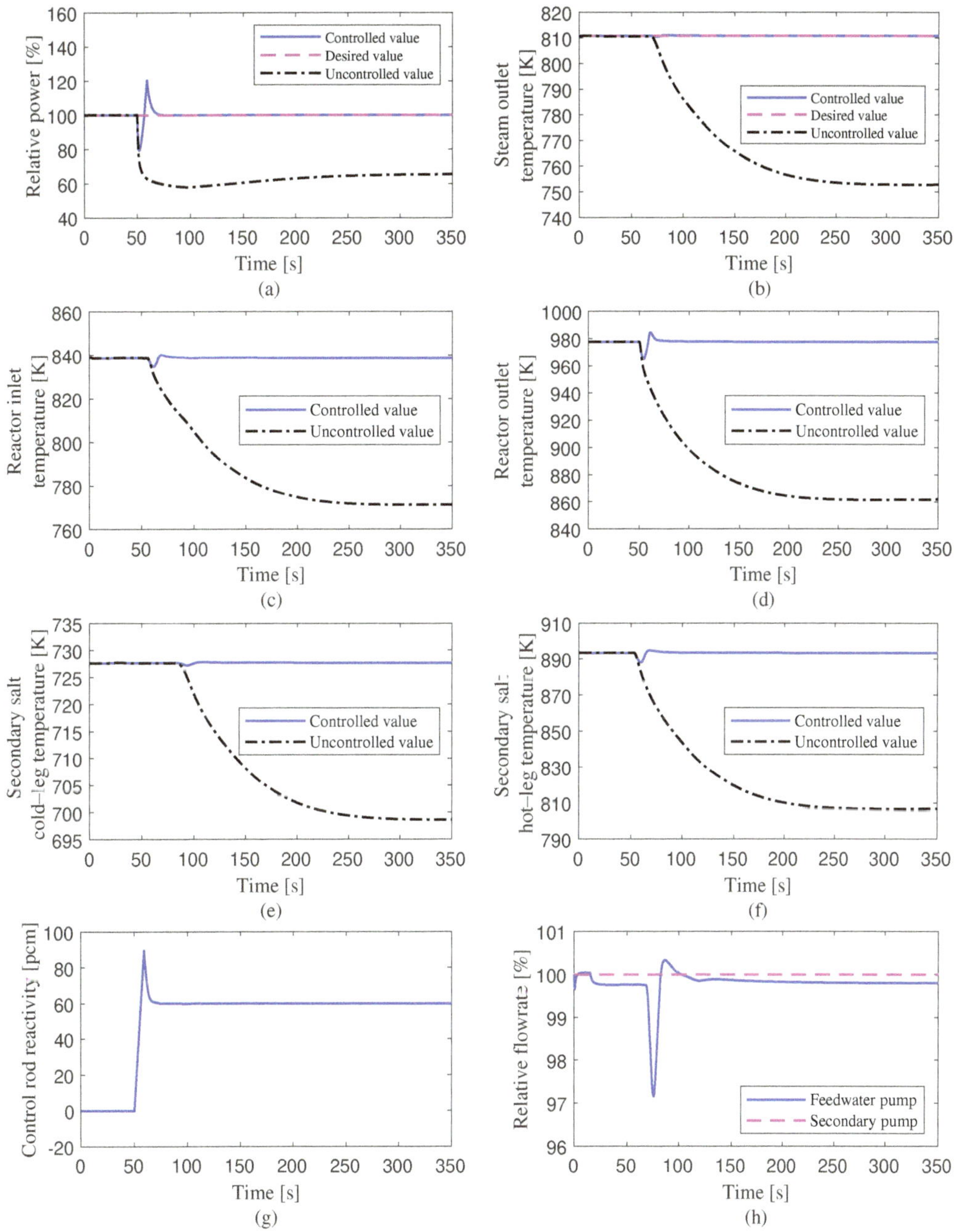

Figure 13. Transient responses of the system parameters under rod drop accident. (**a**) Relative power; (**b**) Steam outlet temperature; (**c**) Reactor inlet temperature; (**d**) Reactor outlet temperature; (**e**) Secondary salt cold–leg temperature; (**f**) Secondary salt hot–leg temperature; (**g**) Control rod reactivity; (**h**) Relative flowrate.

As shown in Figure 13a, the reactor power decreases rapidly to about 79.5% FP level after a step insertion of 60 pcm negative reactivity with the control system. Since the load demand signal keeps at 100% FP, and the setting value of the steam outlet temperature remains at 810.9 K. The power error signal causes the control rod to add reactivity into the reactor at its maximum speed of 10 pcm/s. The sudden reduction of the reactor power also causes a rapid decrease of the reactor outlet temperature.

After a few seconds delay, the reactor inlet temperature reduces with the temperature decrease of the primary fuel. When the reactor inlet and outlet temperatures return to the initial values, the positive reactivity added by the control rod and the negative reactivity introduced by the temperature feedback effects make the reactor power rise by about 121%. Thus, the error signal of the reactor power changes at 59 s, and the control rod starts to introduce negative reactivity to the reactor until the actual power approaches 100% FP, as shown in Figure 13g. Meanwhile, the effect of the reduction in the primary fuel temperature leads to a slight decrease of the steam temperature. The steam temperature controller reduces the feedwater flow rate to return the steam temperature to the desired value. The performance parameters of the control system during this transient are summarized in Table 3, which indicate that both the reactor power and steam temperature are well controlled to their desired values under rod drop accident.

For the uncontrolled case, it can be seen from Figure 13 that the power decreases by about 42% after 60 pcm negative reactivity is inserted into the reactor, and then, the power increases gradually due to the reactivity feedback effects of the fuel and graphite temperatures, finally reaching a new equilibrium value of 66% FP. Meanwhile, the temperatures of the primary fuel, the secondary salt and the steam also decrease with the power drop and then gradually stabilize at a new equilibrium point. The final reactor power and steam temperature are far from the desired values without the control system.

4.3. *Loss of Flow Accident*

The loss of flow accidents occurring in the primary loop and secondary loop, in which the mass flow rate in the loop drops due to the pump trip, are simulated with the developed simulation platform.

4.3.1. Primary Fuel Pump Trip

A sudden loss of one of the four primary fuel pumps at 50 s is simulated with and without the control system, as shown in Figure 14. During this transient, the reduction of load from 100% FP to 75% FP is initiated after a quarter of the primary pumps fail. The main purpose of this simulation is to test the safety characteristics of the control system to avoid overheating of the primary fuel salt.

As shown in Figure 14a, the load demand of the plant is reduced from 100% FP to 75% FP when the loss of flow accident occurs. The error signal between the set-value of the load and its measured value drives the control rod downwards into the reactor for the power adjustment. As the mass flow rate of the primary fuel decreases, both the inlet temperature and outlet temperature rise rapidly. Meanwhile, the mass flow of the secondary salt serves as a feed-forward tuning parameter and it is reduced with the decrease of the demand load. With the rapid decrease of reactor power, the heat transferred to the steam generator reduces rapidly, so that the steam temperature drops. Then, the deviation signal of the steam temperature is transmitted to the feedwater pump in order to reduce the flow rate. In the final steady-state, the reactor power and the steam temperature reach their desired values, as shown in Figure 14a,b. Although there is a short-term large deviation of the steam temperature, which is mainly caused by the simultaneous superposition of the two effects (primary pump trip and load demand reduction). However, this maximum deviation is still within the acceptable range, and the steam temperature quickly returns to the desired value due to the effective adjustment of the control system. The performance parameters of the control system during this transient are summarized in Table 3. The simulation results show that the developed control system not only makes the reactor continue to operate stably under a loss-of-flow accident, but also can produce a rapid adjustment of the reactor power and the steam temperature according to the load demand.

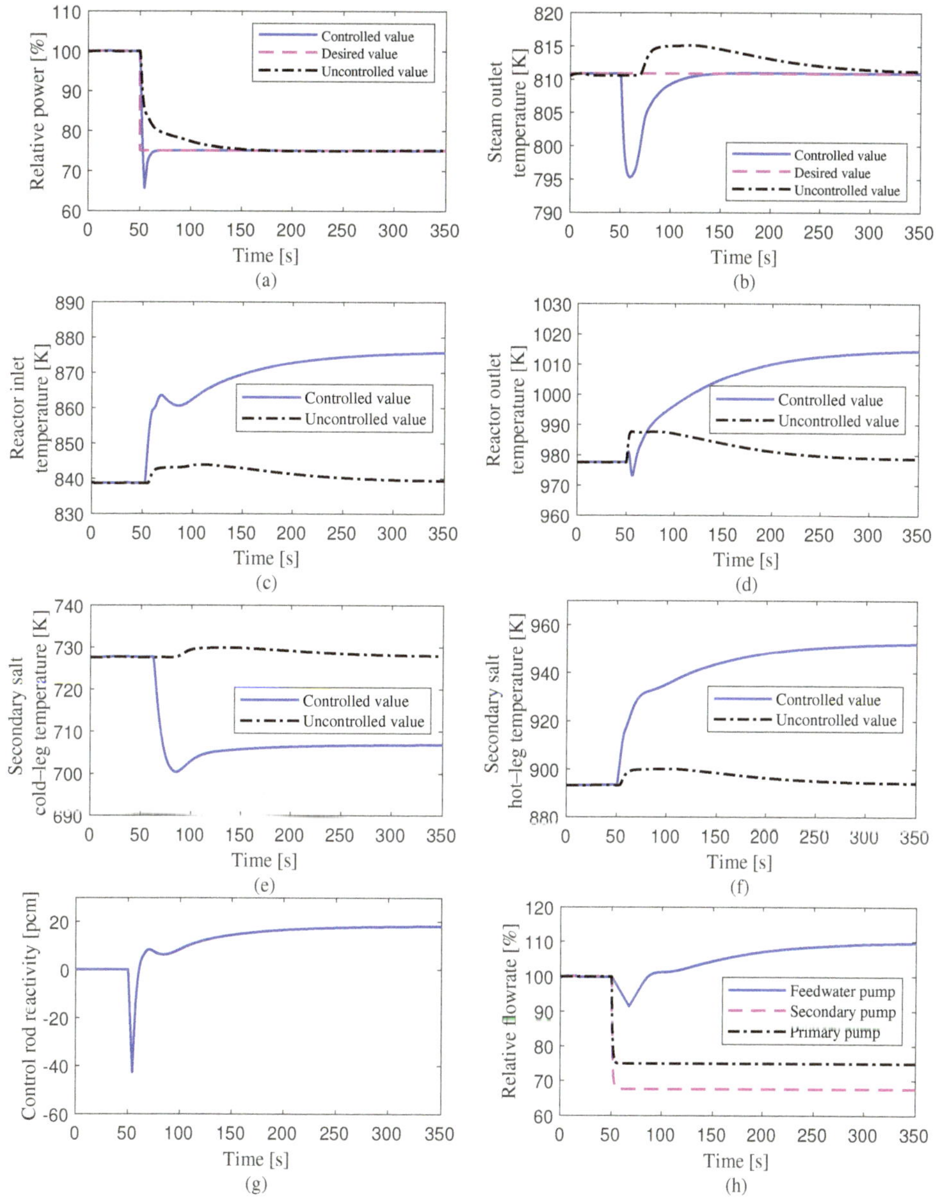

Figure 14. Transient responses of the system parameters under primary fuel pump trip. (**a**) Relative power; (**b**) Steam outlet temperature; (**c**) Reactor inlet temperature; (**d**) Reactor outlet temperature; (**e**) Secondary salt cold–leg temperature; (**f**) Secondary salt hot–leg temperature; (**g**) Control rod reactivity; (**h**) Relative flowrate.

The results of evolution of the system parameters without the control system are also shown in Figure 14. The reactor output power is nearly proportional to the primary-fuel flow rate when the temperature difference remains constant between the reactor outlet and the reactor inlet. Therefore, the uncontrolled actual power automatically drops to about 75% with the step reduction of the primary-fuel flow rate. Meanwhile, the temperatures of the primary fuel, the secondary salt and the steam rise slightly with the decrease of the primary-fuel flow rate and then gradually drop to a new equilibrium point. However, the final values of the reactor power and the steam temperature deviate from their desired values without the control system.

4.3.2. Secondary Salt Pump Trip

In this transient, half of the secondary molten salt pumps are assumed to fail, resulting in a rapid decrease of the mass flow rate to 50% of the steady-state value. The load demand is maintained constant at 100% FP level. Both controlled and uncontrolled cases are simulated and the response results of the system parameters are shown in Figure 15.

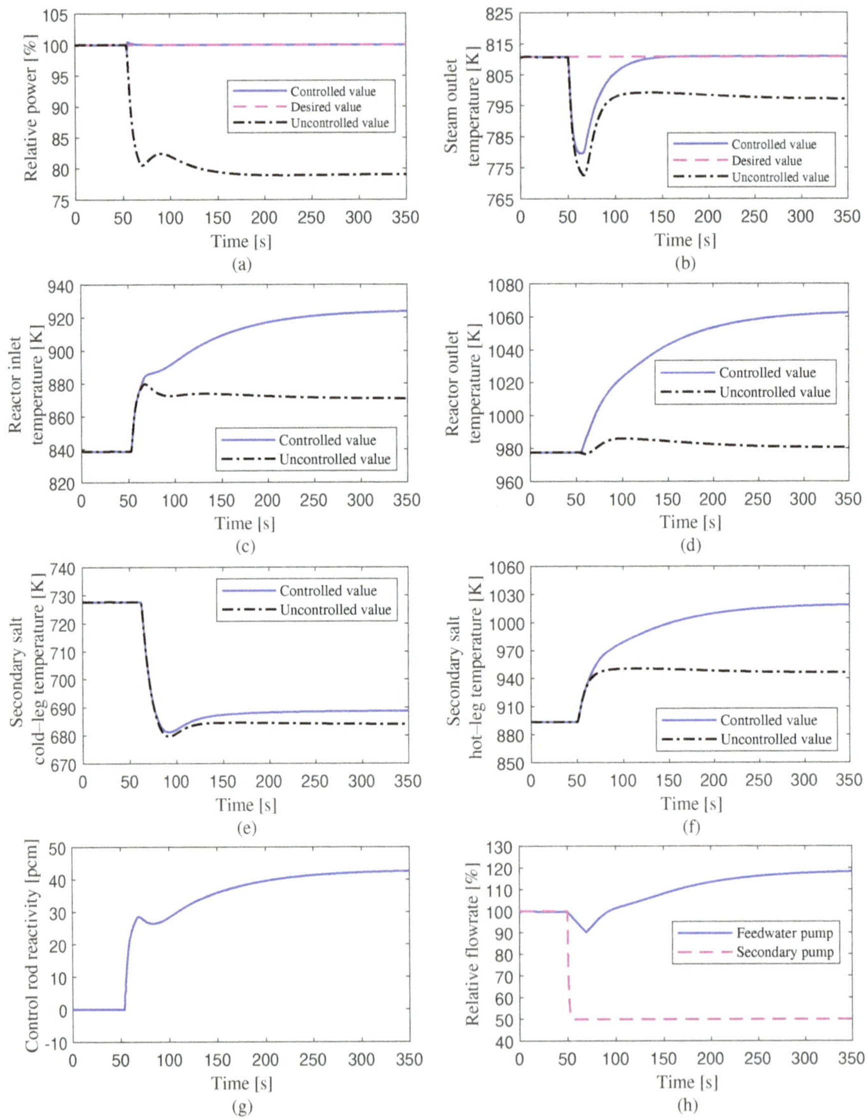

Figure 15. Transient responses of the system parameters under secondary salt pump trip. (**a**) Relative power; (**b**) Steam outlet temperature; (**c**) Reactor inlet temperature; (**d**) Reactor outlet temperature; (**e**) Secondary salt cold–leg temperature; (**f**) Secondary salt hot–leg temperature; (**g**) Control rod reactivity; (**h**) Relative flowrate.

For the controlled case, since half of the secondary pumps stop running, the mass flow rate of the secondary salt drops rapidly in the heat exchanger and steam generator. The flow rate reduction of the secondary salt in the IHX causes that the heat produced by fission reaction in the fuel cannot be taken out in time, and then, the primary fuel temperature increases rapidly from its initial steady-state value, as shown in Figure 15c,d. The reactor power drops due to the negative reactivity introduced by the Doppler effect and the fuel temperature effect with the fuel temperature rising. Because the load demand remains at 100% FP level, the power deviation signal is sent to the control system, and the control rod begins to insert positive reactivity into the reactor to raise the power until it returns to the 100% FP level. The rapid response of the control system makes the maximum power deviation reach only 1.13%. During this transient, the hot-leg temperature of the secondary-salt initially tends to rise and the cold-leg temperature to fall. At the same time, the heat transferred from the secondary salt to the water/steam in the SG is reduced with the rapid decrease of secondary salt flow rate, so that the steam temperature drops rapidly. The error signal between the actual value of the steam temperature and its desired value is transmitted into the feedwater flow-rate controller. It can be seen from Figure 15b that the steam outlet temperature quickly reaches the set-point at 810.9 K. The performance parameters of the control system during this transient are presented in Table 3. The simulation results show that the developed control system could respond quickly to keep the reactor power and the steam temperature at their target values.

For the uncontrolled case, because the mass flow rate of the secondary salt drops rapidly after the loss-of-flow accident occurs, the heat of the primary circuit loop cannot be removed in time. It can be seen from Figure 15c,d that the fuel temperatures of the reactor inlet and outlet rise rapidly. Due to the negative feedback effect of reactivity, the reactor power quickly decreased from 100% FP to about 80% FP, and then stabilizes to a new equilibrium point. Meanwhile, the steam temperature decreases with the power drop and then rises gradually to a new equilibrium point. The final values of the reactor power and the steam temperature deviate from the desired values when the plant control system is not activated.

5. Conclusions

In this study, a simulation platform of the MSR plant was developed to study the control characteristics under normal and accident conditions. The nonlinear dynamic model of the whole system that was built in the platform consisted of a liquid-fuel reactor with a graphite moderator, an intermediate heat exchanger and a steam generator. A new control strategy was presented based on a feed-forward and feedback combined scheme, in which the reactor power control system and steam temperature control system, considering the input saturation and dead zone, were designed to adjust load changes of the plant. In order to investigate the operation and control characteristics of the MSR plant, three different transients were simulated with the developed simulation platform, which include normal load-follow operation, reactivity insertion accident and loss of flow accident. During the load-follow operation, the actual operating value of the reactor power is almost consistent with the desired load, and the range of steam temperature variation does not exceed 1K during the transient. These results show that the designed control system of the MSR plant has a fast load-follow capability and satisfactory control accuracy. For the reactivity insertion transients, the simulation results indicate that the control system could effectively deal with the rod ejection accident, thereby maintaining the stable operation of the MSR plant. In addition, both the reactor power and steam temperature are well controlled to their desired values under rod drop accident. For the loss of flow transients, the simulation results show that the developed control system not only makes the reactor continue to operate stably after one of the four primary fuel pumps fails suddenly, but also could achieve rapid adjustment of the reactor power and steam temperature according to the load demand. Furthermore, the control system could respond quickly to keep the reactor power and the steam temperature at their target values during the transient of secondary salt pump coastdown.

Author Contributions: Conceptualization, R.L. and C.L.; methodology, R.L. and C.L.; software, R.L.; validation, R.L.; formal analysis, R.L.; investigation, R.L. and C.L.; resources, R.L.; data curation, R.L. and C.L.; writing—original draft preparation, R.L.; writing—review and editing, R.L., C.L. and R.M.-J.; visualization, R.L.; supervision, C.L. and R.M.-J.; project administration, R.L.; funding acquisition, R.L. All authors have read and agreed to the published version of the manuscript.

Funding: This research was funded by Scientific Research Foundation of the Education Department of Hunan Province, China (Grant No. 18B265); High-tech Program of the Science and Technology Bureau of Hengyang City, China (Grant No. S2018G9031015321); Fundamental Research Fund for Young Teachers of the University of South China (Grant No. 190XQD064).

Institutional Review Board Statement: Not applicable.

Informed Consent Statement: Not applicable.

Data Availability Statement: Data is contained within the article, further inquiries can be directed to the corresponding authors.

Conflicts of Interest: The authors declare no conflicts of interest.

Abbreviations

The following abbreviations are used in this manuscript:

MSR	Molten Salt Reactor
IHX	Intermediate Heat eXchanger
SG	Steam Generator
PID	Proportion Integration Differentiation
ITAE	Integral Time Absolute Error
MSE	Mean Square Error
MPD	Maximum Percentage Deviation
ST	Settling Time

References

1. Cottrell, W.; Hungerford, H.; Leslie, J.; Meem, J. *Operation of the Aircraft Reactor Experiment*; Technical Report; Oak Ridge National Lab., Tenn.: Oak Ridge, TN, USA, 1955.
2. Bettis, E.; Cottrell, W.; Mann, E.; Meem, J.; Whitman, G. The aircraft reactor experiment—operation. *Nucl. Sci. Eng.* **1957**, *2*, 841–853. [CrossRef]
3. Robertson, R.; Briggs, R.; Smith, O.; Bettis, E. *Two-Fluid Molten-Salt Breeder Reactor Design Study (Status as of 1 January 1968)*; Technical Report; Oak Ridge National Lab.: Oak Ridge, TN, USA, 1970.
4. Taube, M.; Ligou, J. Molten plutonium chlorides fast breeder reactor cooled by molten uranium chloride. *Ann. Nucl. Sci. Eng.* **1974**, *1*, 277–281. [CrossRef]
5. Luzzi, L.; Di Marcello, V.; Cammi, A. *Multi-Physics Approach to the modeling and analysis of Molten Salt Reactors*; Nova Science Publishers, Inc.: Hauppauge, NY, USA, 2012.
6. LeBlanc, D. Molten salt reactors: A new beginning for an old idea. *Nucl. Eng. Des.* **2010**, *240*, 1644–1656. [CrossRef]
7. Locatelli, G.; Mancini, M.; Todeschini, N. Generation IV nuclear reactors: Current status and future prospects. *Energy Policy* **2013**, *61*, 1503–1520. [CrossRef]
8. Varma, V.K.; Holcomb, D.E.; Peretz, F.J.; Bradley, E.C.; Ilas, D.; Qualls, A.; Zaharia, N.M. *AHTR Mechanical, Structural, and Neutronic Preconceptual Design*; Technical Report; Oak Ridge National Lab. (ORNL): Oak Ridge, TN, USA, 2012.
9. Yoder, G.L., Jr.; Aaron, A.; Cunningham, B.; Fugate, D.; Holcomb, D.; Kisner, R.; Peretz, F.; Robb, K.; Wilgen, J.; Wilson, D. An experimental test facility to support development of the fluoride-salt-cooled high-temperature reactor. *Ann. Nucl. Energy* **2014**, *64*, 511–517. [CrossRef]
10. Sheu, R.; Chang, C.; Chao, C.; Liu, Y.W. Depletion analysis on long-term operation of the conceptual Molten Salt Actinide Recycler & Transmuter (MOSART) by using a special sequence based on SCALE6/TRITON. *Ann. Nucl. Energy* **2013**, *53*, 1–8.
11. Fiorina, C.; Aufiero, M.; Cammi, A.; Franceschini, F.; Krepel, J.; Luzzi, L.; Mikityuk, K.; Ricotti, M.E. Investigation of the MSFR core physics and fuel cycle characteristics. *Prog. Nucl. Energy* **2013**, *68*, 153–168. [CrossRef]
12. Uhlíř, J.; Juříček, V. Current czech r&d in thorium molten salt reactor (msr) technology and future directions. In *Thorium Energy for the World*; Springer: Berlin/Heidelberg, Germany, 2016; pp. 139–144.

13. Huke, A.; Ruprecht, G.; Weißbach, D.; Gottlieb, S.; Hussein, A.; Czerski, K. The Dual Fluid Reactor–A novel concept for a fast nuclear reactor of high efficiency. *Ann. Nucl. Energy* **2015**, *80*, 225–235. [CrossRef]
14. Wang, X.; Macian-Juan, R. Steady-state reactor physics of the dual fluid reactor concept. *Int. J. Energy Res.* **2018**, *42*, 4313–4334. [CrossRef]
15. Wang, X.; Liu, C.; Macian-Juan, R. Preliminary hydraulic analysis of the distribution zone in the Dual Fluid Reactor concept. *Prog. Nucl. Energy* **2019**, *110*, 364–373. [CrossRef]
16. Liu, C.; Li, X.; Luo, R.; Macian-Juan, R. Thermal Hydraulics Analysis of the Distribution Zone in Small Modular Dual Fluid Reactor. *Metals* **2020**, *10*, 1065. [CrossRef]
17. Furukawa, K.; Arakawa, K.; Erbay, L.B.; Ito, Y.; Kato, Y.; Kiyavitskaya, H.; Lecocq, A.; Mitachi, K.; Moir, R.; Numata, H.; et al. A road map for the realization of global-scale thorium breeding fuel cycle by single molten-fluoride flow. *Energy Convers. Manag.* **2008**, *49*, 1832–1848. [CrossRef]
18. Jiang, M.; Xu, H.; Dai, Z. Advanced fission energy program-TMSR nuclear energy system. *Bull. Chin. Acad. Sci.* **2012**, *27*, 366–374.
19. Sides, W., Jr. *Control Studies of a 1000-Mw (e) MSBR*; Technical Report; Oak Ridge National Lab.: Oak Ridge, TN, USA, 1970.
20. Tripodo, C.; Di Ronco, A.; Lorenzi, S.; Cammi, A. Development of a control-oriented power plant simulator for the molten salt fast reactor. *EPJ Nucl. Sci. Technol.* **2019**, *5*. [CrossRef]
21. Singh, V.; Lish, M.R.; Chvála, O.; Upadhyaya, B.R. Dynamics and control of molten-salt breeder reactor. *Nucl. Eng. Technol.* **2017**, *49*, 887–895. [CrossRef]
22. Zarei, M. Nonlinear dynamics and control in molten salt reactors. *Nucl. Eng. Des.* **2018**, *332*, 289–296. [CrossRef]
23. Zeng, W.; Jiang, Q.; Xie, J.; Yu, T. A fuzzy-PID composite controller for core power control of liquid molten salt reactor. *Ann. Nucl. Energy* **2020**, *139*, 107234. [CrossRef]
24. Zeng, W.; Zhu, W.; Hui, T.; Chen, L.; Xie, J.; Yu, T. An IMC-PID controller with Particle Swarm Optimization algorithm for MSBR core power control. *Nucl. Eng. Des.* **2020**, *360*, 110513. [CrossRef]
25. Robertson, R.C. *Conceptual Design Study of a Single-Fluid Molten-Salt Breeder Reactor*; Technical Report; Oak Ridge National Lab., Tenn.: Oak Ridge, TN, USA, 1971.
26. Shampine, L.F.; Reichelt, M.W.; Kierzenka, J.A. Solving index-1 DAEs in MATLAB and Simulink. *SIAM Rev.* **1999**, *41*, 538–552. [CrossRef]
27. Burke, O. *Hybrid Computer Simulation of the MSBR*; Technical Report; Oak Ridge National Lab.: Oak Ridge, TN, USA, 1972.
28. Sides, W. *MSBR control studies: Analog simulation program*; Technical Report; Oak Ridge National Lab.: Oak Ridge, TN, USA, **1971**.
29. Wang, P.; Wan, J.; Luo, R.; Zhao, F.; Wei, X. Control parameter optimization for AP1000 reactor using Particle Swarm Optimization. *Ann. Nucl. Energy* **2016**, *87*, 687–695. [CrossRef]

energies

MDPI

Article

The Application of Disturbance-Observer-Based Control in Breath Pressure Control of Aviation Electronic Oxygen Regulator

Rui Pan [1], Guiping Lin [1], Zhigao Shi [2], Yu Zeng [1,*] and Xue Yang [3]

[1] School of Aeronautic Science and Engineering, Beihang University, Beijing 100191, China; panrui@buaa.edu.cn (R.P.); gplin@buaa.edu.cn (G.L.)
[2] Hefei Jianghang Aircraft Equipment Corporation LTD. AVIC, Hefei 230051, China; shizhigao@163.com
[3] Beijing Institute of Spacecraft Environment Engineering, Beijing 100094, China; miemie94@126.com
* Correspondence: zengyu@buaa.edu.cn; Tel.: +86-138-1054-5207

Abstract: The electronic oxygen regulator (EOR) is a new type of aviation oxygen equipment which uses electronic servo control technology to control breathing gas pressure. In this paper, the control method of EOR was studied, and the dynamic model of the aviation oxygen system was established. A disturbance-observer-based controller (DOBC) was designed by the backstepping method to achieve the goal of stable and fast breath pressure control. The sensitivity function was proposed to describe the effect of inspiratory flow on breath pressure. Combined with the frequency domain analysis of the input sensitivity function, the parameters of the DOBC were analyzed and designed. Simulation and experiment studies were carried out to examine the control performance of DOBC in respiratory resistance and positive pressurization process under the influence of noise and time delay in the discrete electronic control system, which could meet the aviation physiology requirements. The research results not only verified the rationality of the application of DOBC in the breath control of EOR, but also proved the effectiveness of the control parameters design method according to the frequency domain analysis, which provided an important design basis for the subsequent study of EOR.

Keywords: electronic oxygen regulator (EOR); pressure control; disturbance observer; sensitivity function; frequency domain analysis

Citation: Pan, R.; Lin, G.; Shi, Z.; Zeng, Y.; Yang, X. The Application of Disturbance-Observer-Based Control in Breath Pressure Control of Aviation Electronic Oxygen Regulator. *Energies* **2021**, *14*, 5189. https://doi.org/10.3390/en14165189

Academic Editors: Ryszard Dindorf, Jakub Takosoglu, Piotr Wos and Adrian Ilinca

Received: 26 July 2021
Accepted: 21 August 2021
Published: 22 August 2021

1. Introduction

Fighter pilots need to be equipped with an aviation oxygen system to avoid the damage caused by low air pressure and oxygen deprivation at high altitude. Oxygen system consists of oxygen source, oxygen regulator, and individual equipment [1–3]. Oxygen regulator is the core component, and its main function is to control the oxygen supply pressure to meet the physiological and protective requirements. Physiological requirement refers to supplying oxygen according to the pilot's breathing demand to reduce respiratory resistance and improve breathing comfort. Protective requirement refers to pressurizing the supply oxygen to prevent pilot from the harsh environment [4,5].

The mechanical oxygen regulator is the traditional type of oxygen regulator, which uses diaphragms for breath control [6]. With the development of microelectronic control technology, the concept of electronic oxygen regulator using sensor and motor was proposed to simplify the structure, improve the pilot's breathing comfort, and provide monitoring capabilities [7]. The French company Airliquide launched the first electronic oxygen regulator for the F-35 which is the only product known to be in use [8]. In the research of civil emergency oxygen regulator, several attempts to combine mechanical control with electronic control have been made. Siska [9] retained the mechanically controlled valve but changed the valve seat from a fixed state to electronically adjustable. Therefore, this electromechanical oxygen regulator could work in two states: with electricity or no electricity. Frederic [10] designed a new kind of electromechanical oxygen regulator by parallel control of mechanical and electronic valves. In general, the aviation oxygen regulator is in the stage of transition from mechanical to electronic. Therefore, it is of great significance to carry out the research on the electronic oxygen regulator (EOR).

The difficulties in breath pressure control were analyzed in references [3,6,11]. Based on these studies and the particularity of pilot breath process, we summarize the following three points about breath pressure control. Firstly, the pilot's breathing demand and the target pressure of control would change with the task load and physiological state, which has strong randomness and uncertainty. Secondly, the breath cycle is very short and the breath flow rate changes fast. In the normal state, the breath cycle is about 6 s, and the peak breath flow rate is about 30 L/min. During flight, the breath cycle may be shortened to about 1 s, and the peak breath flow rate may reach 180 L/min. Thirdly, due to the limitation of space, the volume of the breath chamber is small, so the breath pressure is very easy to change largely. Combined with the above characteristics, EOR needs to have fast response speed, high sensitivity, as well as a stable and robust control law [12].

So far, several studies have been conducted on EORs to improve the control performance of breath pressure. Yu et al. [13] designed an EOR driven by stepping motor and controlled by switch method. Sun et al. [14] improved research in [13] by adjusting the motor speed with fuzzy control theory. Li et al. [15] developed expert PID control for the EOR driven by voice coil motor, which solved the deficiency of fixed parameters. Jiang et al. uses a commercial flow servo valve as the executive structure of EOR to and apply generalized predictive control [16], active disturbance rejection control [17], and adaptive control [18] into breath pressure control. The results show that the intelligent control has better robustness than PID control. All the above studies were exploratory studies on control law schemes of EOR for physiological requirement, but the key to design control parameters did not be pointed out clearly. In addition, only a few of these researches took consideration into protective requirement, so the control of positive pressurization process is still to be studied.

Disturbance-observer-based control (DOBC) is a kind of control method which comprehensively estimates the external disturbance and internal model uncertainty according to the mathematical model and control input, and introduces equivalent compensation in the control to realize the complete control of disturbance. The disturbance observer is extended from the state observer in modern control in which disturbance is extended as a part of the system state. If the estimation error of the disturbance could converge gradually, the system will obtain the ability to compensate the disturbance in time with the estimation value of disturbance, and the control law can achieve great ability of disturbance rejection. DOBC has been usually adapted to supplement the commonly used method of feedback control, making the model-based control design to be applied better to the actual system and improving the robustness of the control system [19–21].

In this paper, a basic structure of EOR driven by hybrid stepping motor was presented, and the mathematical model of breath pressure control was established. The DOBC theory was applied to the breath pressure control of the EOR, and the external load of the pilot's breath flow and the internal model error could be estimated and compensated. The sensitivity function of the input disturbance was used to describe the effect of pilot's breath flow on breath pressure and analyzed in frequency domain, and the control parameters were determined by magnitude-frequency characteristics. Simulation and experiments were carried out to verify the control performance of DOBC on respiratory resistance and positive pressurization.

2. System Principle and Mathematical Model

2.1. System Description

The new type of aviation oxygen system mainly includes four parts: oxygen supply source, electronic oxygen regulator, oxygen mask, and active servo lung, as shown in

Figure 1. High pressure oxygen flow provided from oxygen supply source is adjusted by the EOR so that breath pressure could be keep stable around the target pressure. The oxygen mask has two check valves, one for inhalation and the other for exhalation. During breathing, the two check valves open alternately, not only saving oxygen supply, but also relief the exhaled air out of the mask. It is ensured that pilots would not repeatedly inhale the exhaled air with low oxygen concentration. Active servo lung was designed to simulate human lungs, which could intermittently inhale and exhale gas [22].

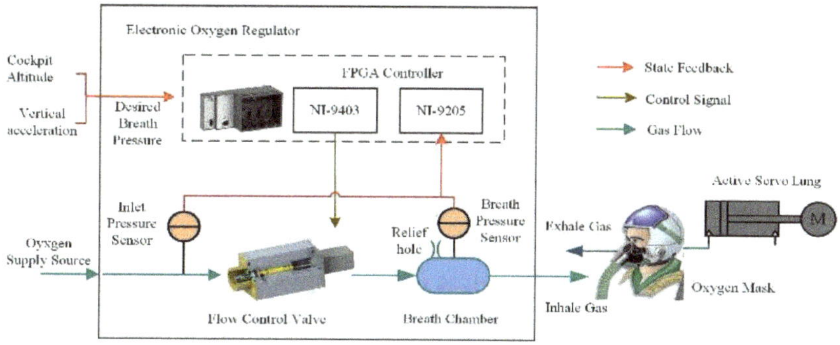

Figure 1. The structure and working principle of aviation oxygen system.

The electronic oxygen regulator mainly includes inlet pressure sensor, flow control valve, breath chamber, breath pressure sensor, and controller. The desired breath pressure was set in accordance with the cockpit height and vertical acceleration signal from external sensors, to achieve the purpose of positive breath pressure for high altitude and anti-G. When the pilot inhales, the check valve for inhalation on mask opens. Gas in the breath chamber would be breathed into the mask, and the breath pressure would drop below the desired pressure. The controller drives the flow control valve open to provide sufficient oxygen flow to the pilot to maintain breath pressure stable. When the pilot exhales, the check valve for inhalation on mask turns to close and the check valve for exhalation opens, the controller closes the flow control valve immediately to avoid excessive gas supply, affecting the pilot's exhalation process. Relief hole on the breath chamber is used to relief excessive breath pressure.

The pressure difference between the breath pressure and the desired pressure is usually used to describe the working performance of EOR. The negative pressure difference indicates the inspiratory resistance, and the positive pressure difference indicates the expiratory resistance. In order to avoid the phenomenon of breathing fatigue, the pressure difference under control of the EOR should keep within the respiratory resistance threshold in Table 1. The control accuracy of breath pressure in positive pressurization process should reach $\pm$ 0.1 kPa in the pressurized range of 0~10 kPa to accurately realize the protective effect of pressurized oxygen supply.

Table 1. Respiratory resistance threshold under different pulmonary ventilation volumes.

Pulmonary Ventilation Volume (L/min)	Inspiratory Resistance (kPa)	Expiratory Resistance (kPa)
10	$\leq$0.49	$\leq$0.25
20	$\leq$0.64	$\leq$0.39
30	$\leq$0.78	$\leq$0.59
45	$\leq$0.88	$\leq$1.08

2.2. Mathematical Model

According to the characteristics of the thermal process, the following reasonable assumptions were made to simplify the model:

(1) The gas could be descried with the ideal gas state equation and the thermal process of gas in each chamber accords with the polytropic thermal process;

(2) The gas distribution in each chamber is even and pressure could be expressed by a uniform value;

(3) The flow of gas through the hole and valves was considered as the one-dimensional steady adiabatic flow.

During the working process, there will be continuous flow into and out of the breath chamber, and the exchange of mass, heat, and work exists between the system and the outside world. Therefore, the breath chamber was modeled as an open system. The control volume was selected according to the edge of the chamber, and the mathematical model of the state parameters such as pressure, temperature, and gas mass in the breath chamber were established with the mass conservation equation and the polytropic thermal process equation. According to the law of mass conservation, the mass increment in the control volume is equal to the difference between the mass into and out of the control volume. The differential form is written as

$$dm = \delta m_{in} - \delta m_{out}, \tag{1}$$

where dm represents the gas mass increment in the breath chamber, δm_{in} and δm_{out} are the mass entering and leaving the system. Based on the ideal gas state equation $pV = mRT$, the gas mass increment could be written as

$$\frac{dm}{m} = \frac{dp}{p} + \frac{dV}{V} - \frac{dT}{T}, \tag{2}$$

where m, p, ρ, and T represent the mass, pressure, density, and temperature of the gas in the breath chamber, respectively; V is volume of breath chamber; and R is the gas constant. Since the volume of the breath chamber is fixed, the volume differential in the mass equation above can be ignored. According to the polytropic thermal process equation $p/\rho^n = const$, relation between temperature and pressure could be descried as

$$\frac{dT}{T} = \frac{n-1}{n}\frac{dp}{p}, \tag{3}$$

where n is the polytropic index, which was chosen as 1.35, because the polytropic process in the breath chamber is close to the adiabatic process. Combined with the Equations (1)–(3), the differential function of breath pressure could be given by

$$\frac{dp}{dt} = \frac{nRT}{V}\left(\frac{\delta m_{in}}{dt} - \frac{\delta m_{out}}{dt}\right). \tag{4}$$

According to the relationship between the downstream and upstream pressure ratio and the critical pressure ratio, the mass flow through the flow control valve and relief hole can be divided into critical flow and subcritical flow. The mass flow rate of the critical flow is only determined by the upstream pressure, while the mass flow rate of the subcritical flow is affected by both upstream and downstream pressure. The mass flow model is as follows:

$$\begin{cases} q_m = \mu A p_{in}\sqrt{\frac{2k}{(k-1)RT_{in}}\left(\varepsilon^{2/k} - \varepsilon^{(k+1)/k}\right)} \\ \varepsilon = \max(p_{out}/p_{in}, \varepsilon_{cr}) \end{cases}, \tag{5}$$

where q_m is the mass flow rate, kg/s; μ is the flow coefficient; A is the flow area, m^2; p_{in} and p_{out} indicate upstream and downstream pressure, Pa; k is the adiabatic constant, with the value of 1.4. ε Refers to the pressure ratio, and ε_{cr} refers to the critical pressure ratio, with the value of 0.528. According to the condition of upstream and downstream pressure, the mass flow of the flow control valve should be calculated as critical flow, and the mass flow of the relief hole should be calculated as subcritical flow.

The human lung is innervated by the nervous system and periodically expands and contracts by the respiratory muscles. It is generally believed that the instantaneous breath flow is close to the sinusoidal process, which can be expressed as

$$Q_B(t) = \pi V_t N sin\left(\frac{2\pi}{60/N}t\right), \tag{6}$$

where Q_B is the instantaneous volume flow of breath, L/min. Tidal volume V_t is the amount of air inhaled or exhaled in each breath cycle, with the value of 1 L for pilots. The respiratory rate N is the number of breath cycle completed per minute. The product of tidal volume V_t and respiratory rate N becomes pulmonary ventilation volume, L/min.

Due to the presence of two check valves on the mask, only the inspiratory flow would affect the pressure in the breath chamber. Therefore, the mass flow of breath is given by

$$q_{m,B}(t) = \begin{cases} \frac{\rho}{60} \times Q_B(t) \; during\ inhalation \\ 0 \qquad\qquad\quad during\ exhalation \end{cases}, \tag{7}$$

where $q_{m,B}$ is the instantaneous mass flow of breath from the breath chamber, g/s.

According to the above model, the mass flow variations in the breathing chamber include: control input—oxygen supply flow of the flow control valve, load disturbance—mass flow of respiration, state variable—leakage flow through the relief hole. The mathematical model of breath pressure control is

$$\frac{dp}{dt} = \frac{nRT}{V}\left(q_{m,in}(A(t), p_{in}) - q_{m,leakage}(p) - q_{m,B}(t)\right). \tag{8}$$

3. DOBC Design

In order to maintain the control stability and achieve better control accuracy of breath pressure, a disturbance-observer-based controller was designed to quickly adjust the oxygen supply flow and accurately compensate the inspiration flow and leakage.

3.1. Design of Feedback Control

Let $b = nRT/V$, and according to the definition of mass flow variation, the differential function of breath pressure can be written as

$$\dot{p} = b(u - f(p) - d(t)), \tag{9}$$

where $f(p)$ represents the leakage of the system, u and d indicate input flow and inspiration flow, respectively. There may be some uncertainties in the system, such as parameter error between the model and the actual system. The input flow may also be not equal to the expected flow. Considering these uncertainties, the differential equation of the actual breath pressure is given by

$$\dot{p} = (b + \Delta b)(u + \Delta u - f(p) - \Delta f(p) - d(t)), \tag{10}$$

where Δb reflects the influence of temperature variation and the uncertainty of chamber volume and polytropic index. Δu represents the uncertainty of the input flow, which may be caused by the flow control error of the flow control valve. $\Delta f(p)$ represents the uncertainty of the leakage, which may be caused by the inaccurate leakage area or the variational flow coefficient. Let D_i represent the disturbance caused by internal model uncertainty, D_o represent the disturbance caused by external load, and D represent the total disturbance of the system. These disturbances could be given by

$$\begin{cases} D_i = (b + \Delta b)(\Delta u - \Delta f(p)) \\ D_o = (b + \Delta b)d(t) \\ D = D_i - D_o \end{cases}. \tag{11}$$

The differential equation of breath pressure with uncertainty is rewritten as follows:

$$\dot{p} = b(u - f(p)) + D. \tag{12}$$

Let p_d represent the desired breath pressure, which is set according to the cockpit height and vertical acceleration. For control system, the target is to make breath pressure stable around the desired breath pressure and the control error was expected to the 0. The control error of the system e and its change rate is

$$\begin{cases} e = p_d - p \\ \dot{e} = \dot{p_d} - \dot{p} = \dot{p_d} - b(u - f(p)) - D \end{cases}. \tag{13}$$

Define the Lyapunov function V_1 as $V_1 = e^2/2$, then the derivative of V_1 is

$$\dot{V_1} = e\dot{e} = e(\dot{p_d} - b(u - f(p)) - D). \tag{14}$$

The control input is design by back-stepping method as follows:

$$u = \frac{1}{b}(\dot{p_d} - D + k_1 e) + f(p), \tag{15}$$

where k_1 is the state feedback gain. Then the derivative of Lyapunov function V_1 can be rewritten as $\dot{V_1} = -k_1 e^2$. According to Lyapunov stability criterion, if Lyapunov function satisfies the conditions below:

$$\begin{cases} V(e) > 0 \\ \dot{V}(e) \leq 0 \\ e \neq 0, \dot{V}(e) \neq 0 \end{cases}, \tag{16}$$

the system has Lyapunov stability and the control error tends to 0. Therefore, under this feedback control law, the control stability of breath pressure could be guaranteed. The convergence speed of control error is related to the state feedback gain k_1. The control error converged faster with larger k_1, but it also might cause overshoot of the system. There is disturbance term of the system in the control input of the feedback control, so the second part is going to design an appropriate disturbance observer and use the estimated disturbance to calculate control input.

3.2. *Design of Disturbance Observer*

In the high frequency control system, the uncertainty of the system parameters and the external low frequency disturbance caused by the inspiratory flow can be regarded as a slow time-varying process, so it can be assumed that $D(t) \approx const, \dot{D}(t) \approx const$. Let $\hat{D}$ represent the disturbance estimated by the disturbance observe, and the disturbance estimation error $\tilde{D}$ is $\tilde{D} = D - \hat{D}$, the derivative of disturbance estimation error is

$$\dot{\tilde{D}} = \dot{D} - \dot{\hat{D}} \approx -\dot{\hat{D}}. \tag{17}$$

Define Lyapunov function V_2 as $V_2 = e^2/2 + \tilde{D}^2/2\gamma$, where γ is the disturbance estimation coefficient. The derivative of V_2 is $\dot{V_2} = e\dot{e} + \tilde{D}\dot{\tilde{D}}/\gamma$. By substituting the control input into the derivative of V_2, we can get:

$$\dot{V_2} = -k_1 e^2 - \tilde{D}(\dot{\hat{D}}/\gamma + e). \tag{18}$$

Let the estimation function of the disturbance observer be $\dot{\hat{D}} = -\gamma e$, then the derivative of V_2 can be rewritten as $\dot{V_2} = -k_1 e^2$. According to Lyapunov stability criterion, the system controlled by state feedback and disturbance observer has Lyapunov stability. The control input of DOBC can be expressed as

$$\begin{cases} \hat{D} = \int_0^t -\gamma e(\tau)d\tau \\ u = \frac{1}{b}(\dot{p_d} - D + k_1 e) + f(p) \end{cases}. \tag{19}$$

3.3. Design of Anti-Windup Mechanism

Since the input flow rate is always positive, the disturbance observer may appear integral saturation. In order to reduce its influence, a clamping anti-windup mechanism was added into DOBC. After each completion of the control input calculation, the clamping anti-windup mechanism will examine whether the calculated control input has reached the limitation. If the control input exceeds the limitation and the sign of the total input is the same as the control error, the disturbance observer will stop working. Under any other condition, the disturbance observer would work normally. The DOBC with clamping anti-windup mechanism is as follows:

$$\hat{D}(k) = \hat{D}(k) - \alpha \times \gamma e(k) \times T_s, \tag{20}$$

where T_s is the sampling period and α is the anti-windup coefficient:

$$\alpha = \begin{cases} 0 \; if\{u(k-1) \notin [u_{min}, u_{max}] \parallel u(k-1) \times e(k) \geq 0\} \\ 1 \qquad\qquad else \end{cases}. \tag{21}$$

The block diagram of DOBC with clamping anti-windup mechanism is shown in Figure 2.

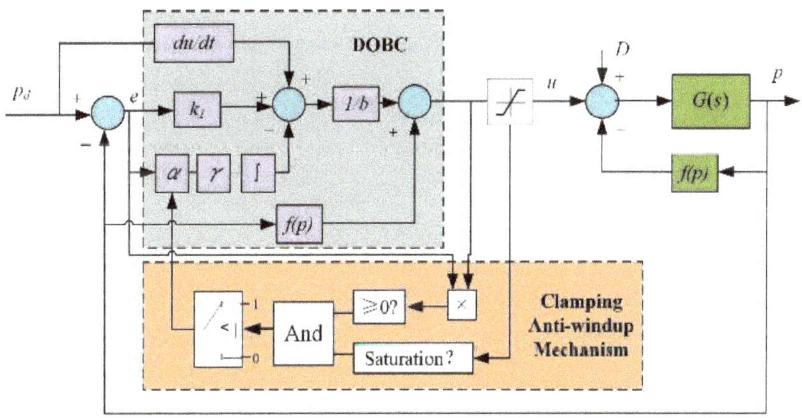

Figure 2. The block diagram of DOBC with clamping anti-windup mechanism.

3.4. Design of Control Parameters

Laplace transform is applied to the DOBC in normal working mode to obtain the transfer function of the control input:

$$U(s) = \frac{1}{b}\left(sP_d(s) + \gamma\frac{E(s)}{s} + k_1 E(s)\right) + f(P(s)). \tag{22}$$

Assuming that the desired pressure remains constant, the leakage is relatively small compared to the inspiratory flow and the supply flow. The transfer function of control law can be simplified as $C(s) = U(s)/E(s) = \frac{1}{b}(k_1 + \frac{\gamma}{s})$. Similarly, the transfer function of breath pressure could be obtained by Laplace transform, $G(s) = b/s$. According to the transfer function of control law and the plant, the influence of the inspiratory flow on the breath pressure can be expressed by the input sensitivity function:

$$S_i(s) = \frac{G(s)}{1 + G(s)C(s)} = \frac{bs}{s^2 + k_1 s + \gamma}. \tag{23}$$

Let the feedback gain and disturbance observer coefficient as $k_1 = 2\xi\omega_n, \gamma = {\omega_n}^2$. Then, the characteristic equation of the close-loop system can be expressed with the natural frequency ω_n and damping coefficient ξ in the standard form of the second-order system. The magnitude-frequency characteristics of the input sensitivity function can be expressed as

$$M(\omega) = \left| \frac{bj\omega}{-\omega^2 + 2\xi\omega_n s + {\omega_n}^2} \right| = \frac{b/\omega_n}{\sqrt{(2\xi)^2 + \left(\frac{\omega_n}{\omega} - \frac{\omega}{\omega_n}\right)^2}}. \tag{24}$$

It can be analyzed that the magnitude-frequency characteristic decreases monotonically with the increase of damping coefficient. When natural frequency ω_n is greater than input disturbance frequency ω, the magnitude-frequency characteristic decreases monotonically with the increase of natural frequency, and increases monotonically with the increase of input disturbance frequency.

In order to make the respiratory resistance within the threshold in Table 1 it is necessary to limit the magnitude-frequency characteristics of the input sensitivity function under different input disturbance frequencies. Combined with the sinusoidal formula of inspiratory flow, the design objective of magnitude-frequency characteristics could be given by

$$When\ \omega = 2\pi f = \frac{2\pi}{60/N}, M(\omega) \leq \frac{bj\omega}{-\omega^2 + 2\xi\omega_n s + {\omega_n}^2}. \tag{25}$$

According to the above equitation, the respiratory resistance threshold in Table 1 could be converted into the desired magnitude-frequency characteristic in Table 2.

Table 2. Desired magnitude-frequency characteristics of the input sensitivity function.

Pulmonary Ventilation Volume (L/min)	Input Disturbance Frequency (Hz)	Maximum of Inspiratory Flow (g/s)	Inspiratory Resistance Threshold (kPa)	Desired Magnitude-Frequency Characteristics
10	1/6	0.65	≤0.49	≤0.754
20	1/3	1.3	≤0.64	≤0.492
30	1/2	1.95	≤0.78	≤0.4
45	3/4	2.9	≤0.88	≤0.303

Combined with the monotonically relationship analyzed above, it can be concluded from the Table 2 that if the magnitude-frequency characteristic of the input sensitivity function drops below 0.303 when the input disturbance frequency is 3/4 Hz, the control performance of DOBC could meet the inspiratory resistance threshold. The volume of breath chamber was set as 0.3 L, and the temperature was set as 293.15 K, so the coefficient b was calculated as 378.6 kPa/(g/s). The damping coefficient was selected as 0.707, and the natural frequencies were set as 10 Hz, 15 Hz, and 20 Hz, respectively. The Bode diagram of input sensitivity function was drawn in frequency range less than 1 Hz, as shown in Figure 3.

It can be seen from the magnitude-frequency characteristic diagram (upper panel) of the Bode diagram that when the natural frequency reaches 15 Hz, the magnitude-frequency characteristic can be maintained below the desired magnitude-frequency characteristic. At this time, the respiratory resistance control can theoretically achieve the control target. On the other hand, it can be seen from the phase-frequency characteristic diagram (lower panel) of Bode Diagram that the phase lag is between 80° and 90° at each natural frequency, which means that the response of breath pressure is ahead of the inspiratory flow in phase.

Because of the sinusoidal wave of the inspiratory flow, the phase lead feature makes the peak of the inspiratory resistance appear at the beginning of the inhalation. Moreover, at the end of the inhalation, pressure error will become positive, which may lead to a higher expiratory resistance.

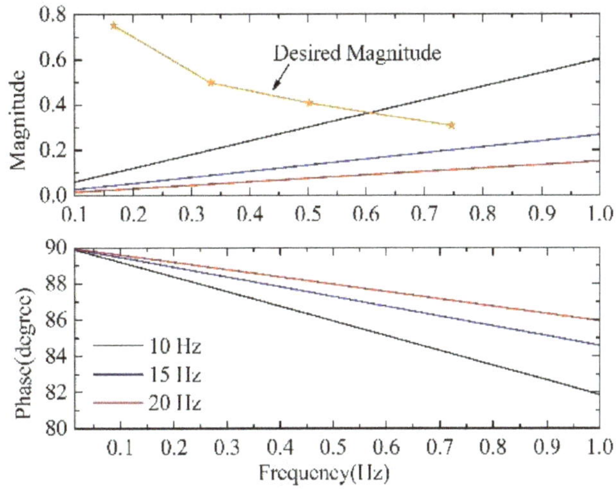

Figure 3. The Bode diagram of input sensitivity function, upper panel: magnitude-frequency characteristic, lower panel: phase-frequency characteristic.

4. Results and Discussion

4.1. Simulation Research

Simulation Researches were carried out to simulate the breath pressure during the whole breath process. Simulation parameters and conditions were set as follows. The natural frequency was set as 15 Hz and the damping coefficient was set as 0.707. Therefore, the state feedback gain k_1 was calculated as 133, and disturbance observer coefficient was calculated as 8883. In order to simulate the discrete characteristics in the electronic control system, the sampling period of the controller T_s was set as 1 ms, and the time delay of system T_d was set as 3 ms. White noise signal with the standard deviation of 0.005 kPa was added to the feedback loop to simulate the measurement noise of breath pressure signal.

- Simulation of normal breath

The desired pressure was set as 0 kPa to simulate the normal breath process. Figure 4 shows the dynamic curves of breath pressure under the control of DOBC for four different pulmonary ventilation rates of 10 L/min, 20 L/min, 30 L/min, and 45 L/min, respectively.

It can be seen from the simulation results that DOBC can provide appropriate supply flow according to the feedback breath pressure to meet the control target of respiratory resistance. In the stage of inhalation, when the pulmonary ventilation volume is small, the control process was affected significantly by the measurement noise, and the pressure fluctuates frequently near the desired pressure; when the pulmonary ventilation volume gets larger, the influence of measurement noise reduced gradually, and the phase response of breath pressure is basically consistent with the phase frequency characteristics in theoretical analysis, with the peak of inspiratory resistance at the initial of inhalation, and positive pressure at the end of inhalation. During exhalation, the breath pressure maintains a positive pressure of 0.1 kPa, result in a small expiratory resistance. This phenomenon is caused by the influence of measurement noise on control input of DOBC, making the supply flow fluctuate in a very small range.

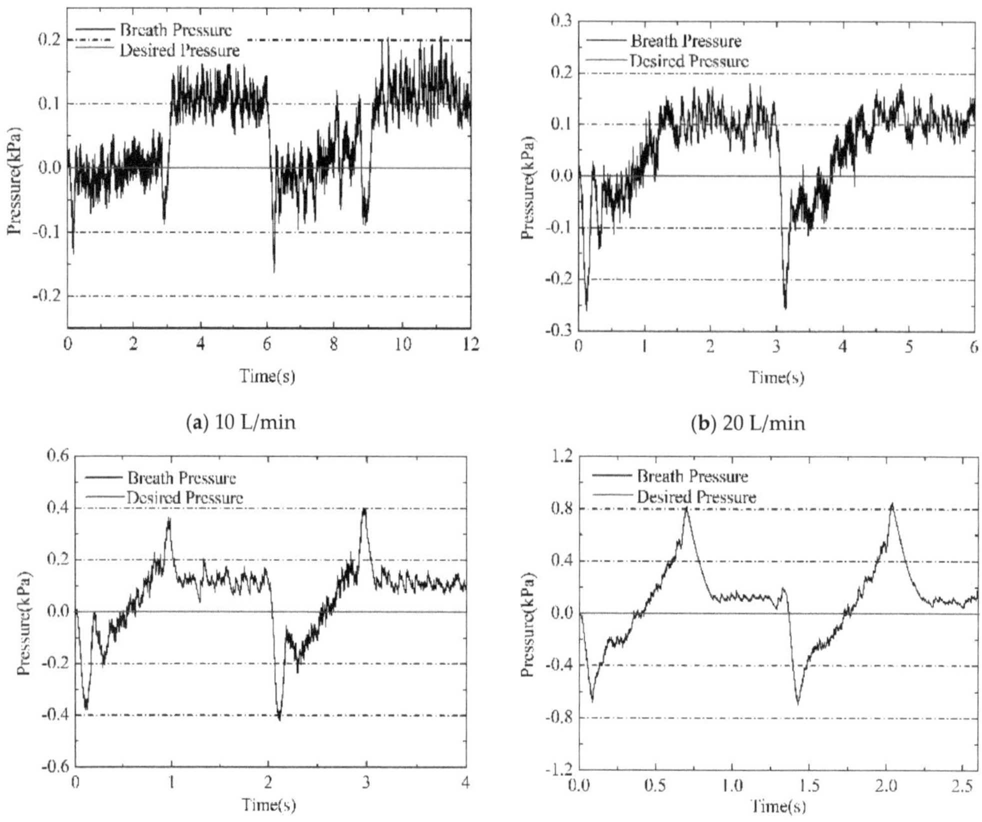

(**a**) 10 L/min (**b**) 20 L/min (**c**) 30 L/min (**d**) 45 L/min

Figure 4. Simulation results of breath pressure during breath under different pulmonary ventilation volume, (**a**) 10 L/min, (**b**) 20 L/min, (**c**) 30 L/min, and (**d**) 45 L/min.

- Simulation of positive pressurization

The desired pressure was increased from 0 kPa to 10 kPa in a step process of 1 kPa to simulate the positive pressurization process. The curve of controlled breath pressure and the control error of DOBC were shown in Figure 5. In order to compare with the accuracy requirement, the control error was also amplified in part.

The simulation results show that the DOBC can reduce the control error in a very short time after the step of desired pressure signal and the overshoot could be limited to about 0.4 kPa. The control system has good ability of reference following and small overshoot characteristic. Furthermore, DOBC can achieve pressure control stably in the small breath chamber, and the steady-state error of control is basically maintained in the range of ±0.1 kPa, which basically meets the control accuracy requirement. According to the characteristics of the sensitivity function, the small input sensitivity function means that the complementary sensitivity function is close to 1. Therefore, the large control parameters in DOBC make close-loop system have good reference following performance and disturbance rejection at the same time. As a sacrifice, the negative effect of measurement noise was inevitably amplified. It can be deduced that the frequent fluctuation of control error is caused by measurement noise.

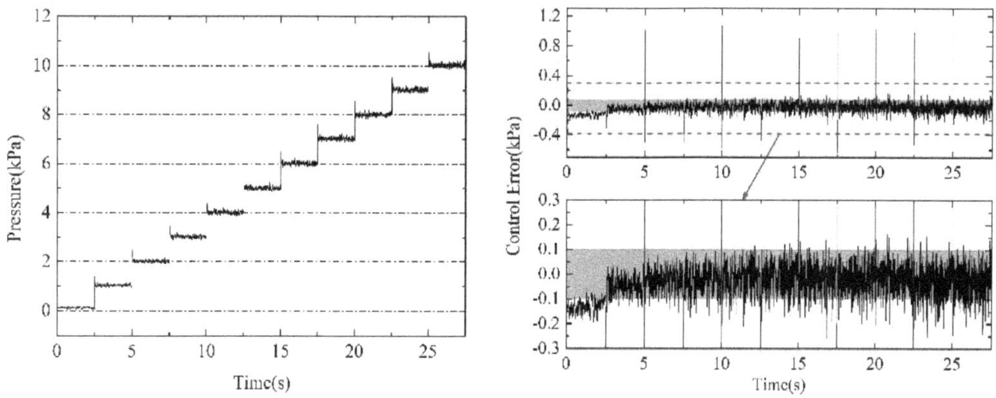

(**a**) Breath Pressure (**b**) Control error (amplification in lower panel)

Figure 5. Simulation results of breath pressure and control error under DOBC during positive pressurization process: (**a**) breath pressure curve; (**b**) control error curve and its amplification in lower panel.

4.2. Experimental Research

The experimental platform of EOR is shown in Figure 6. The function of the host computer is to communicate with the FPGA controller, setting the desired pressure and record experimental data. Active servo lung can simulate breathing movements of a variety of pulmonary ventilation volume. The pressure of oxygen supply source is adjusted in the range of 0.08 MPa~0.6 MPa working pressure in advance. When the breath pressure is 10 kPa, the leakage flow rate of the relief hole has been calibrated as 0.062 g/s, which is equivalent to the volume flow rate of 3 L/min. The mask pressure sensor was added to evaluate the pilot's breath comfort. The YM-6 pressurized oxygen mask was selected as the mask. The NI-9146 chassis with onboard FPGA chip is selected as the controller, and the acquisition and control functions of the controller are realized by NI-9205 and NI-9403, respectively. The operating frequency of the controller was set at 1 kHz. The control parameters of DOBC used in the experiment were set in accordance with the theory analysis, which has been used in the simulation. The flow control valve is driven by a high-precision, linear motion hybrid stepping motor. The flow curves under different inlet pressures are shown in Figure 7. Although the flow control valve has a hysteresis characteristic, the flow rate is generally linear with the output step. The output step of the flow control valve is adjusted according to the required flow calculated by DOBC, inlet pressure signal, and flow characteristic curve.

- Experiment of normal breath

The normal breathing processes of four pulmonary ventilation volume were simulated by using the active servo lung. The recorded experimental results of breath pressure and mask pressure were shown in Figure 8. In addition, the maximum values of inspiratory resistance (IR) and expiratory resistance (ER) obtained from simulation and experiment were compared in Figure 9 with the thresholds in Table 1. It could be found that the control parameters of DOBC designed based on the frequency domain analysis could make the respiratory resistance under each pulmonary ventilation volume meet the threshold requirements of IR and ER. Combined with the mask pressure curve in Figure 8, the breath pressure was kept relatively stable and the pressure variation in mask was mainly affected by the flow resistance of the check valve on mask. This indicates that the pilot will have a good breath experience and not suffer from breathing fatigue.

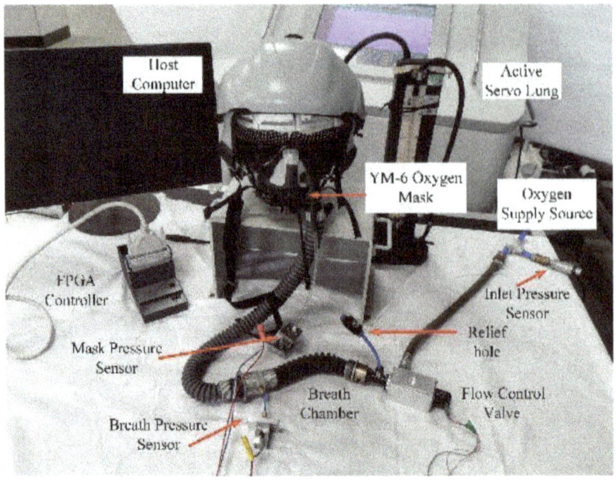

Figure 6. Experimental platform.

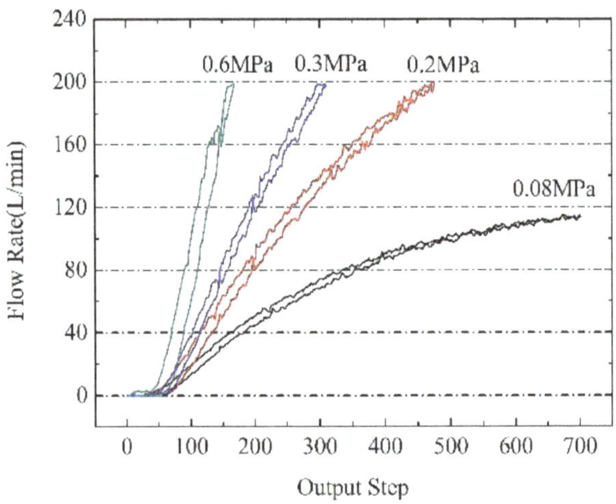

Figure 7. Characteristics of flow control valve under different inlet pressure.

- Experiment of positive pressurization

The desired pressure in experiment of positive pressurization process was set as same as the simulation condition. The curve of controlled breath pressure and the control error were shown in Figure 10.

The experimental results also verified the breath pressure control performance of DOBC in positive pressurization of the EOR. When the desired pressure was increased in the step of 1 kPa, the rise time of breath pressure was about 0.2 s, without significant overshoot. The steady-state error was kept within the control accuracy requirements of ±0.1 kPa, and only exceeded the requirements at few time points, which is acceptable in practical use. The continuous fluctuation of pressure was caused by the measurement noise amplified by large control parameters, and the reason has been explained in the discussion of simulation results.

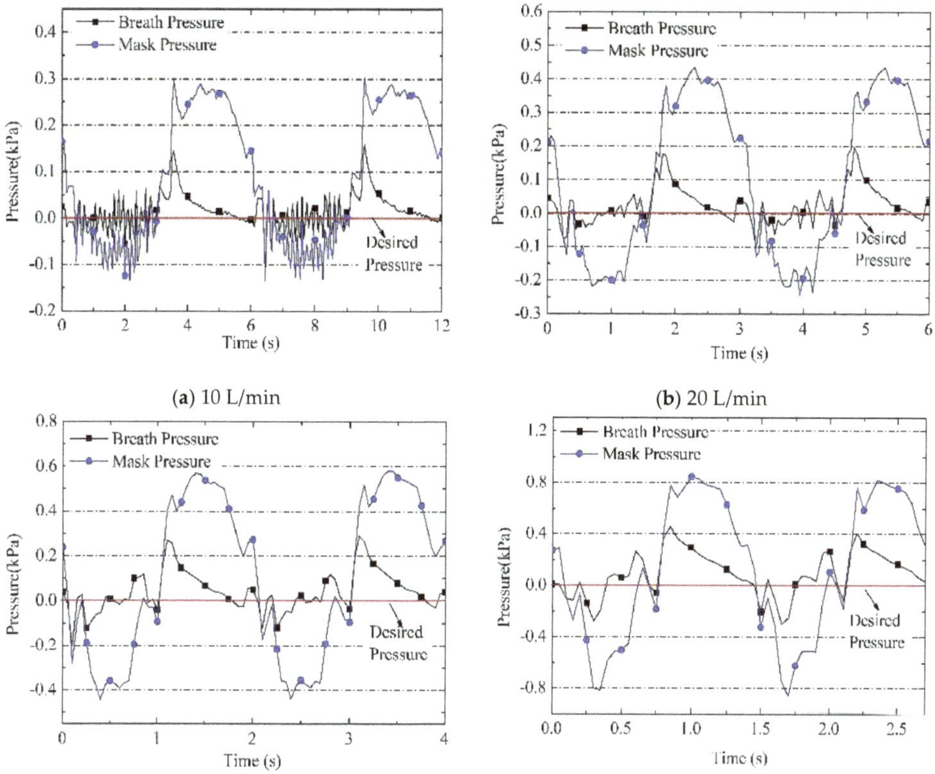

(**a**) 10 L/min (**b**) 20 L/min

(**c**) 30 L/min (**d**) 45 L/min

Figure 8. Experiment results of breath pressure during breath under different pulmonary ventilation volume, (**a**) 10 L/min, (**b**) 20 L/min, (**c**) 30 L/min, and (**d**) 45 L/min.

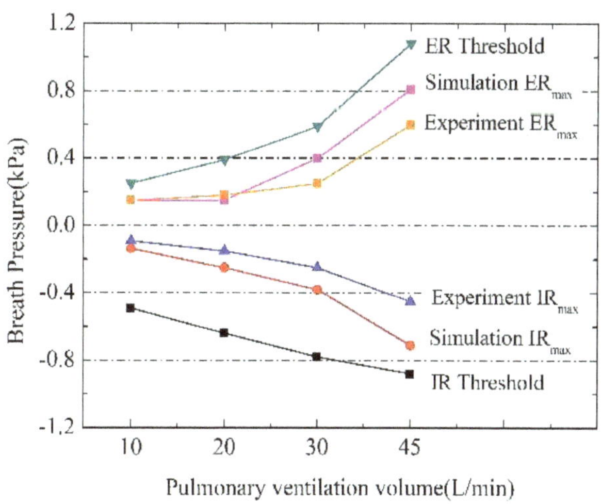

Figure 9. Comparison of respiratory resistance between the threshold values in Table 1, simulation results, and experiment results.

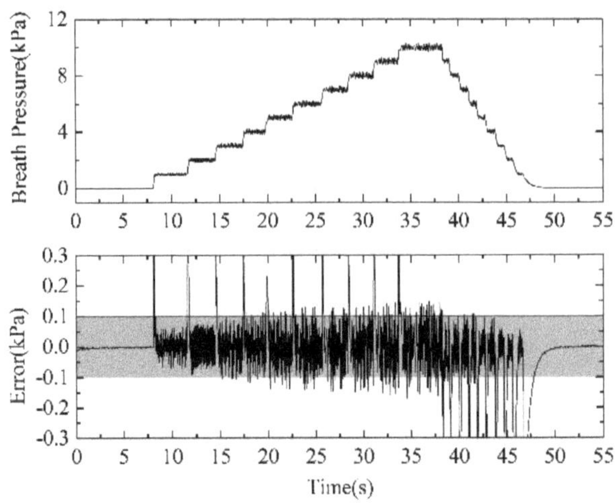

Figure 10. Experiment results of breath pressure under DOBC during positive pressurization process in upper panel and control error in lower panel.

Comparing the simulation results with the experimental results, it was found that the simulation research can nearly reflect the actual DOBC control performance. The experimental results are slightly better than the simulation results. The error between simulation and experiment might come from model errors and parameter uncertainties. First, the actual physical processes were simplified in mathematical model. Furthermore, the simulated white noise signal could not be identical to the measurement noise in experiment. Secondly, although some physical parameters can be measured directly, some other parameters might not be set accurately, such as standard deviation of noise and delay time of system. In addition, hysteresis characteristics and the fitting of linear gain of flow control valve will also cause errors between simulation and experiment. Overall, the simulation accuracy is acceptable and the mathematical model could be use in further study about the EOR.

5. Conclusions and Perspectives

In this paper, the mathematical model of the electronic oxygen regulator was established, and the main factors affecting the breath pressure control were analyzed. Based on the Lyapunov stability and backstepping design method, the DOBC with better stability and robustness has been designed, which can compensate the model error and the parameter uncertainty in the system. To avoid integral saturation, the clapping anti-windup mechanism was designed and adopted. The influence of the inspiratory flow on the breath pressure was described by the input sensitivity function, and the frequency domain analysis was applied to design the control parameter of DOBC for the first time, which could provide important theoretical guidance for the design of controller and control parameters of EOR.

The study shows that when the natural frequency of the close-loop control system reaches 15 Hz, the inspiratory resistance can meet the threshold requirements in theory. Through simulation and experimental researches, performance of the respiratory resistance and positive pressurization of the EOR under DOBC were obtained, which meets the physiological and protective requirements. It can be concluded that the application of DOBC in breath pressure control of EOR was reasonable and effective. However, it was found that although larger control parameters could reduce the respiratory resistance range, it will inevitably lead to the amplification of measurement noise, so that the control steady-state error cannot be completely eliminated and the pressure fluctuate continually. Further studies will be conducted to solve the contradiction.

Although there are certain errors between simulation and experiment, the main trend and approximate range of breath pressure can still be estimated by simulation, which proved the validity of the proposed model and provides a basis for further research on breath pressure control of EOR through simulation technology.

In the future study, the design method of control parameter based on frequency domain analysis will be applied to some other control laws. Then, we could compare the advantages and disadvantages of various control methods in the breath pressure control of EOR fairly. It is also possible to conduct physiological experiments in low-pressure-chamber to simulate performance of EOR during flight more realistically.

Author Contributions: Conceptualization, R.P., G.L., Z.S. and Y.Z.; methodology and investigation, R.P., G.L., Y.Z. and X.Y.; formal analysis, R.P., Y.Z. and X.Y.; software, R.P.; validation and resources, R.P. and Z.S.; writing—original draft preparation, R.P.; writing—review and editing, R.P. and Y.Z.; supervision and project administration, G.L. All authors have read and agreed to the published version of the manuscript.

Funding: This research received no external funding.

Institutional Review Board Statement: Not applicable.

Informed Consent Statement: Not applicable.

Data Availability Statement: Not applicable.

Conflicts of Interest: The authors declare no conflict of interest.

References

1. Xiao, H.J. *Application Physiology of Aircraft Oxygen Protective Equipment*; Military Medical Science Press: Beijing, China, 2005; pp. 34–46. (In Chinese)
2. Rainford, D.J.; Gradwell, D.P. *Ernsting's Aviation Medicine*, 4th ed.; Oxford University Press: London, UK, 2006; pp. 105–127.
3. Yan, S.H.; Yixuan, W.A.; Maolin, C.A.; Zhang, B.; Jian, Z.H. An aviation oxygen supply system based on a mechanical ventilation model. *Chin. J. Aeronaut.* **2018**, *31*, 197–204. [CrossRef]
4. Ackles, K.N.; Porlier JA, G.; Holness, D.E. Protection against the physiological effects of positive pressure breathing. *Aviat. Space Environ. Med.* **1978**, *49*, 753–758. [PubMed]
5. Lauritzsen, L.P.; Pfitzner, J. Pressure breathing in fighter aircraft for G accelerations and loss of cabin pressurization at altitude—A brief review. *J. Can. Danesthésie* **2003**, *50*, 415. [CrossRef]
6. Pan, R.; Lin, G.P.; Zeng, Y.; Yang, X.; Shi, Z.G. Modeling and Simulation of Diaphragm Oxygen Regulator Pressure Control system. In Proceedings of the 2019 IEEE 10th International Conference on Mechanical and Aerospace Engineering (ICMAE), Brussels, Belgium, 22–25 July 2019; pp. 6–10. [CrossRef]
7. Beaumont, M.; Lejeune, D.; Isabey, D.; Marotte, H.; Harf, A.; Lofaso, F. Positive pressure generation by pneumatic and electronic O2 regulators: A bench experimental evaluation. *Aviat. Space Environ. Med.* **1999**, *70*, 812–818. [CrossRef] [PubMed]
8. Xiao, H.J. Molecular sieve oxygen systems on French military aircraft. *Aeronaut. Astronaut.* **1996**, *1*, 38–42. (In Chinese)
9. Siska, W.D., Jr.; Collins, R.; Carleton Technologies Inc. Electromechanical Oxygen Valve and Regulator. United States Patent US 7677529 B2, 10 March 2010.
10. Frampton, R.; B/E Intellectual Property. Electromechanical Regulator with Primary and Backup Modes of Operation for Regulating Passenger Oxygen. United States Patent US 7604019 B2, 20 October 2010.
11. Yu, Z.; Zhao, J. Simulation of pressurized oxygen supply performance of air oxygen supply system based on Simulink. *Microcomput. Appl.* **2010**, *31*, 1–6. (In Chinese)
12. Yu, Z.; Bing, S.; Weidong, W.; Guiping, L. Simulation and optimization of a new pulmonary mechanism in electronic oxygen regulator. In Proceedings of the 2016 IEEE International Conference on Aircraft Utility Systems (AUS), Beijing, China, 10–12 October 2016; pp. 719–722. [CrossRef]
13. Yu, X.; Sun, B.; Lin, G.; Wang, H. Application of ATmega128 microcontroller in electronic oxygen regulator. *Micro-Comput. Appl.* **2009**, *12*, 50–56. (In Chinese)
14. Sun, C.; Cai, Y.; Long, H. Research on stepping motor fuzzy control technology application in the aircraft electronic oxygen regulator. *Meas. Control Technol.* **2013**, *4*, 82–87. (In Chinese)
15. Li, X.; Lin, G.; Zeng, Y.; Wu, F. Design of electronic oxygen regulator PID control system based on LabVIEW. *Comput. Meas. Control.* **2016**, *3*, 80–83. (In Chinese)
16. Yuxin, J.; Qinglin, S.; Zengqiang, C.; Sanpeng, D. Modeling and simulation of an electronic oxygen regulator based on generalized predictive control algorithm. In Proceedings of the 34th China Control Conference (CCC), Hangzhou, China, 28–30 July 2015; pp. 4067–4072. [CrossRef]

17. Jiang, Y.; Sun, Q.; Zhang, X.; Chen, Z. Pressure Regulation for Oxygen Mask Based on Active Disturbance Rejection Control. *IEEE Trans. Ind. Electron.* **2017**, 6402–6411. [CrossRef]
18. Jiang, Y.; Sun, Q.; Tan, P.; Chen, Z. Modeling and Simulation of an Electronic Oxygen Regulator Based on All-Coefficient Adaptive Control. *ASME J. Dyn. Sys. Meas. Control* **2016**, *138*, 081010. [CrossRef]
19. Amare, N.D.; Kim, D.H.; Yang, S.J.; Son, Y.I. Boundary Conditions for Transient and Robust Performance of a Reduced-Order Model-Based State Feedback Controller with PI Observer. *Energies* **2021**, *14*, 2881. [CrossRef]
20. Gao, N.; Lin, X.; Wu, W.; Blaabjerg, F. Grid Current Feedback Active Damping Control Based on Disturbance Observer for Battery Energy Storage Power Conversion System with LCL Filter. *Energies* **2021**, *14*, 1482. [CrossRef]
21. Baran, J.; Jąderko, A. An MPPT Control of a PMSG-Based WECS with Disturbance Compensation and Wind Speed Estimation. *Energies* **2020**, *13*, 6344. [CrossRef]
22. Rui, P.A.; Guiping, L.I.; Zhigao, S.H.; Yu, Z.E.; Xue, Y.A. Analysis and control optimization of positive pressure fluctuation in electromechanical oxygen regulator. *Chin. J. Aeronaut.* **2021**, *34*, 205–213. [CrossRef]

energies

Article

Determination of the Theoretical and Actual Working Volume of a Hydraulic Motor—Part II (The Method Based on the Characteristics of Effective Absorbency of the Motor)

Pawel Sliwinski

Faculty of Mechanical Engineering and Ship Technology, Gdansk University of Technology, 80-233 Gdansk, Poland; pawel.sliwinski@pg.edu.pl

Abstract: In this article, the second method of determination of the theoretical and actual working volume of a hydraulic motor is described. The proposed new method is based on the characteristics of effective absorbency of the motor. The effective absorbency has been defined as the ratio of flow rate in a motor to the rotational speed of the motor's shaft. It has been shown that the effective absorbency is a nonlinear function of the rotational speed and nonlinear function of the pressure drop in the motor's working chambers. Furthermore, it has been proven that the actual working volume of a motor is a function of a third degree of pressure drop in the motor's working chamber. The actual working volume should be taken to assess the mechanical and volumetric energy losses in the motor. Furthermore, the influence of the flowmeter location in the measurement system and the compressibility of liquid on the result of the theoretical and actual working volume calculation was also taken into account and is described in this article. The differences in the assessment of the volumetric efficiency assuming the theoretical and actual working volume was also shown.

Keywords: hydraulic motor; effective absorbency; theoretical working volume; actual working volume; volumetric efficiency; mechanical efficiency

Citation: Sliwinski, P. Determination of the Theoretical and Actual Working Volume of a Hydraulic Motor—Part II (The Method Based on the Characteristics of Effective Absorbency of the Motor). *Energies* **2021**, *14*, 1648. https://doi.org/10.3390/en14061648

Academic Editors: Ryszard Dindorf and Federico Barrero

Received: 4 February 2021
Accepted: 11 March 2021
Published: 16 March 2021

1. Introduction

For users and designers of a hydraulic system, the volumetric and pressure-mechanical efficiency of the hydraulic positive displacement machine (motor or pump) is very important [1–6]. So far, these efficiencies are calculated based on the theoretical working volume q_t. The theoretical working volume q_t is a parameter that corresponds to the volume of liquid flowing without energy losses through the positive displacement machine during one full rotation of the shaft. The q_t is determined from experimental data. The value of q_t is constant in the whole range of the rotational speed n and the pressure drop Δp in a positive displacement machine [7–14].

It happens that the geometric working volume q_g is used instead of the q_t to calculate the energy losses in positive displacement machines. The q_g is calculated from geometrical dimensions of the working mechanism using special mathematical formulas which are different for various types of positive displacement machines [15]. The theoretical working volume q_t does not have to be equal to the geometrical working volume q_g and is usually not equal. The differences can be up to several percent [7,8,15]. The main reasons for these differences are:

- Looseness in the working mechanism,
- errors in the manufacturing of the components,
- assembly errors,
- changes in ambient temperature,
- changes in working liquid temperature, etc.

If the q_t corresponds to the volume of liquid flowing without energy losses through the motor during one full revolution of the motor shaft, then the q_t should be determined by taking into consideration the pressure drop Δp_i in the working mechanism and the liquid compressibility β. If the liquid compressibility β is taken into account, then the location of a flow meter in the measuring system is of great important [12]. However, in practice, it is assumed that the pressure drop Δp in a motor is equal to the pressure drop Δp_i in the motor working chambers, and the liquid compressibility β is neglected [7,8,13,14].

In a loaded hydraulic motor, the working chambers are resiliently deformed (due to the pressure difference Δp_i). The working chamber under the pressure will have a bigger volume. Then, the working volume can differ from the geometric working volume q_g [8,12]. The working volume under the pressure difference Δp_i was called the actual working volume q_r [12]. Therefore, q_r is a function of the pressure drop Δp_i. It will be better to assume q_r instead of q_t and q_g to assess the losses in a hydraulic motor. Furthermore, in [12], it was shown that q_t and q_r should be determined by taking into account the liquid compressibility β. The precision determination q_r allows us to more precisely assess energetic losses and more precisely assess the clearances in the hydraulic motor working mechanism.

So far, there have been only a few publications on the methodology of determination of the theoretical working volume q_t of a displacement machine. The first method was described by Wilson in 1950 [16] and Schlosser and Hilbrands in 1963 [9–11]. This method was corrected and described by Toet in 1970 [13] and in 2019 [14], and also by Balawender in 1974 [8]. These authors described in their publications the method of determining the q_t based on the characteristics ($Q = f(n)_{\Delta p = const.}$) of flow rate Q in a motor as a function of rotational speed n at constant pressure drop Δp in this motor. The Toet and Balawender method have been referred in literature, like in [17–20].

The latest article on the methodology of determination the theoretical working volume was published in 2020 [12]. In this publication, the influence of liquid compressibility β and the influence of pressure drop Δp_{ich} in the motor internal channels on the q_t are described.

The second method of determining the q_t was proposed by Balawender [8]. This method is based on the characteristics $q_e = f(n^{-1})_{\Delta p = const.}$ of effective absorbency q_e as a function of rotational speed n at constant pressure drop Δp in the motor. Thus, the influence of pressure drop in the motor internal channel and liquid compressibility β were neglected.

According to ISO 8426, the theoretical working volume q_t is defined as a ratio of flow rate increment ΔQ to the corresponding rotational speed increment Δn [21]. The method according to ISO standard is the simplest of all the known methods and gives an inaccurate result.

There were also attempts to calculate the q_t by analyzing the flow rate for a single pump chamber [22]. However, this method becomes problematic to use in practice.

For this purpose, in the following sections are described:

(a). Known methods for determining the theoretical working volume q_t;
(b). the proposed new method of determining the theoretical q_t and actual working volume q_r based on the characteristics $q_e = f(n)_{\Delta pi = const.}$ (taking into account the compressibility β of liquid and the pressure drop Δp_{ich} in the internal channel of a motor);
(c). the practical implementation of the proposed method;
(d). the experimental tests of the hydraulic satellite motor for confirming the correctness of the proposed new method;
(e). comparing the result of new method with the result of Balawender's method.

2. The Analysis of Flow Rate in a Hydraulic Motor

In order to determine the theoretical working volume q_t of the motor, it is necessary to measure and analyze the flow rate in this motor. Balawender recommends using two flow meters for this—one located in the inflow line of the motor and second in the outflow line (Figure 1) [7,8], whereas Toet indicates only one flowmeter located in the inflow line of the motor [13,14].

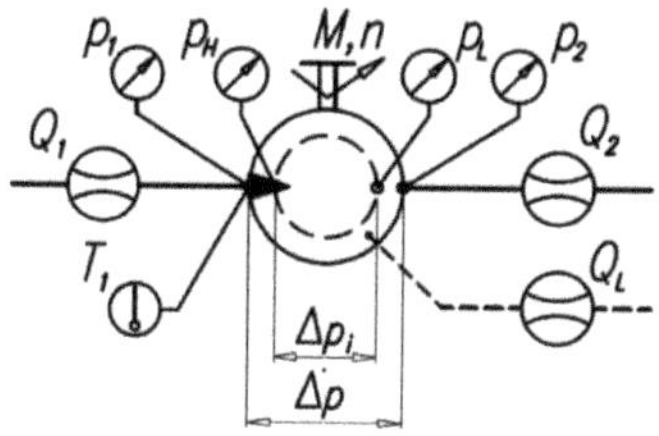

Figure 1. Parameters measured in a hydraulic motor: Q_1, Q_2 and Q_L—flowrate, p_1 and p_2—pressure in the inflow and outflow port, respectively, p_H and p_L—pressure in the high-pressure and low-pressure working chamber respectively, M—load (torque), n—rotational speed, T_1—temperature, Δp—pressure drop in the motor, Δp_i—pressure drop in the motor working chambers.

2.1. Balawender's Model

The flow rate in a hydraulic motor is described by the following formulas [7,8]:

$$Q_1 = Q_g + Q_u + Q_k + Q_{C1} + \underbrace{Q_{Li} - Q_{Le1}}_{Q_{L1}} \quad (1)$$

$$Q_2 = Q_g + Q_u + Q_k + Q_{C2} + \underbrace{Q_{Li} - Q_{Le2}}_{Q_{L2}} \quad (2)$$

where [7,8,12]:

- Q_g, Q_u, Q_k, Q_C—the components of the flow rate Q lineary dependent on the geometric volume q_g, the pressure drop Δp, and rotational speed n;
- Q_{Li}—the internal leakage (from the high-pressure working chambers to the low-pressure working chambers):
- Q_{Le}—the external leakage linearly dependent on the theoretical working volume q_t and the pressure drop Δp (leakage from the working chambers on the outside of the motor—leakage from the motor body by the third hose):

$$Q_{Le1} + Q_{Le2} = Q_{Le} \quad (3)$$

$$Q_{Le1} \approx Q_{Le2} \approx 0.5 \cdot Q_{Le} \quad (4)$$

2.2. Sliwinski's Model

The Balawender's model of flow rate in a hydraulic motor is a general model. This model does not take into account the construction parameters of the motor. In this way, this model can try to adapt to any type of hydraulic motor, which is his advantage. On the other hand, the small accuracy is a disadvantage of this model. However, Sliwinski recommends using the model of flow rate in the motor based on analysis of construction and analysis of phenomena in the motor. Therefore, the type of motor and its construction is important. In this way, the basis for development of the model of flow rate in the motor is development of the model of volumetric losses in this motor. The mathematical model of volumetric losses, dedicated to the satellite motor (described in Section 6.1), is following [23,24]:

$$Q_L = \Delta Q_L + Q_{Lfg} + Q_C + Q_{Le} \quad (5)$$

where:

- Q_{Lfg}—flow rate in flat clearances of working mechanism [24,25]:

$$Q_{Lfg} = A_1 \cdot m^{\left(\frac{1-\beta}{2-\beta}\right)} \cdot \left(\frac{1}{\nu}\right)^{\left(\frac{\beta}{2-\beta}\right)} \cdot \left(\frac{2 \cdot h^3}{K \cdot A_2 \cdot \rho}\right)^{\left(\frac{1}{2-\beta}\right)} \cdot \Delta p_i^{\left(\frac{1}{2-\beta}\right)} \quad (6)$$

- Q_C—the flow rate in commutation unit clearances [24,26]:

$$Q_C = C_1 \cdot (D_o \cdot h_s)^{C_2} \cdot \left(\frac{1}{\rho \cdot \nu^{C_3}}\right)^{\gamma} \cdot \Delta p_i^{\gamma} \tag{7}$$

- ΔQ_L—the component of volumetric losses depends on the pressure drop Δp_i in the working chambers and rotational speed n of the motor [23,24]:

$$\Delta Q_L = \Delta q_L \cdot n \tag{8}$$

where:

$$\Delta q_L = \left(C_q \cdot \Delta p_i + \frac{C_{id}}{n^{0.5}}\right) \cdot m^2 \cdot H \tag{9}$$

- Q_{Le}—external leakage;
- $C_1, C_2, C_3, C_q, C_{id}, A_1, A_2, K$, and γ—coefficients;
- β—the degree of laminarity of the flow;
- ν—kinematic viscosity of liquid;
- m—teeth module;
- h—equivalent axial clearances of the rotor and satellites;
- h_S—axial clearance of the satellites;
- H—height of the working mechanism (equal to height of curvature);
- D_O—the diameter of inlet and outlet holes in the commutation plate.

The flow rate Q in the motor is:

$$Q = Q_t + Q_L \tag{10}$$

where Q_t is the theoretical flow rate in the motor:

$$Q_t = q_t \cdot n \tag{11}$$

If the influence of liquid compressibility β on the flow rate measurement result is omitted, then:

$$Q = Q_1 = Q_2 + Q_{Le} \tag{12}$$

If $Q_{Le} = 0$ then:

$$Q = Q_1 = Q_2 \tag{13}$$

3. The Pressure Drop in the Motor Working Chambers

The pressure difference Δp_i in the motor working chambers has an influence on the volumetric losses in the motor—as indicated above. This Δp_i is also taken into account during the determination of the theoretical working volume q_t of a motor. However, direct measurements of this pressure are very difficult or impossible. It is much easier to determine the Δp_i by measuring the pressure drop in Δp_{ich} in the internal channel of the motor. Then [12,24,27]:

$$\Delta p_i = \Delta p - \Delta p_{ich} \tag{14}$$

In most hydraulic motors, the inflow and outflow internal channels have the same shape and dimensions. Then the pressure drop Δp_{ich} can be calculated according to the formula [12,24,27]:

$$\Delta p_{ich} = 2 \cdot (p_1 - p_H) \tag{15}$$

The methodology of determination of the pressure drop Δp_{ich} in the internal channel of a motor is widely described in [24,27].

4. Known Method to Determine the Theoretical Working Volume

4.1. Balawender's and Toet's Method—Based on Characteristics $Q = f(n)$

Balawender and Toet claim that the flow rate Q (Q_1 or Q_2) in a hydraulic motor is described by a linear function [7,8,12,14]:

$$Q_{i,(\Delta pi=const)} = A_i \cdot n + B_i \quad (16)$$

where:

$$A_i = \lim_{\Delta p_i \to 0}\left(\frac{\partial Q_i}{\partial n}\right)_{(\Delta p_i)} \approx \lim_{\Delta p \to 0}\left(\frac{\partial Q_i}{\partial n}\right)_{(\Delta p)} \quad (17)$$

$$B_{i(\Delta pi=const)} = Q_{Li,i} + Q_{Le,i} \quad (18)$$

Balawender suggests testing the motor in the range of speed $n_{max} \le 3n_{min}$. The n_{min} is the minimum speed at which the motor can work stably. The n_{max} is the maximum speed for which the assumption $\Delta p_i \approx \Delta p$ is true [7,8,12]. If $\Delta p_i \approx \Delta p$, then Equations (16) and (17) take the form:

$$Q_{i,(\Delta p=const)} = A_i \cdot n + B_i \quad (19)$$

$$B_{i(\Delta p=const)} = Q_{Li,i} + Q_{Le,i} \quad (20)$$

The flow rate characteristics $Q_1 = f(n)_{\Delta p=const.}$ are shown in 0 (Figure 2).

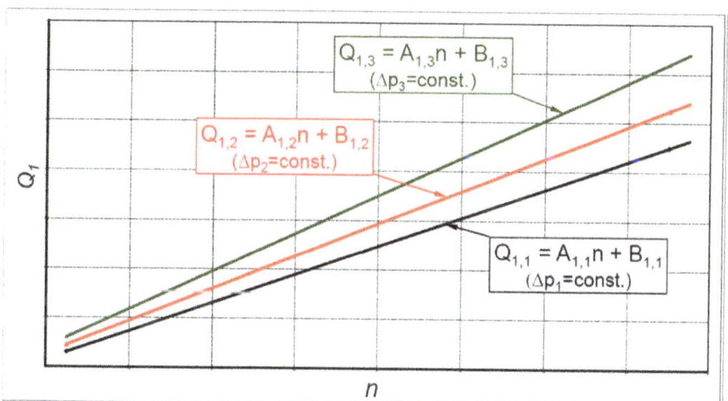

Figure 2. Flow rate Q_1 vs. rotational speed n [7,8,12,14]. Similar characteristics are obtained for Q_2.

According to Balawender, between A and Δp is the following relationship (Figure 3) [7,8]:

$$A = a \cdot \Delta p + q_a \quad (21)$$

Similarly between B and Δp is (Figure 3) [7,8]:

$$B = q_{Le} = b \cdot \Delta p + q_{Le,0} \quad (22)$$

The theoretical working volume is described as [7,8,12]:

$$q_{t1} = \lim_{\Delta p \to 0} A_{1(\Delta p)} - \frac{1}{2} \cdot \lim_{\Delta p \to 0} q_{Le(\Delta p)} \quad (23)$$

and

$$q_{t2} = \lim_{\Delta p \to 0} A_{2(\Delta p)} + \frac{1}{2} \cdot \lim_{\Delta p \to 0} q_{Le(\Delta p)} \quad (24)$$

In practice, only one flowmeter is applied in measuring system. Then $q_t = q_{t1}$ or $q_t = q_{t2}$.

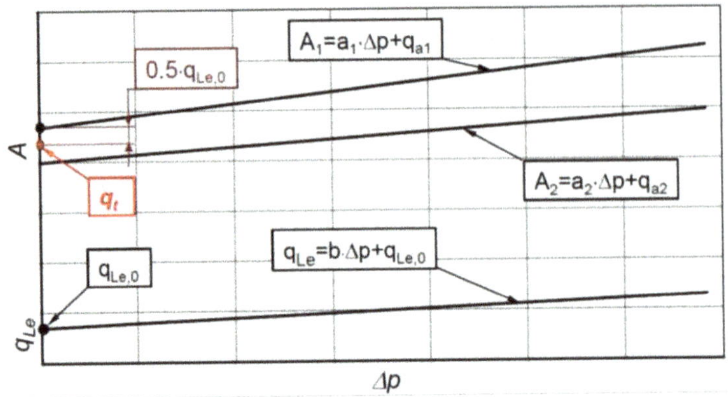

Figure 3. Linear relationship between working volume A and leakage q_{Le} and pressure difference Δp [7,8,12].

4.2. The Influence of Working Pressure on Working Volume

According to Balawender and Toet, the geometric working volume q_g is a linear function of pressure [7,8,14]:

$$q_{g(\Delta p_i)} = q_{g(\Delta p_i=0)} \cdot (1 \pm K_6 \cdot \Delta p_i) \tag{25}$$

This is a simplification, of course. Similarly, Sliwinski assumed that the increase in the positive displacement machine's working volume is a function of pressure drop in the motor working chambers and can be described by the Formula (9).

The influence of working pressure on the working volume was also noticed by Osiecki during the test of prototype of axial piston pump with cam-driven unit [6]. With increasing the load of the pump, the working volume of the pump also increased. Thus, it can be seen that the adoption of the theoretical working volume q_t can introduce a significant error in the evaluation of the losses in displacement machines. Therefore, in [12], it has been shown that is not advisable to assume q_t for the assessment of losses in the motor in the whole range of its load. However, it is better and more precise to use the actual working volume q_r defined as [12]:

$$q_r = q_{g(\Delta p_i=0)} \underbrace{\pm f\left(\Delta p_i^{\,3}\right) \pm f\left(\Delta p_i^{\,2}\right) \pm f(\Delta p_i)}_{\Delta q_{g(\Delta p_i)}} \tag{26}$$

4.3. Balawender's Method—Based on Characteristics $q_e = f(n)_{\Delta p = const.}$

The effective absorbency q_e of the motor is defined as [7,8]:

$$q_e = \frac{Q}{n} \tag{27}$$

In relations to the flowmeters Q_1 and Q_2 (Figure 1):

$$q_{e1} = \frac{Q_1}{n} \tag{28}$$

$$q_{e2} = \frac{Q_2}{n} \tag{29}$$

In his considerations, Balawender omits the pressure drop Δp_{ich} in internal channels of the motor and assumes: $\Delta p_i \approx \Delta p$.

Therefore, taking into consideration the Formulas (1) and (2) [7,8]:

$$q_{e1} = \underbrace{q_g + q_u + q_k + q_{C1}}_{q_{w1}} + \underbrace{\frac{Q_{L1}}{n}}_{q_{Le1}} \tag{30}$$

$$q_{e2} = \underbrace{q_g + q_u + q_k + q_{C2}}_{q_{w2}} + \underbrace{\frac{Q_{L2}}{n}}_{q_{Le2}} \tag{31}$$

According to Balawender, the q_e is the linear function of $1/n$ (Figure 4).

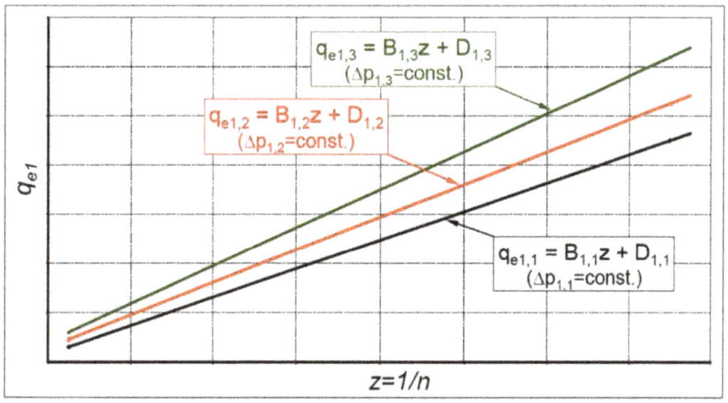

Figure 4. Linear interrelationship between q_{e1} and $1/n$ for different constant Δp in the motor for flowmeter Q_1 (Figure 1) [8].

If the rotational speed n increases, then the leakage Q_L has a smaller share in the effective flow rate in the motor. In effect [8]:

$$\lim_{n\to\infty}(q_{e1})_{(\Delta p)} = (q_{w1} + q_{Le1})_{(\Delta p)} \approx (q_{w1})_{(\Delta p)} - 0.5\cdot q_{Le1} = D_1 \tag{32}$$

$$\lim_{n\to\infty}(q_{e2})_{(\Delta p)} = (q_{w2} - q_{Le2})_{(\Delta p)} \approx (q_{w2})_{(\Delta p)} + 0.5\cdot q_{Le2} = D_2 \tag{33}$$

According to Balawender, the q_{w1} and q_{w2} are the linear function of Δp (Figure 5).

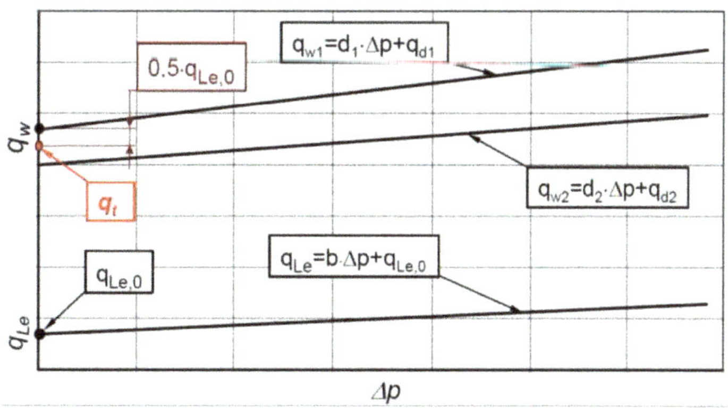

Figure 5. Linear relationship between working volume q_w and leakage q_{Le} and pressure drop [7,8].

Similarly, if the Δp decreases, then the leakages Q_L also decreases. Then the q_t is calculated from the following formulas [6,8]:

$$\lim_{\Delta p \to 0}(D_1) = q_g + q_u - 0.5 \cdot q_{Le0} = q_{t1} \quad (34)$$

$$\lim_{\Delta p \to 0}(D_2) = q_g + q_u + 0.5 \cdot q_{Le0} = q_{t2} \quad (35)$$

In practice, only one flowmeter is applied in the measuring system. Then $q_t = q_{t1}$ or $q_t = q_{t2}$.

4.4. The Influence of the Flow Meter Location in the Measurement System on the Flow Rate in the Hydraulic Motor

4.4.1. Toet's Method

Toet recommends placing a flow meter in the inlet line to a hydraulic motor. Then the flow meter measures total flow rate, including the external leakage. However, if the flow meter is located in the low-pressure line, the influence of the liquid compressibility on the flow rate should be taken into account. Then [14]:

$$Q_{(p_1)} = Q_2 + \underbrace{q_{g(\Delta p=0,T_1)} \cdot K_3 \cdot \Delta p - K_4 \cdot (\Delta p)^2 + K_5 \cdot (\Delta p)^3 \cdot n}_{Q_C} \quad (36)$$

where K_1, K_2, K_3, K_4, K_5 are constants. That is, the effect of liquid compressibility is related to the geometric working volume q_g. Furthermore, the influence of liquid compressibility on the leakage Q_L is neglected. Therefore, this assumption introduces some error in calculations of the theoretical working volume q_t.

4.4.2. Balawender's Method

Balawender recommends locating the flow meter in two places of hydraulic circuit. The first place is the high-pressure line of the motor. Then $Q_{(p1)} \approx Q_1$ [8]. The second place is the low-pressure line of the motor. For this location of flow meter [8]:

$$Q_{(p_1)} = \kappa \cdot Q_2 \quad (37)$$

where κ is the correction coefficient. For oil, for t = 40 °C, the value of the coefficient is [8]:

$$\kappa = 0.97 - 75 \cdot 10^{-5} \cdot \Delta p \quad (38)$$

and Δp is expressed in [MPa].

Finally, in a practical approach to determine the theoretical working volume, both Toet and Balawender simplify the problem, and the effect of the flow meter location in the measuring system is finally neglected.

4.4.3. Sliwinski's Method

The flow rate $Q_{(pH)}$ related to the pressure p_H in the high-pressure working chamber is possible to calculate depending on the location of a flow meter in measuring system. Thus:

(a). For the flow meter located in the high-pressure line to the motor (Figure 1), the flow rate is [12]:

$$Q_{1(p_H)} = Q_1 \cdot \left(1 + \int_{p_H}^{p_1} \frac{1}{K_{Z(p)}} \cdot \partial p\right) \quad (39)$$

(b). For the flow meter located in the low-pressure line to the motor (Figure 1), the flow rate is:

$$Q_{2(p_H)} = Q_2 \cdot \left(1 - \int_{p_2}^{p_H} \frac{1}{K_{Z(p)}} \cdot \partial p\right) \quad (40)$$

where:

- $K_{Z(p)}$—the tangential isentropic bulk modulus [28]:

$$K_{Z(p)} = \frac{e^{\left(\frac{p_o - p}{K}\right)} \cdot \left\lfloor 1 + m \cdot \frac{p - p_o}{K} \right\rfloor^{\left(-\frac{1}{m}\right)} + \frac{X_o}{1 - X_o} \cdot \left(\frac{p_o}{p}\right)^{\left(\frac{1}{n}\right)}}{\frac{e^{\left(\frac{p_o - p}{K}\right)}}{K} \cdot \left\lfloor 1 + m \cdot \frac{p - p_o}{K} \right\rfloor^{\left(\frac{1-m}{m}\right)} + \frac{X_o}{n \cdot p_o \cdot (1 - X_o)} \cdot \left(\frac{p_o}{p}\right)^{\left(\frac{1+n}{n}\right)}} \tag{41}$$

- K—the bulk modulus of non-aerated oil at atmospheric pressure p_o;
- X_o—the amount of non-dissolved air in the oil at atmospheric pressure p_o;
- p—the absolute pressure;
- n—the polytrophic exponent;
- m—the coefficient of the influence of pressure p on the bulk modulus K.

5. New Method of Determining the Theoretical and Actual Working Volume Based on the Characteristics of Effective Absorbency of the Motor

In the above described Balawender's method of determining theoretical working volume, it is assumed that $\Delta p_i = \Delta p$. That is, the pressure drop Δp_{ich} in internal channels of a motor is omitted. This simplification can have the influence on the value of theoretical working volume q_t of the motor. Furthermore, Balawender assumed that the volumetric losses Q_L in a motor are linear function of rotational speed n (Section 2.1). Whereas, the analysis of volumetric losses in the motor, carried out by Sliwinski, show that these losses are a nonlinear function of speed n (Formula (9)). Furthermore, various researchers argue that the increase in working volume of a motor is a linear function of pressure drop in the motor (Formulas (9) and (25)) [7,8,13,14]. This is a simplification, of course. In fact, the motor's working chamber is a spatial element, and each of its dimensions changes linearly under the influence of pressure. Therefore, the change in working volume of a motor is a nonlinear function of the pressure drop Δp_{ich} in its working chamber.

In the below described new method of determining the theoretical working volume $q_{t,}$ it is assumed:

- Δp_i instead Δp;
- nonlinear dependence of volumetric losses Q_L on rotational speed n;
- nonlinear dependence of increasing in working volume Δq on pressure drop Δp_i;
- the compressibility β of the liquid, that is the measured flow rate is compensated to the pressure in the filled chamber of the motor (the location of flowmeter in the test stand is included).

5.1. The Increase in Working Volume

In publications [7,8,23,24] and in Section 2.2, it is written that the rotational speed n has an influence on the increase in volumetric losses Δq (Formulas (8) and (9)). This is because of the movement of working mechanism elements within the slack. This movement causes an additional increase in working volume. In effect, additional volumetric loss is created depending on the speed. Furthermore, the increase in working volume Δq is a nonlinear function of pressure drop Δp_i [23,24]. So, the increase in working volume Δq_g caused by the pressure drop Δp_i in the working chamber cannot be the reason of the volumetric losses in a motor. Therefore, Formula (9) is inaccurate, and instead of it, the following is proposed:

$$\Delta q = \Delta q_g + \Delta q_L \tag{42}$$

where:

$$\Delta q_g = m^2 \cdot H \cdot \left(C_{q1} \cdot \Delta p_i + C_{q2} \cdot \Delta p_i^{\ 2} + C_{q3} \cdot \Delta p_i^{\ 3} \right) \tag{43}$$

$$\Delta q_L = \frac{C_{id}}{n^{0,5}} \cdot m^2 \cdot H \tag{44}$$

5.2. The Actual Working Volume of a Loaded Motor

In [12], it was shown that with a change in pressure drop Δp_i in working chambers the working volume also is in change according to formula:

$$q_r = q_t + \underbrace{\left(C_{q1}\cdot\Delta p_i + C_{q2}\cdot\Delta p_i{}^2 + C_{q3}\cdot\Delta p_i{}^3\right)\cdot m^2\cdot H}_{\Delta q_g} \tag{45}$$

The q_r was called the actual working volume. For $\Delta p_i = 0$ is $qr = q_t$. The value of q_r is obtained experimentally. The determination of q_r is possible only when the component of flow rate Q_C (caused by liquid compressibility) is eliminated (see Section 4.4).

5.3. Flow Rate in a Motor

Taking into account Formula (45), the flow rate in a motor should be described by:

$$Q = \underbrace{q_r\cdot n}_{Q_r} + \underbrace{\Delta q_L\cdot n + Q_{Lfg} + Q_C + Q_{Le}}_{Q_L} \tag{46}$$

where Δq_L is described by Formula (44).

It can be assumed that the flow rate Q is a linear function of rotational speed n of a motor. The flow rate Q is measured by flowmeter and in this way Formulas (12) and (13) are applied.

5.4. The Effective Absorbency

The effective absorbency q_e of the motor should be calculated according to the following formulas:

$$q_{e1} = \frac{Q_{1(p_H)}}{n} \tag{47}$$

$$q_{e2} = \frac{Q_{2(p_H)}}{n} \tag{48}$$

Thus, after substitution (5), (8), (9), and (10) into (47) and (48), respectively, the following are obtained:

$$q_{e1} = q_t + \left(C_{q1}\cdot\Delta p_i + C_{q2}\cdot\Delta p_i{}^2 + C_{q3}\cdot\Delta p_i{}^3\right)\cdot m^2\cdot H + \frac{C_{id}}{n^{0.5}}\cdot m^2\cdot H + \frac{Q_{Lfg} + Q_C + Q_{Le}}{n} \tag{49}$$

$$q_{e2} = q_t + \left(C_{q1}\cdot\Delta p_i + C_{q2}\cdot\Delta p_i{}^2 + C_{q3}\cdot\Delta p_i{}^3\right)\cdot m^2\cdot H + \frac{C_{id}}{n^{0.5}}\cdot m^2\cdot H + \frac{Q_{Lfg} + Q_C}{n} \tag{50}$$

Therefore, the effective absorbency q_e is a linear function of Δp_i and nonlinear function of the inverse of rotational speed n of the motor. That is:

$$q_{e1}(or\ q_{e2}) = f\left(\Delta p_i, n^{-0.5}\right) \tag{51}$$

In addition, it is worth noting that in the case of flow meter located in the motor outflow line the component Q_{Le} is omitted (because is not measured).

5.5. Theoretical Working Volume

If rotational speed n of a motor increases, then the share of leakages Q_L in the general balance of flow rate Q in the motor decreases. Therefore, the value of q_e, described by Formulas (49) and (50), strives to a certain value q_r, namely:

$$\lim_{n\to\infty}(q_{e1})_{(\Delta p_i)} = q_{r1} = q_{t1} + \Delta q_{g1} \tag{52}$$

$$\lim_{n\to\infty}(q_{e2})_{(\Delta p_i)} = q_{r2} = q_{t2} + \Delta q_{g2} \tag{53}$$

If $\Delta p_i = 0$, then $\Delta q_g = 0$. Thus, the theoretical working volume can be defined as follows:

$$\lim_{\Delta p_i \to 0} (q_{r1}) = q_{t1} \tag{54}$$

$$\lim_{\Delta p_i \to 0} (q_{r2}) = q_{t2} \tag{55}$$

If the measurement of flow rate Q_1 and Q_2 are made using flow meters of the same type and class and the compressibility of liquid is taken into account, then the theoretical working volumes (expressed by Formulas (54) and (55)) should have the same value:

$$q_{t1} = q_{t2} = q_t \tag{56}$$

In practice, only one flowmeter is applied in measuring system. Then $q_t = q_{t1}$ or $q_t = q_{t2}$.

5.6. Practical Implementation of the New Method

The practical implementation of the methodology of determining the theoretical working volume q_t based on the characteristics of $Q = f(n)_{\Delta pi=const.}$ were described in [12]. In this place it is worth once again present this implementation and adjust it to the proposed new method. So, in order to determine the theoretical q_t and actual working volume q_r it is necessary [12]:

(a). To measure the flow Q_1 or Q_2 at a constant inlet temperature T_1 and a constant pressure drop Δp for several values of n (no less than five). The measurement has to be taken once again using at least five different pressure drops Δp;
(b). to determine the pressure drop Δp_{ich} in the internal channel of the motor;
(c). to calculate the pressure drop Δp_i in the working chamber of the motor for data from item (a) and item (b);
(d). if the flow meter is located in the low-pressure line, the influence of the liquid compressibility should be taken into account, and the flow rate $Q_{(pH)}$ corresponding to the value of high-pressure p_H in the working chamber should be calculated;
(e). to plot the characteristics of $q_e = f(1/n)_{\Delta pi=const.}$ and calculate the q_r according to Formulas (52) or (53) for each Δp_i;
(f). to plot the characteristics of $q_r = f(\Delta p_i)$ and calculate the trend line of q_r with equation (equation in the form (45)). Finally, from the obtained the theoretical working volume q_t should be read.

6. Results of Experimental Research

6.1. Tested Motor

A prototype of a hydraulics satellite motor was selected for experimental tests. The construction of this motor is presented in Figure 6. The working mechanism of the satellite motor consists of internally toothed curvature C, externally toothed rotor R, and satellites S (Figure 7). The principle of operation of this working mechanism was widely described in other author's publications, especially in [23–27,29,30].

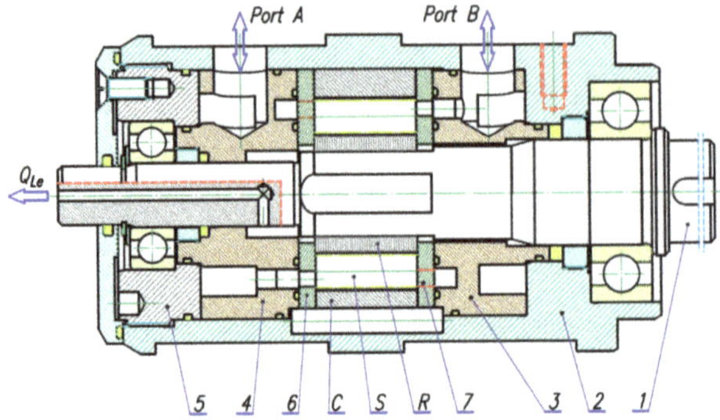

Figure 6. Hydraulic satellite motor [12,24]: *C*—curvature (stator); *S*—satellite; *R*—rotor; 1—shaft; 2—case; 3 and 4—inflow and outflow manifolds; 5—rear body; 6 and 7—distribution (compensation) plates.

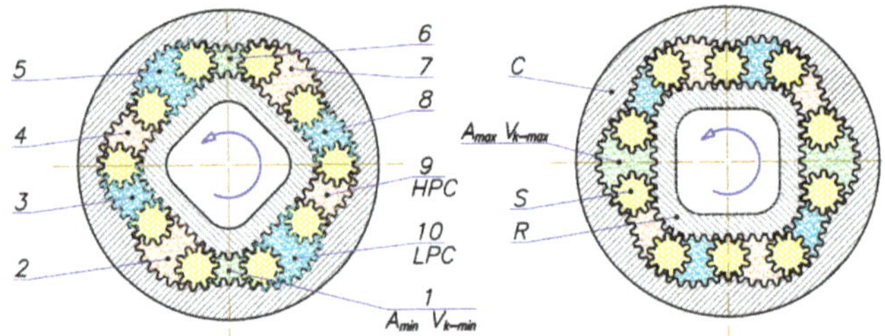

Figure 7. Satellite mechanism of type II [12,23–27,30]: *C*—curvature, *R*—rotor, *S*—satellite, 1÷10—working chambers, *HPC*—high pressure chambers, *LPC*—low pressure chambers, $V_{k\text{-}min}$—working chamber with minimum volume (dead chamber) and with minimum area A_{min}, $V_{k\text{-}max}$—working chamber with maximum volume and with maximum area A_{max}.

The number of filling and emptying cycles of the working chambers per one rotation of the shaft is the product of the number n_R of humps of the rotor and the number n_C of humps of the stator. Thus, 24 cycles correspond to one shaft revolution [12,24].

The geometric working volume q_g of the satellite mechanism depends on the number n_R of humps of the rotor and the number n_C of humps of the stator and is calculated according to the following formula [12,24]:

$$q_g = n_C \cdot n_R \cdot H \cdot (A_{max} - A_{min}) \tag{57}$$

The following geometrical parameters characterized the satellite motor [12]:

- The height of the working mechanism H = 25 mm;
- the minimum area of the working chamber A_{min} = 26.11 mm^2;
- the maximum area of the working chamber A_{max} = 83.51 mm^2.

Thus, the geometric working volume of the satellite mechanism is q_g = 34.44 cm^3/rev.

6.2. The Test Stand and Measuring Apparatus

The satellite motor was tested on a test stand with power recuperation. The diagram of the measurement system of this test stand is shown in Figure 8.

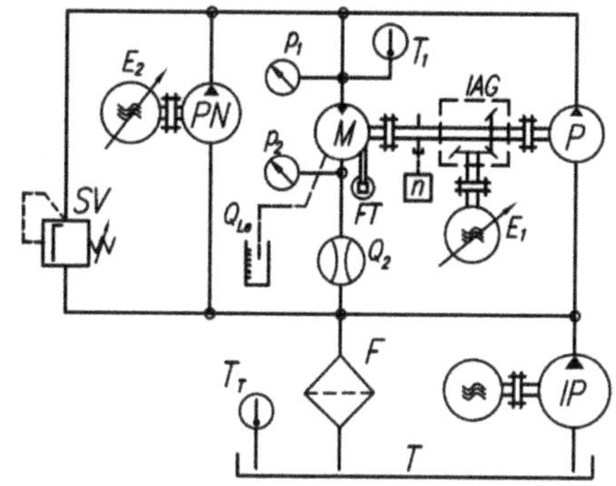

Figure 8. Diagram of the test stand measurement system [12,29]: P—pump, M—tested hydraulic motor, PN—pump for filling leaks in P and M, IP—impeller pump (pre-supply pomp), SV—safety valve, F—filter, T—reservoir, IAG—intersecting axis gear, E_1 and E_2—electric motors with frequency converters, T_1, T_T—temperature sensors, Q_2—flowmeter, Q_{Le}—leakage measurement, FT—force transducer (torque measurement), n—inductive sensor (rotational speed measurement).

Parameters of the motor measured during the test were summarized in Table 1.

Table 1. Parameters of the tested motor.

Parameter	Device	Range	Class
Pressure p_1 (in the inflow port)	Strain gauge pressure transducer	0–10 MPa and 0–40 MPa	0.3
Pressure p_2 (in the outflow port)	Strain gauge pressure transducer	0–2.5 MPa	0.3
Flow rate Q (motor absorption)	Piston flowmeter (the flowmeter chamber volume 0.63 dm^3)	0–200 L/min	0.2
Torque M	Strain gauge force transducer mounted on the arm 0.5 m	0–100 N	0.1
Rotational speed n	inductive sensor	Accuracy of measurement ±0.01 rpm	
temperature T_1 (in the inflow port)	RTD temperature sensor	Max. meas. error 0.5 °C	A

In order to determine the working volume of the motor with the smallest possible error, it is important to maintain the setting of speed n, pressure drop Δp, and liquid temperature T_1 with the least possible deviation. Thus [12]:

- For rotational speed n the deviation was ±0.1 rpm;
- for the pressure drop Δp the deviation was ±0.05 MPa;
- for the temperature in inflow port T_1 the deviation was ±1.0 °C.

6.3. Working Liquid Parameters

The oil Total Azolla 46 was the working liquid during the satellite motor tests. The temperature in the inflow port of the motor was T_1 = 43 °C which corresponds to the kinematic viscosity ν = 40 cSt and oil density ρ = 873 kg/m^3.

The characteristic of tangential isentropic bulk modulus $K_{Z(p)}$ of Total Azolla 46 oil is shown in Figure 9. The $K_{Z(p)}$ was calculated according to the Formula (41) for the data given in Table 2.

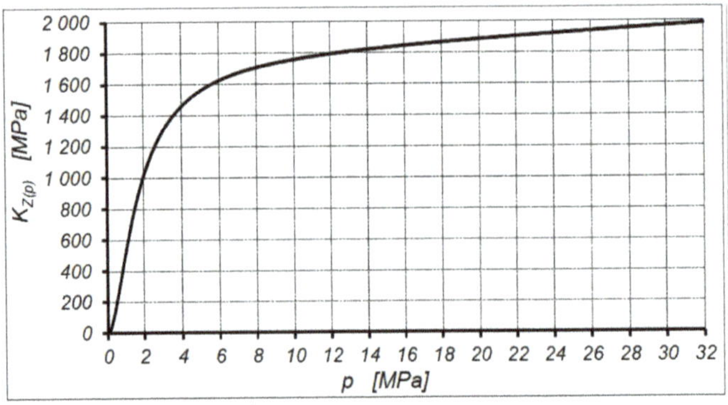

Figure 9. Characteristic $K_{Z(p)} = f(p)$ of oil Total Azolla 46 [12,28].

Table 2. Liquid parameters for calculation the $K_{Z(p)}$ [12,30].

Parameter	Value
K	1775 MPa
X_o	0.01
n	1.4
m	9.19
p_o	0.1 MPa

6.4. Pressure Drop in Motor Internal Channels

Refs. [24,27] precisely described the method of determining the pressure drop Δp_{ich} in a motor's internal channels. For the tested motor, the Δp_{ich} can be described by the following empirical formula [12]:

$$\Delta p_{ich} = 0.003224 \cdot Q_2{}^2 + 0.02183 \cdot Q_2 \tag{58}$$

where Q_2 in [L/min] and Δp_{ich} in [MPa].

6.5. Motor Output Flow Rate Characteristics

The satellite motor was tested in the range of rotational speed $n = 50 \div 1500$ rpm and in this speed range the characteristics $Q_2 = f(n)$ at $\Delta p = const.$ were determined (Figure 10). In order to maintain the clarity of the graphs the characteristics are shown only for two extreme pressure drops in the motor, that is, for $\Delta p = 2$ MPa and $\Delta p = 32$ MPa.

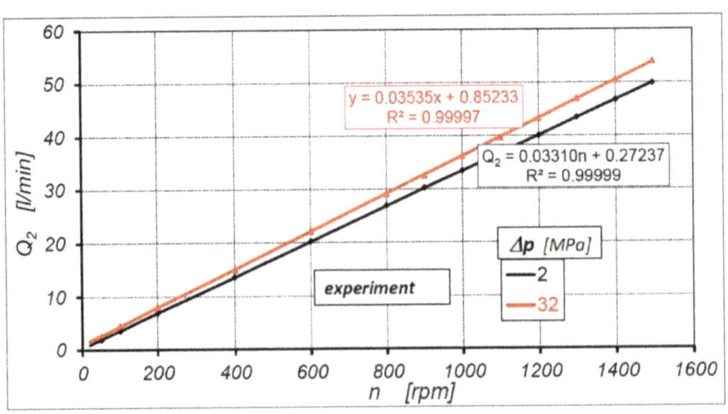

Figure 10. Characteristics of motor output flow rate Q_2 as a function of rotational speed n at $\Delta p = const.$ [12].

The results of the experiment show that the characteristics of flow rate Q_2 (at Δp = const.) are a linear function of rotational speed n, and they are compatible with theoretical characteristics presented in Figure 2. Therefore, the characteristics of Q_2 can be described by Equation (19), as shown in Figure 10.

The value of external leakage Q_{Le} in the tested motor is no more than 0.06 L/min in the whole range of the motor load [12]. Due to a very small value, the leakage Q_{Le} was omitted.

In Figure 11, the characteristics of output flow rate Q_2 at constant pressure drop Δp_i in working chambers (for Δp_i = 2 MPa and Δp_i = 32 MPa) are shown. The Δp_i was calculated according to Formulas (14) and (58) (that is was taken into account the pressure drop Δp_{ich} in the internal channels of the motor). The influence of liquid compressibility was omitted.

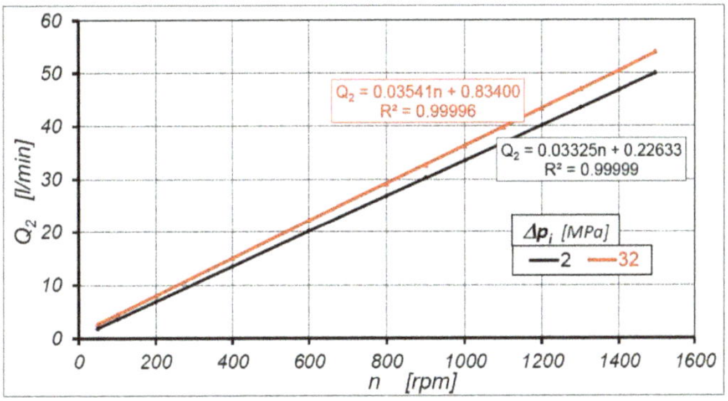

Figure 11. Characteristics of motor output flow rate Q_2 as a function of rotational speed n at Δp_i = *const*. The influence of liquid compressibility is omitted [12].

The flow rate $Q_{(pH)}$ related to the pressure p_H in the high-pressure chamber of the motor was calculated (using the method described in Section 4.4.3—that is the liquid compressibility was taken into account) and shown in Figure 12 (for Δp_i = 2 MPa and Δp_i = 32 MPa).

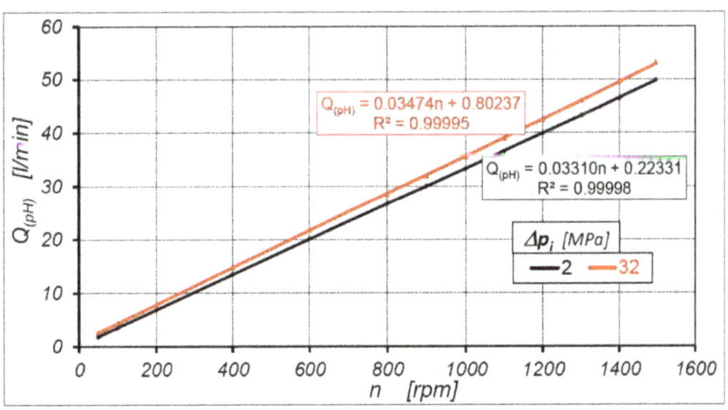

Figure 12. Characteristics of $Q_{(pH)} = f(n)$ at Δpi = *const.* related to the pressure in the high-pressure working chamber [12].

6.6. Theoretical Working Volume According to Balawender's Method

The characteristics of $q_e = f(n^{-1})$ determined according to Balawender's method are shown in Figure 13.

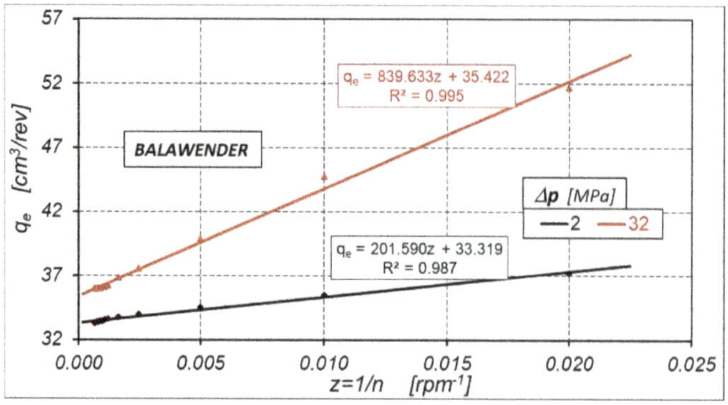

Figure 13. Characteristics of $q_e = f(n^{-1})$ at Δp=*const*.—according to Balawender's method.

According to the Balawender method, the theoretical working volume q_t of the satellite motor is q_t = 33.312 cm^3/rev. (Figure 14).

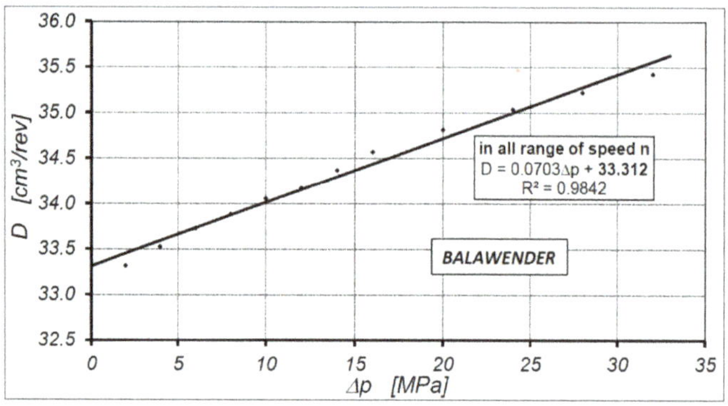

Figure 14. Characteristics of $D = f$ (Δp) (Balawender's method).

6.7. Motor Theoretical Working Volume According to Proposed New Method

The characteristics of $q_e = f(n^{-0.5})$ determined according to the proposed new method are shown in Figure 15.

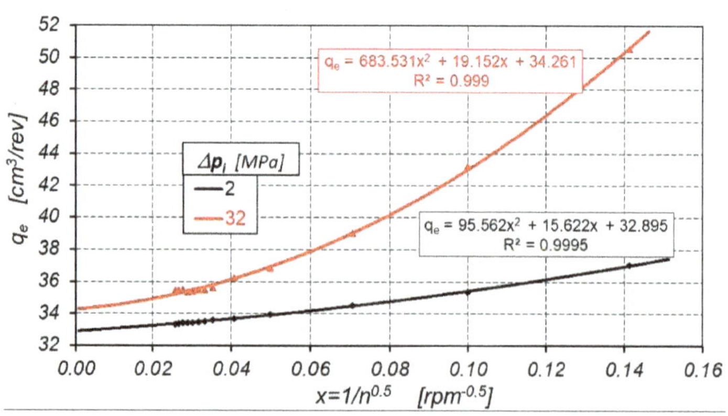

Figure 15. Characteristics $q_e = f(n^{-0.5})$ at Δp = *const*.—according to the new method.

According to the new method, the theoretical working volume of the satellite motor is q_t = 32.559 cm^3/rev. (Figure 16).

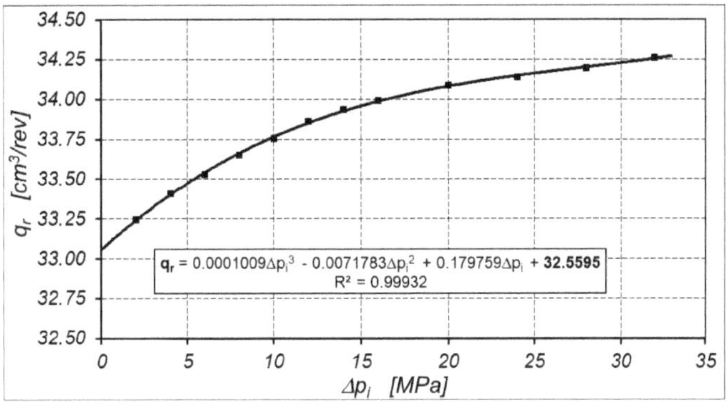

Figure 16. Characteristics of working volume q_r as a function of pressure drop Δp_i in the motor working chambers (according to the new method).

7. Discussion

The theoretical analysis indicates that the flow rate in a hydraulic motor is a linear function of rotational speed (see Equations (10), (11), and (46)). All experimental data (see characteristics $Q_2 = f(n)_{\Delta p=const.}$ in Figure 10, $Q_2 = f(n)_{\Delta pi=const.}$ in Figure 11, and the characteristics of $Q_{(pH)} = f(n)_{\Delta pi=const.}$ in Figure 12) were described by a linear trend line. In all trend line the correlation coefficient R^2 between flow rate and rotational speed is close to one. A very high value of the correlation coefficient R^2 is an effect of maintaining very precise parameters during the tests (described in Section 6.2). Furthermore the use of piston flow meter also contributed to the small scatter of the measurement results (the piston flow meter measured the average flow rate corresponding to more than 18 revolutions of the motor shaft). Thus, the results of the experimental research confirm the theoretical consideration and the flow rate in a hydraulic motor can be described by a linear function of rotational speed of a motor.

The value of the geometric working volume q_g of the motor (34.44 cm^3/rev) is greater than the theoretical working volume q_t (32.5595 cm^3/rev—calculated according to the proposed new method) by 1.8805 cm^3/rev. That is, the q_g is as much as 5.5% larger than the q_t calculated according to the new method. The main reason of the difference between the q_t and q_g is technology of manufacturing of working mechanism elements. These elements are electrically cutting with wire with a certain allowance for finishing treatment (lapping). In this way, the q_g is definitely smaller than q_t.

The results of the experimental research confirm that:

(a). It is possible to determine the pressure drop Δp_{ich} in internal channels and to calculate the pressure drop Δp_i in working chambers;
(b). the effective absorbency q_e is a linear function of Δp_i and non-linear function of the inverse of rotational speed n of the motor (formula (51) and Figure 15);
(c). the rotational speed n of a hydraulic motor has no influence on the theoretical working volume q_t;
(d). the actual working volume q_r is a third order polynomial function of pressure drop Δp_i in working chambers of the motor (Figure 16). That is, Formula (45) is true. In the tested satellite motor is:

$$q_r = \underbrace{32.5595}_{q_t} + \underbrace{0.179759 \cdot \Delta p_i - 0.0071783 \cdot \Delta p_i^{\,2} + 0.0001009 \cdot \Delta p_i^{\,3}}_{\Delta q_g} \quad (59)$$

Furthermore it has been shown that is possible to determine the q_r and q_t including the influence of oil compressibility β.

The results of the experimental research show some difference (about 2.2%) in the theoretical working volume q_t calculated using the new method (q_t = 32.5595 cm^3/rev.) and calculated using Balawender's method (q_t = 33.312 cm^3/rev.). As the final result of the theoretical working volume should be adopted q_t = 32.5595 cm^3/rev. calculated according to the proposed new method. This decision can be justified by the fact that the Balawender method is a simplified method (adopted simplification: $\Delta p_i \approx \Delta p$, linear relationship between q_e and $1/n$ (Figure 13) and linear relationship between D and Δp (Figure 14)). For users of hydraulic motor, the difference of 2.2% is not large and is usually imperceptible. However, for researchers of positive displacement machines, this difference can be significant because it directly influences on assessment of energetic losses in the motor.

Due to the fact the working volume increases under the pressure (Figure 16), the values of q_r according to Formula (45) should be taken into account to calculate the volumetric and mechanical losses in a hydraulic motor.

In the publicly available literature, the volumetric efficiency in a hydraulic motor is defined as:

$$\eta_v = \frac{q_t \cdot n}{q_t \cdot n + Q_L} \tag{60}$$

For theoretical working volume calculated according to a simplified Balawender's method, the volumetric efficiency η_v of the motor for Δp < 2 MPa (calculated according to the above formula) is larger than one (η_v > 1) (Figure 17). It has no physical sense, of course.

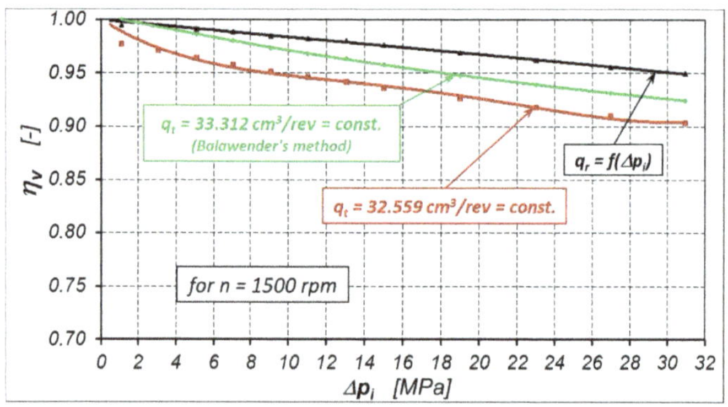

Figure 17. Volumetric efficiency $\eta_v = f(\Delta p_i)$ of satellite motor at n = 1500 rpm for q_r and q_t calculated according to the Balawender's method and the new method.

The theoretical working volume q_t in all range of pressure drop Δp in a motor is adopted for calculation of losses in this motor and its efficiencies. In this way:

- The volumetric losses are overestimated;
- the mechanical losses are underestimated;
- the volumetric efficiency η_v is underestimated (Figure 17);
- the mechanical efficiency η_m is overestimated.

In the context of the aforementioned theoretical q_t and actual working volume q_r, instead of the theoretical flow rate Q_t in a hydraulic motor, the actual flow rate Q_r should be used, defined as:

$$Q_r = q_r \cdot n \tag{61}$$

where for $\Delta p_i = 0$ is $q_r = q_t$ and $Q_r = Q_t$. Whereas the volumetric efficiency for any Δp_i is:

$$\eta_v = \frac{q_r \cdot n}{q_r \cdot n + \Delta q_L \cdot n + Q_{Lfg} + Q_C + Q_{Le}} \tag{62}$$

An example of the volumetric efficiency characteristics of the satellite motor for $n = 1500$ rpm, calculated according to Formulas (60) and (62), is shown in Figure 17.

Thus, it is necessary to take the actual working volume q_r (as a function of pressure drop Δp_i in working chambers) to calculate the losses in the hydraulic motor.

8. Summary

The above article presents a new methodology for determining the theoretical q_t and actual q_r working volume of a hydraulic motor. The proposed new method takes into account the pressure drop Δp_{ich} in the motor's internal channels and the liquid compressibility (the location of the flow meter in the measurement system). The actual working volume q_r is a non-linear function of the pressure drop Δp_i in the working chambers. Therefore, the actual working volume q_r should be taken to calculate the volumetric and mechanical losses, and also the volumetric and mechanical efficiency of the motor. In the tested satellite motor, the q_r is changing about 3% in all range of Δp_i. Is it a lot or a little? It is difficult to assess. Therefore, in order to be convinced of this, a satellite motor with a greater stiffness of the working mechanism should be built. The greater stiffness of the working mechanism is possible to obtain in two ways (separately or together):

- Mainly by increasing the diameter of the bypass,
- by changing the construction of the teeth in working mechanism elements.

The geometrical working volume of a satellite motor is definitely bigger than the theoretical working volume (about 5.5%). To explain this difference, it would be worthwhile to test the motor with reduced tip clearances and backlash clearances in the satellite mechanism. In this way, it will reduce the volume between the mating teeth of the satellite mechanism. This volume also has an impact on the process of pumping liquid through the working mechanism. The smaller this volume, the smaller the volumetric losses? The smaller an influence of rotational speed on the volumetric losses in a hydraulic motor? This thesis would require proof.

Funding: This research was funded by National Centre for Research and Development in Poland, grant number LIDER/35/102/L-2/10/NCBiR.

Conflicts of Interest: The author declares no conflict of interest.

References

1. Antoniak, P.; Stosiak, M.; Towarnicki, K. Preliminary testing of the gear pump with internal gearing with modification of the sickle insert. In Proceedings of the Engineering Mechanics 2019, Svratka, Czech Republic, 13–16 May 2019.
2. Kollek, W.; Osiński, P.; Warzyńska, U. The influence of gear micropump body asymmetry on stress distribution. *Pol. Marit. Res.* **2017**, *24*, 60–65. [CrossRef]
3. Kollek, W.; Osinski, P.; Stosiak, M.; Wilczyński, A.; Cichon, P. Problems relating to high-pressure gear micropumps. *Arch. Civ. Mech. Eng.* **2014**, *14*, 88–95. [CrossRef]
4. Patrosz, P. Influence of Gaps' geometry change on leakage flow in axial piston pumps. In *Lecture Notes in Mechanical Engineering*; Cavas-Martínez, F., Chaari, F., Gherardini, F., Haddar, M., Ivanov, V., Kwon, Y.W., Trojanowska, J., di Mare, F., Eds.; Springer International Publishing: Geneva, Switzerland, 2020; pp. 76–89.
5. Załuski, P. Experimental research of an axial piston pump with displaced swash plate axis of rotation. In *Lecture Notes in Mechanical Engineering*; Cavas-Martínez, F., Chaari, F., Gherardini, F., Haddar, M., Ivanov, V., Kwon, Y.W., Trojanowska, J., di Mare, F., Eds.; Springer International Publishing: Geneva, Switzerland, 2020; pp. 135–145.
6. Osiecki, L. *Commutation Units of Hydraulics Axial Piston Machines*; Gdansk University of Technology Publishers: Gdansk, Poland, 2006.
7. Balawender, A. *Analiza Energetyczna i Metodyka Badań Silników Hydraulicznych Wolnoobrotowych (Energy Analysis and Methods of Testing of Low-Speed Hydraulic Motors)*; Scientific Book of the Gdansk University of Technology; Mechanika No. 54; Gdansk University of Technology Publishing House: Gdansk, Poland, 1988.

8. Balawender, A. Opracowanie Metodyki Wyznaczania Teoretycznej Objętości Roboczej Pomp i Silników Hydraulicznych Wyporowych. (The Development of the Methodology for the Determination of the Theoretical Working Volume of Positive Displacement Pumps and Hydraulic Motors). Ph.D. Thesis, Gdansk University of Technology, Gdansk, Poland, 1974.
9. Schlosser, W.M.J. Meten aan Verdringerpompen. Ph.D. Thesis, Technical Hogeschool Delft, Delft, The Netherlands, 1957.
10. Schlosser, W.M.J.; Hilbrands, J.W. Das theoretische Hubvolumen von Verdrangerpumpen. *Olhydraul. Pneum.* **1963**, *7*, 133–138. (In German)
11. Schlosser, W.M.J.; Hilbrands, J.W. Das volumetrische Wirkungsgrad von Verdrongerpumpen. *Olhydraul. Pneum.* **1963**, *7*, 469–476.
12. Sliwinski, P. Determination of the theoretical and actual working volume of a hydraulic motor. *Energies* **2020**, *13*, 5933. [CrossRef]
13. Toet, G. Die Bestimmung des theoretischen Hubvolumens von hydrostatischen Verdrangerpumen und Motoren aus volumetrischen Messungen. *Olhydraul. Pnaum.* **1970**, *14*, 185–190.
14. Toet, G.; Johnson, J.; Montague, J.; Torres, K.; Garcia-Bravo, J. The Determination of the theoretical stroke volume of hydrostatic positive displacement pumps and motors from volumetric measurements. *Energies* **2019**, *12*, 415. [CrossRef]
15. Stryczek, S. *Napęd Hydrostatyczny (Hydrostatic Drive)*; WNT: Warszawa, Poland, 2005.
16. Wilson, W.E. *Positive Displacement Pumps and Fluid Motors*; Publication Corporation: New York, NY, USA, 1950.
17. Garcia-Bravo, J.; Nicholson, J. What is the real size of that pump? *Fluid Power J. Manuf. Dir.* **2018**, *25*, 20–21. Available online: https://fluidpowerjournal.com/real-size-pump/ (accessed on 4 January 2021).
18. Manring, N.; Williamson, C. The theoretical volumetric displacement of a check-valve type, digital displacement pump. *J. Dyn. Syst. Meas. Control* **2018**, *141*, 12–14. [CrossRef]
19. Myszkowski, A. Energy analysis of an ideal suction-pressure unit. *Arch. Mech. Technol. Mater.* **2015**, *35*, 51–60.
20. Post, W. Models for steady-state performance of hydraulic pumps: Determination of displacement. In Proceedings of the 9th Bath International Fluid Power Workshop, University of Bath, Bath, UK, 9–11 September 1996; Volume 9.
21. International Organisation for Standardization. *Hydraulic Fluid Power. Positive Displacement*; International Organization for Standardization: Geneva, Switzerland, 2008.
22. Kim, T.; Kalbfleisch, P.; Ivantysynova, M. The effect of cross porting on derived displacement volume. *Int. J. Fluid Power* **2014**, *15*, 77–85. [CrossRef]
23. Sliwinski, P. The influence of water and mineral oil on volumetric losses in a hydraulic motor. *Pol. Marit. Res.* **2017**, *24*, 213–223. [CrossRef]
24. Sliwinski, P. *Satelitowe Maszyny Wyporowe. Podstawy Projektowania i Analiza Strat Energetycznych (Satellite Displacement Machines. Basic of Design and Analysis of Power Loss)*; Gdansk University of Technology Publishers: Gdansk, Poland, 2016.
25. Śliwiński, P. Flow of liquid in flat gaps of the satellite motor working mechanism. *Pol. Marit. Res.* **2014**, *21*, 50–57. [CrossRef]
26. Śliwiński, P. Influence of water and mineral oil on the leaks in satellite motor commutation unit clearances. *Pol. Marit. Res.* **2017**, *24*, 58–67. [CrossRef]
27. Sliwinski, P.; Patrosz, P. The influence of water and mineral oil on pressure losses in hydraulic motor. In *Lecture Notes in Mechanical Engineering*; Cavas-Martínez, F., Chaari, F., Gherardini, F., Haddar, M., Ivanov, V., Kwon, Y.W., Trojanowska, J., di Mare, F., Eds.; Springer International Publishing: Geneva, Switzerland, 2020; pp. 112–122.
28. Zaluski, P. Wpływy Położenia Osi Obrotu Tarczy Wychylnej na Sprawność Objętościową Pomp Wielotłoczkowych Osiowych (Influence of the Position of the Swash Plate Rotation Axis on the Volumetric Efficiency of Axial Piston Pumps). Ph.D. Thesis, Gdansk University of Technology, Gdansk, Poland, 2017.
29. Śliwiński, P. The influence of water and mineral oil on mechanical losses in a hydraulic motor for offshore and marine applications. *Pol. Marit. Res.* **2020**, *27*, 125–135. [CrossRef]
30. Sliwinski, P.; Patrosz, P. Satellite Operating Mechanism of the Hydraulic Displacement Machine. Patent PL218888, 27 February 2015.

energies

MDPI

Article

Methods of Determining Pressure Drop in Internal Channels of a Hydraulic Motor

Pawel Sliwinski * and **Piotr Patrosz**

Faculty of Mechanical Engineering and Ship Technology, Gdansk University of Technology, 80-233 Gdansk, Poland; piotr.patrosz@pg.edu.plnd
* Correspondence: pawel.sliwinski@pg.edu.pl

Abstract: In this paper, new methods for determining the pressure drop in internal channels of a hydraulic motor are proposed and described. Mathematical models of pressure losses in internal channels have also been described. Experimental tests of the satellite motor were carried out according to one of the proposed methods. The tests were carried out for two liquids, i.e., water and mineral oil. Experimental studies have shown that at a high flow rate in the motor supplied with water the pressure losses are a dozen or so percent greater than in the motor supplied with oil. However, at low flow rates is the inverse, that is, the pressure losses in the motor supplied with water are about ten percent lower than in the motor supplied with oil. The CFD calculation of the pressure drop in the internal channel of the motor was also conducted. It was assumed that holes in the commutation unit plate are placed face to face and that the liquid did not cause changes in the working chambers' volume. In this way, it has been proven that those simplified assumptions can have up to a 50% difference in relation to the experimental tests.

Keywords: k pressure losses; satellite motor; water; oil; CFD calculation; internal channels in motor; pressure efficiency

Citation: Sliwinski, P.; Patrosz, P. Methods of Determining Pressure Drop in Internal Channels of a Hydraulic Motor. *Energies* **2021**, *14*, 5669. https://doi.org/10.3390/en14185669

Academic Editor: Ryszard Dindorf

Received: 4 August 2021
Accepted: 3 September 2021
Published: 9 September 2021

1. Introduction

The task of a hydraulic motor is to convert hydraulic energy into mechanical energy. Liquid is an energy carrier in the hydraulic system. Currently, mineral oil is the most commonly used liquid in hydraulic systems [1]. However, in some industrial sectors, a liquid that is non-flammable (mining, steel mills, etc.) or non-toxic for the environment and human health (food industry) is required [2].

Water is a liquid which is non-flammable, non-toxic and certainly suitable for energy transfer in hydraulic systems. Furthermore, water is generally available as a working liquid. There is a growing trend throughout the world towards researching and developing components and hydraulic systems that are supplied with water [1–4]. The development of water hydraulics is important for various industrial sectors, e.g., for the food industry, the mining industry and marine technology [5]. In comparison to mineral oil, water has a very low viscosity and low lubricating properties [6]. These features adversely influence the efficiency of energy conversion in hydraulic systems [7]. Despite this, attempts are being made to develop innovative components and hydraulic systems supplied with water [2].

Each hydraulic element (pump, motor and valves) is recommended to a specific type of working liquid. For example, a hydraulic motor recommended to oil circuits should not be used in water systems. In other cases, this motor has very little durability and a low efficiency [7]. Both the pump and the hydraulic motor are components of the hydraulic system that have large energy losses, including volumetric, mechanical and pressure losses [1,7–16].

The main sources of volumetric losses in hydraulic motors are the leakage in the clearances of the working mechanism, the leakage in the clearances of the commutation unit and the compressibility of liquid in the death chamber of the working mechanism [7–11,17–25]. However, the main sources of mechanical losses are friction between moving parts of the working mechanism and friction in bearings and seals [22,25]. In the case of pressure losses in hydraulic motors, this pressure loss results from internal channels in the motor. The internal channels are defined as channels inside the motor body. Their task is to supply liquid from the motor inflow port to the working chambers (inflow channels) and remove the liquid from the working chambers to the outflow port (outflow channels). The pressure drop in internal channels of the motor mainly depends on the geometric dimensions of those channels, the liquid parameters (viscosity and density) and the flow rate of the liquid. The pressure drop in internal channels of the motor has a significant impact on the energy conversion efficiency in the motor and on the parameters of hydraulic systems [8,11,17,22,26–28].

The results of this research proved that all of the above-mentioned types of losses were influenced by the type of liquid. In general, a motor supplied with a low viscosity liquid generates larger energy losses than a motor supplied with oil [20–23]. Thus far, there are no research results comparing the influence of water and mineral oil on the pressure losses in hydraulic motors. There is also no specific information about the designs of motors that can be supplied with both water and mineral oil. There is a rich literature on flows in orifices and other simple openings; however, there are not any prominent studies in the literature that compare the influence of the type of liquid on the flow characteristics in these internal channels.

Therefore, from the scientific and cognitive point of view, researching and describing the influence of water and mineral oil on pressure losses in a hydraulic motor is appropriate and justified. The issue of the influence of the type of liquid on the pressure losses in the hydraulic motor is a new issue, represents an important scientific problem and is the subject of this article. Consequently, the following objectives have been defined for this article:

(a) Describe a mathematical model of the pressure losses;
(b) Results of experimental tests of pressure in working chambers and pressure losses in motor supplied with mineral oil and water and their comparison;
(c) Results of CFD calculations of pressure losses in motor supplied with mineral oil and water and their comparison;
(d) Compare the results of experimental research with CFD calculations.

The experimental research on the influence of the type of liquid on pressure losses was carried out using a prototype of a hydraulic satellite motor presented in Section 6. The satellite motor was chosen to test because the authors of this publication are the co-creators of this motor and have conducted extensive research on the development of this motor. This motor is already produced but not yet widely known in the world.

2. Pressure Drop in the Internal Channels of the Hydraulic Motor

In the hydraulic motor the pressure drop Δp_{ich} in internal channels increases the value of the pressure drop Δp. The pressure drop Δp is measured in motor ports (Figure 1), according to the formula [29]:

$$\Delta p = \Delta p_i + \Delta p_{ich} \quad (1)$$

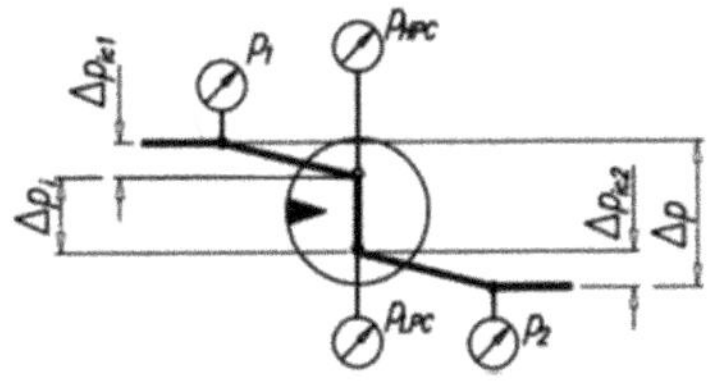

Figure 1. Pressure drop in hydraulic motor [23]: Δp_{ic1}—pressure drop in inflow internal channel, Δp_{ic2}—pressure drop in outflow internal channel, p_{HPC}—pressure in high-pressure working chamber, p_{LPC}—pressure in low-pressure working chamber, p_1—pressure in inflow motor port, p_2—pressure in outflow motor port.

Pressure drop Δp_{ich} in internal channels is the following sum [29]:

$$\Delta p_{ich} = \Delta p_{ic1} + \Delta p_{ic2} \tag{2}$$

where:

- Δp_{ic1}—the pressure drop in inflow internal channel;
- Δp_{ic2}—the pressure drop in outflow internal channel.

In a hydraulic motor with variable shaft rotation directions, if the inflow and outflow internal channels in a hydraulic motor have the same shape and dimensions then is only necessary to measure Δp_{ic1} and the Δp_{ich} is calculated according to the formula:

$$\Delta p_{ich} = 2 \cdot \Delta p_{ic1} \tag{3}$$

3. Known Method for Measuring the Pressure Drop in Internal Channels—Method 1

The measurement of pressure drop in internal channels of hydraulic motor during his normal work is problematic. The main problem is the pressure measurement in working chambers; it is technologically difficult to implement or sometimes impossible. Therefore, it is easier to determine the pressure drop in internal channels of a hydraulic motor if this motor works as a pump.

In known method the test stand is equipped with a drive motor *DM* with constant speed and does not include an electronic measurement data recording system (Figure 2). Changes in the setting of the throttle valve *TV* in a low-pressure line are made and the pressure p_1 in the low-pressure port of motor (suction port—motor works as a pump) is measured. The flow rate Q is measured at the moment of cavitation in the motor. Then, during cavitation, a certain value of pressure p_{LPC} in the low-pressure working chambers of the motor should be assumed. Usually, for simplicity, a vacuum of 1 bar is assumed (p_{LPC} = –1 bar) [9,15,17].

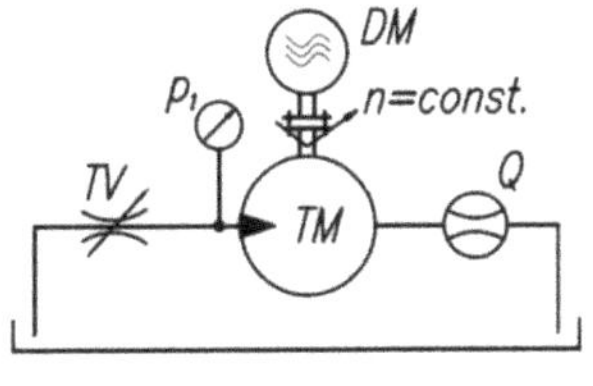

Figure 2. Scheme of the hydraulic circuit of the test stand for the measurement of the pressure drop in the internal channel of the hydraulic motor [23]: *TM*—tested motor; *DM*—drive motor; *TV*—throttle valve; *Q*—flow meter; p_1—manometer.

The experiment execution time is relatively short. The appearance of cavitation in the engine's working chambers does not cause damage in such a short time.

4. Proposed Methods for Measuring the Pressure Drop in Internal Channels

The disadvantages of method 1 are:

- The difficulty to catch the beginning of cavitation;
- The constant speed of motor *DM* (Figure 2) prevents the full characteristics of the pressure drop in the internal channel from being obtained;
- Unknown values of pressure in working chambers. Thus, the value of −1 bar in low-pressure working chamber was assumed.

Therefore, new methods that avoid this inconvenience are proposed below.

4.1. Method 2

Method 2 is the experimental method. In this method, similar to method 1, the hydraulic motor 1 works as a pump and is driven by an electric motor *DM* (Figure 3). The speed n of electric motor *DM* and machine 1 is set via a frequency converter. The displacement machine 1 is supplied by a pump 3 through throttle valve *TV*. At a particular setting of the valve 4, the speed n is increased from its minimum value. A recording data system collects the values of the speed n, delivery Q and pressure p_1. In this way, the characteristics of $Q = f(n)$ and $p_1 = f(n)$ are created (Figure 4). After some characteristic point 1 (Figure 4), the delivery Q does not change despite the increase in the rotational speed n. Point 1 is necessary to register the flow rate Q_1 and the pressure $p_{1\text{-}1}$. In the next step, the rotational speed n is decreased to its minimum, the setting of valve *TV* is changed, and, at the end, the speed n is increased again and the values of n, Q and p_1 are registered. In this way, the point 2 (parameters Q_2 and $p_{1\text{-}2}$) are obtained. In order to obtain the full characteristics of the pressure drop in the internal channel of the motor, this procedure should be carried out several times with different settings of valve *TV*.

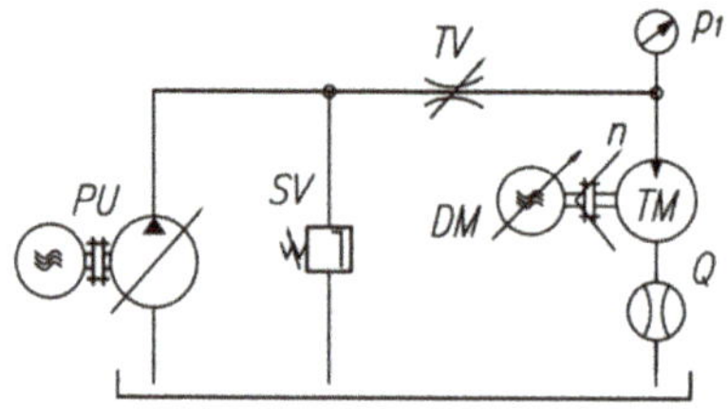

Figure 3. Scheme of hydraulic circuit of the test stand for the measurement of the pressure drop in the internal channel of the hydraulic motor according to method 2: *TM*—tested motor, *DM*—drive motor, *PU*—pump unit, *TV*—throttle valve, *SV*—safety valve, *Q*—flow meter, p_1—manometer (pressure in inflow port of motor), n—rotational speed.

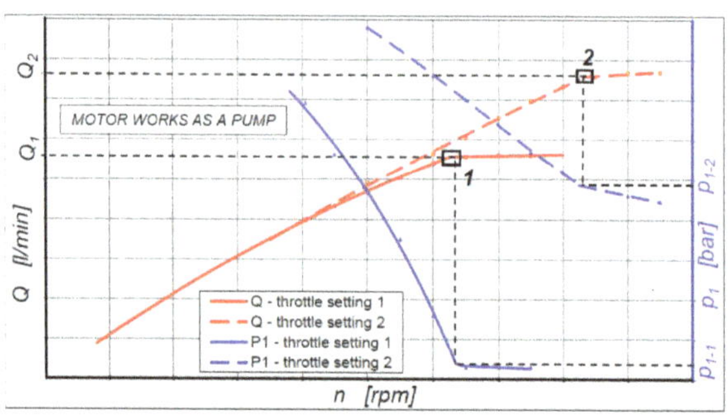

Figure 4. Delivery Q of motor 1 working as a pump and the pressure p_1 in the suction port of motor 1 for two different settings of throttle valve *TV*—according to the scheme in Figure 3 [23].

For points 1,2, … ,*n* (Figure 4) in the suction working chambers of tested motor TM, there exists pressure p_{LPC} close to the vacuum. Therefore, the pressure drop Δp_{ic1} in the internal channel (from the inflow (suction) port of the motor, which is p_1, to the suction working chambers) is:

$$\Delta p_{ic1} = p_1 - p_{LPC} \tag{4}$$

For a hydraulic motor with the same internal channels on the inflow and the outflow side, the total pressure drop Δp_{ich} is calculated according to formula (3). Theoretical characteristics $\Delta p_{ich} = f(Q)$ are shown in Figure 5.

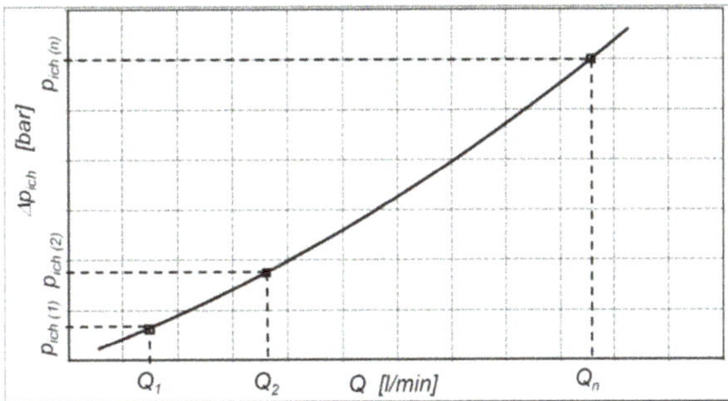

Figure 5. Theoretical characteristics of $\Delta p_{ich} = f(Q)$ [23].

4.2. Method 3

Method 3 is the experimental method. In this method, similar to methods 1 and 2, the hydraulic motor *TM* works as a pump and is driven by an electric motor *DM*. The speed n of the electric motor *DM* and tested motor *TM* is set via a frequency converter (Figure 6). In this method, a measurement data acquisition system at the same time records:

(a) The pressure p_1 and p_2 in motor ports;
(b) The pressure p_{LPC} and p_{HPC} in motor working chambers;
(c) The delivery *Q*;
(d) The rotational speed *n* of the motor shaft.

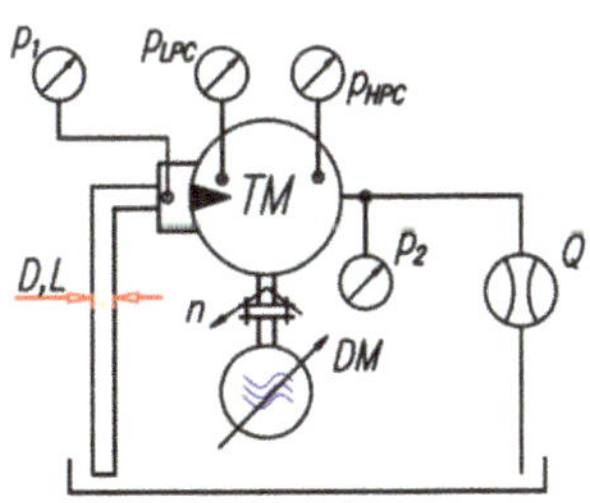

Figure 6. Hydraulic circuit of the test stand for the measurement of the pressure drop in the internal channel of the hydraulic motor according to method 3: *D*—diameter of suction tube, *L*—length of suction tube, *TM*—tested motor, *DM*—electric motor (controlled by frequency converter), Q—flow meter, *n*—rotational speed, p_{HPC}—pressure in high-pressure working chamber, p_{HPC}—pressure in low-pressure working chamber, p_1—pressure in inflow motor port, p_2—pressure in outflow motor port.

If cavitation occurs in the low-pressure working chamber, then the increase in rotational speed n of the tested motor does not increase the flow rate *Q* in this motor. This is because the pressure drop between the tank and low-pressure chamber is constant (in tank is zero and in the low-pressure chamber during full cavitation the pressure is also constant). If the pressure drop is constant, then the flow rate Q also must to be constant. Therefore, for Q = const. is incomplete filling of the working chamber.

In all described methods the flow meter was located in the outflow hose of the motor. This flow meter creates considerable resistance to flow (the pressure p_2 in the drain connection is greater than zero). So, a throttle valve to increase pressure p_2 was not needed and the cavitation in the low-pressure working chamber has no influence on the measurement of flow rate Q.

At the stage of theoretical considerations, it can be assumed that the pressure drop in the internal channels of the motor depends on the type of overlap in the commutation unit of the working mechanism. Therefore, in a motor with the negative overlap in the commutation unit (especially on the high-pressure side) there is no additional pressure increase caused by liquid compression in the closed working chamber. Then, the pressure drop is calculated according to Formula (1) considering (Figure 7):

$$\Delta p_{ic1} = p_1 - p_{LPC} \tag{5}$$

$$\Delta p_{ic2} = p_{HPC} - p_2 \tag{6}$$

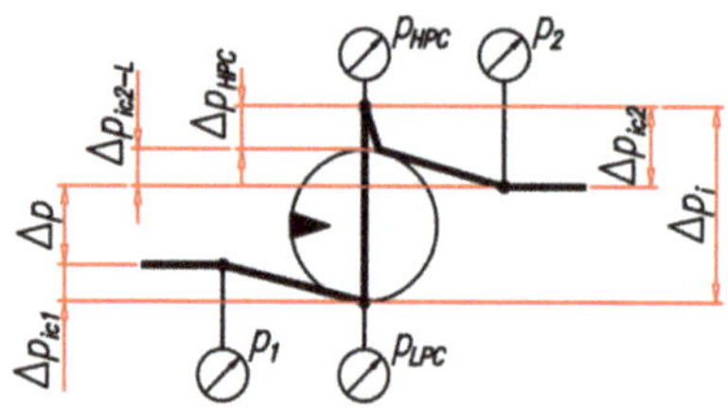

Figure 7. The progress of pressure in a hydraulic motor (working as a pump) including the increase of pressure Δp_{HPC} in a high-pressure working chamber.

However, in a motor with the positive overlap or zero overlap in the closed working chamber additional increase in pressure is created. Furthermore, the higher the rotational speed of the shaft, the higher the p_{HPC} in the working chamber. Therefore, the pressure drop is calculated as (Figure 9):

$$\Delta p_{ich} = \Delta p_{ic1} + \Delta p_{ic2-L} \tag{7}$$

where:

$$\Delta p_{ic2-L} = p_{2-L} - p_{HPC-L} \tag{8}$$

- $p_{2\text{-}L}$—the pressure in port *B* measured for the opposite direction of the shaft rotation (rotation to the left);
- $p_{HPC\text{-}L}$—the pressure in the suction chamber measured for the opposite direction of the shaft rotation (rotation to the left).

In order to be abundantly clear, the suction working chamber with pressure p_{LPC} becomes a high-pressure working chamber with pressure $p_{LPC\text{-}L}$ after changing the direction of rotation. However, a high-pressure working chamber with pressure p_{HPC} becomes a low-pressure working chamber with pressure $p_{HPC\text{-}L}$ after changing the direction of rotation.

Theoretical characteristics of the pressure drop in internal channels of a hydraulic motor are represented exactly as Figure 5 shows.

An increase of pressure Δp_{HPC} in a high-pressure working chamber caused by compression of liquid is:

$$\Delta p_{HPC} = \Delta p_{ic2} - \Delta p_{ic2-L} \tag{9}$$

Thus, the change of pressure in the hydraulic motor should progress as in Figure 7. The increase of pressure Δp_{HPC} has a direct influence on the mechanical losses in the working mechanism. Therefore, the Δp_{HPC} is not the component of the pressure losses in the internal channels of the motor [23,29].

In a hydraulic motor with a variable direction of shaft rotation the internal channels on the supply side and on the outflow side have the same dimensions. Then, the total pressure drop Δp_{ich} can be calculated according to Formula (3).

4.3. Method 4

Method 4 is also the experimental method. Is this method, the pressure drop is measured in a loaded hydraulic motor equipped with pressure sensors like in Figure 1. In this method, the test stand must be able to set the motor speed. The measurement data recording system is not required.

Due to the fact that method 4 is analogous to methods 2 and 3, it will not be described.

4.4. Practical Aspects of Applying the Methods 2 and 3

From the practical point of view the biggest disadvantages of method 3 is necessity of measurement of the pressure in the working chamber of the tested machine. Therefore, this method rather will not be used in some types of positive displacement machines, like in an axial piston pump or motor. However, in other displacement machines (like in satellite displacement machines) the possibility of measurement of the pressure in the working chamber is a very important advantage because it is easy to obtain the characteristics of pressure drop Δp_{ic1} and Δp_{ic2} in internal channels. The measurement of pressure p_1 and p_{HPC} (and similarly p_2 and p_{LPC}) gives higher accuracy of the characteristic $\Delta p_{ic1} = f(Q)$ (or $\Delta p_{ic2} = f(Q)$) in all ranges of Q. This is not possible in method 2, where the pressure p_{LPC} has to be assumed.

Method 2 requires a more advanced test stand (additional pump and throttle valve) and larger number of measurement series than method 3. Furthermore, method 3 makes it possible to designate an increase of pressure Δp_{HPC} in a high-pressure working chamber caused by compression of liquid.

5. Mathematical Models of the Pressure Drop in the Internal Channels of the Hydraulic Motor

The pressure drop Δp_{ich} in the internal channels of the hydraulic motor is energetic losses and is commonly called "pressure losses". The geometry of the internal channels (shape and dimensions) has influence on the value of the pressure drop. Any changes in the channel geometry has a large influence on the direction of the liquid flow and on the average liquid flow speed c. In this way, the pressure drop Δp_{ich} is treated as a local pressure loss. The local pressure loss is described by the well-known formula:

$$\Delta p_{ich} = \xi \cdot \rho \cdot \frac{c^2}{2} \tag{10}$$

where:

- ξ—the coefficient of pressure loss;
- ρ—the density of liquid.

In the case of hydraulic machines, the coefficient of pressure loss is expressed by the formula [17]:

$$\xi = \frac{C_1}{Re} + C_2 \tag{11}$$

where C_1 and C_2 are constants.

In a hydraulic motor, the internal channels are short and have irregular geometry, so it is not possible to calculate Reynolds number Re and the average speed c in this way.

However, it is known that $Re = f(c)$ and $c = f(Q)$. Therefore, the pressure losses in internal channels of the motor can described by:

$$\Delta p_{ich} = \underbrace{C_t \cdot \rho \cdot Q^2}_{turbulent\ flow\ component} + \underbrace{C_l \cdot \nu \cdot \rho \cdot Q}_{laminar\ flow\ component} \tag{12}$$

where:

- C_t—the constant of turbulent flow component;
- C_l—the constant of laminar flow component;
- ν—the kinematic viscosity.

The advantages of the above model are [29]:

(a) The geometry of the motor's internal channels is omitted;
(b) The pressure losses are expressed as a function of density and viscosity of the liquid and the liquid flow rate (these are the parameters having a direct impact on the pressure losses);
(c) The values of constants C_t and C_l can be calculated based on the equation of the trend line of the characteristic shown in Figure 5.

In the literature, the pressure losses Δp_{ich} are also described by the following formula [17,27]:

$$\Delta p_{ich} = C_{ich} \cdot \rho \cdot \omega^2 \cdot \sqrt[3]{\left(\frac{V_t}{2 \cdot \pi}\right)^2} \tag{13}$$

where:

- ω—the angular speed of the motor shaft;
- V_t—the theoretical displacement;
- C_{ich}—the coefficient.

The disadvantages of the above model are as follows:

(a) The angular speed ω has an indirect influence on the pressure drop Δp_{ich} in the internal channels because angular speed is a function of theoretical absorbency Q_t of the motor and a function of volumetric losses Q_{vl} in the motor. Thus, only the absorbency Q (flow rate consumed by the motor) has influence on Δp_{ich}. The absorbency Q is defined as:

$$Q = Q_t + Q_{vl} \tag{14}$$

(b) Theoretical displacement V_t has influence only on the theoretical absorbency Q_t;
(c) Does not take into account the influence of liquid viscosity on the pressure drop. The flow in internal channels is assumed as turbulent flow only.

Z. Paszota describes the pressure losses in the internal channels of the motor in the following way [11]:

$$\Delta p_{ich} = k_8 \cdot p_n \cdot \left(\frac{Q}{Q_{pt}}\right)^{a_{Qp}} \cdot \left(\frac{v}{v_n}\right)^{a_{vp}} \tag{15}$$

where:

- k_8—the coefficient determining the pressure loss in the internal channels and in the commutation unit of the motor;
- a_{Qp}—the exponent of the influence of the liquid flow rate in the internal channels on pressure losses;
- a_{vp}—the exponent of the influence of liquid viscosity on pressure losses;
- p_n—the nominal pressure of the pump;
- Q_{Pt}—the theoretical delivery of the pump;
- v_n—the reference viscosity of the liquid.

The inconvenience of the above model is the dependence of the pressure drop Δp_{ich} on the parameters of the pump supplying the motor. Furthermore, it is necessary to assume the reference viscosity ν_n of the liquid. Therefore, the exponents a_{Qp} and $a_{\nu p}$ and the coefficient k_8 depend on the parameters of the pump and on the reference viscosity of the liquid.

6. Tested Motor

The experimental research was carried out with the use of the prototype of a satellite motor (Figures 8 and 9). The working mechanism of this motor is a satellite mechanism (Figure 10). On both sides of this mechanism are distribution plates (Figures 9 and 11). The principle of operation of this mechanism and principle of operation of a satellite motor are widely described in many publications, such as in [8,20–23,29–32].

Figure 8. General view of hydraulic satellite motor (SM type) [30].

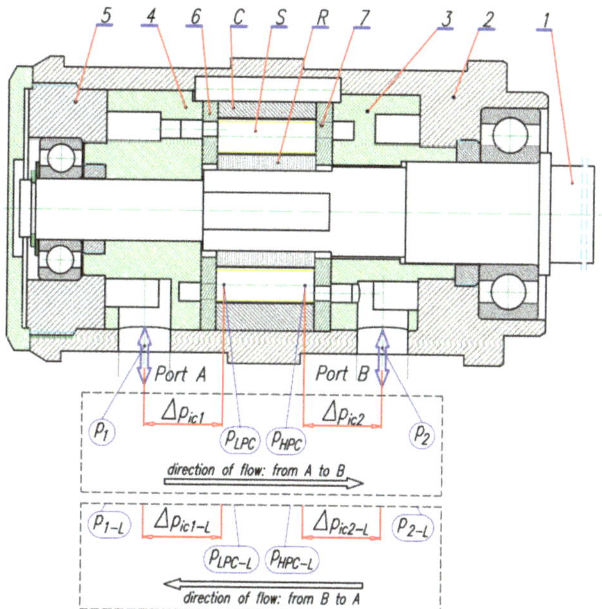

Figure 9. Construction of satellite motor and pressure drops in internal channels: *C*—curvature (stator), *S*—satellite, *R*—rotor, 1—shaft, 2—casing, 3 and 4—inflow and outflow manifolds, 5—rear plate, 6 and 7—distribution plates.

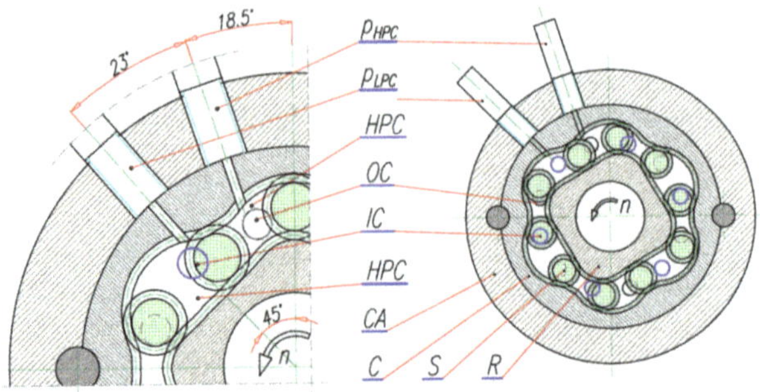

Figure 10. The cross section of the motor through the working mechanism and the location of pressure sensors p_{LPC} and p_{HPC} [23,32]: *C*—curvature (stator), *R*—rotor, *S*—satellite, *CA*—casing, *LPC*—low-pressure working chamber, *HPC*—high-pressure working chamber, *IC*—inflow hole, *OC*—outflow hole.

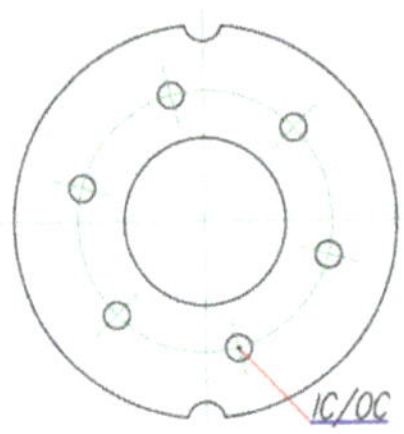

Figure 11. Distribution plate: IC/OC—inflow/outflow hole [23].

The toothed unit, shown in Figure 10, is the satellite working mechanism of the motor. It consists of a toothed rotor R (4 humps), toothed stator C (6 humps) and ten wheels S (satellite).

The principle of operation of satellite motor was widely described in [20–23].

The satellite motor used for the test had the following geometrical parameters [30,31]:

- The theoretical displacement q_t = 32.94 cm^3/rev.;
- The teeth module m = 0.75 mm;
- The height of working mechanism H = 25 mm.

7. Results of the Research on the Pressure Drop in the Internal Channels of the Motor

In Section 4 it was written that method 3 makes it possible, besides the pressure drop Δp_{ich}, to designate an increase of pressure Δp_{HPC} caused by compression of liquid in a high-pressure working chamber. Therefore, the research on the pressure drop in the internal channels of the satellite motor was carried out according to method 3.

During the experiment the following parameters were recorded:

(a) p_1 and p_2—pressures in motor ports;
(b) p_{LPC} and p_{HPC}—pressures in motor working chambers;
(c) Q—liquid stream feeding the motor;
(d) n—the rotational speed of the motor shaft.

All pressures were measured by a strain gauge pressure transducer with range $-1 \div 3$ bar and class 0.3. The flow rate Q was measured by mass flow meter with range 33 lpm and class 0.1 and the rotational speed n was measured using the inductive sensor.

During the test the temperature of the liquid in the test stand was kept constant. The temperature of liquid in the inflow port of motor was controlled. In this way viscosity of the liquid was controlled. This temperature was not registered.

7.1. Working Liquids

Laboratory research on the pressure drop in the internal channels of the satellite motor was carried out using the following liquids:

(a) Tap water (ν = 0.9 cSt and ρ = 999.8 kg/m^3);
(b) Mineral oil (ν = 40 cSt and ρ = 862.0 kg/m^3).

7.2. Experimental Data—Motor Supplied with Water

In the tested motor (Figures 8 and 9), the internal channels on the supply side have the same dimensions as the internal channels on the outflow side. Therefore, it is enough to measure only Δp_{ic1}. Then, the total pressure drop Δp_{ich} in the internal channels of the motor can be calculated according to Formula (3).

The recorded results of pressure p_1, p_2, p_{HPC}, p_{LPC} and flow rate Q in the motor as a function of the shaft speed n are shown in Figure 12. In addition, characteristics Δp_{ic1} and Δp_{ic2}, whose values were calculated according to Formulas (5) and (6), are shown in Figure 13. Cavitation occurs (the flow rate Q = const.) at rotational speeds above 630 rpm. In Figure 14, characteristics Δp_{ic1} and Δp_{ic2} are shown at speeds up to 630 rpm—cavitation does not occur at these speeds.

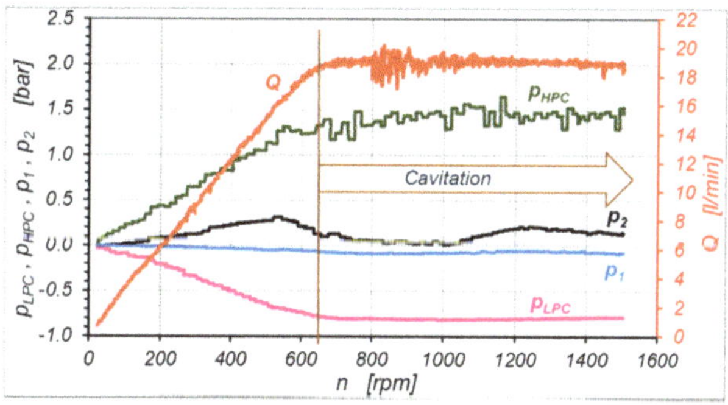

Figure 12. Characteristics of $p_1 = f(n)$, $p_2 = f(n)$, $p_{HPC} = f(n)$, $p_{LPC} = f(n)$ and $Q = f(n)$ over the entire range of the motor shaft speed n—motor supplied with water [22,23].

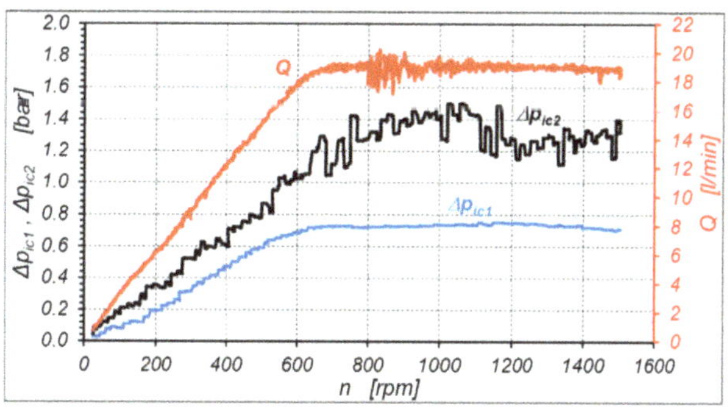

Figure 13. Characteristics of pressure drop $\Delta p_{ic1} = f(n)$ and $\Delta p_{ic2} = f(n)$ in the internal channels of the motor over the entire range of the motor shaft speed n—motor supplied with water [23].

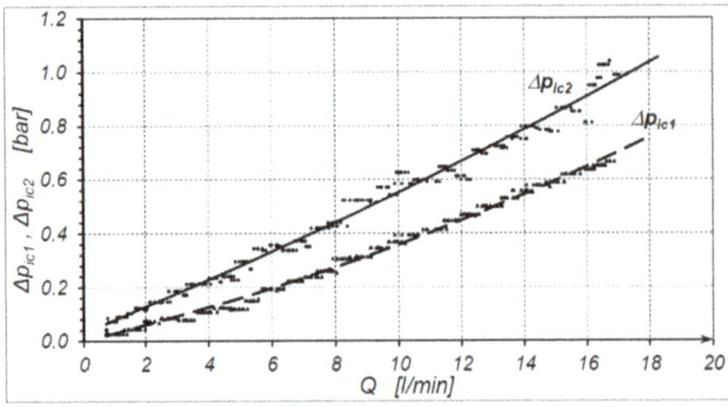

Figure 14. Characteristics of the pressure drop $\Delta p_{ic1} = f(Q)$ and $\Delta p_{ic2} = f(Q)$ in internal channels of the motor supplied with water (no cavitation) [23].

The results of the research on the motor supplied with water show (Figure 12) that in a low-pressure working chamber LPC (Figure 10), during cavitation, the minimum value of the pressure p_{LPC} was –0.85 bar.

7.3. Experimental Data—Motor Supplied with Oil

The recorded results of pressure p_1, p_2, p_{HPC}, p_{LPC} and flow rate Q of oil in the motor as a function of the shaft speed n are shown in Figure 15. Characteristics Δp_{ic1} and Δp_{ic2}, whose values were calculated according to Formulas (5) and (6) are shown in Figure 16. At rotational speeds above 550 rpm, cavitation occurs (the flow rate Q = const.). In Figure 17, characteristics Δp_{ic1} and Δp_{ic2} are shown at speeds up to 580 rpm—cavitation does not occur at these speeds.

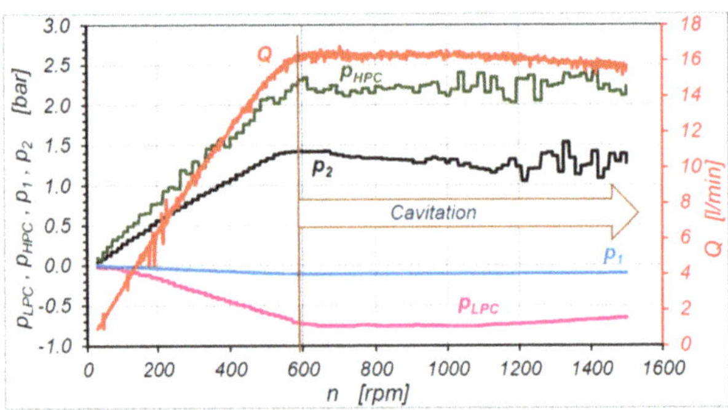

Figure 15. Characteristics of $p_1 = f(n)$, $p_2 = f(n)$, $p_{HPC} = f(n)$, $p_{LPC} = f(n)$ and $Q = f(n)$ over the entire range of the motor shaft speed n—motor supplied with oil [23].

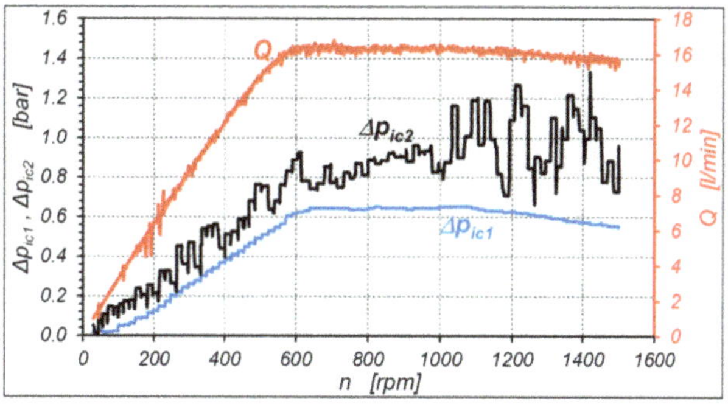

Figure 16. Characteristics of pressure drop $\Delta p_{ic1} = f(n)$ and $\Delta p_{ic2} = f(n)$ in the internal channels of the motor over the entire range of the motor shaft speed, n—motor supplied with oil [23].

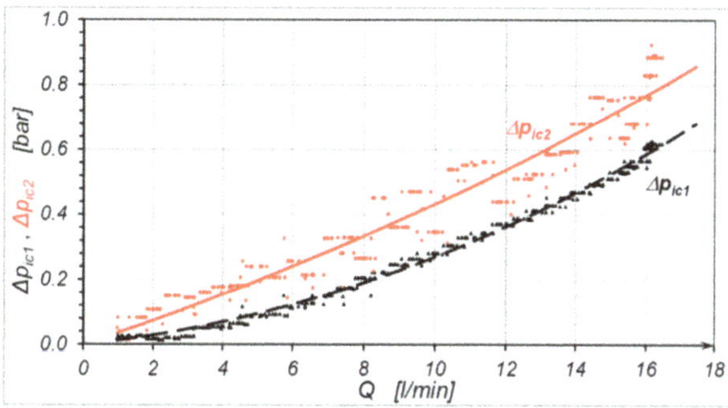

Figure 17. Characteristics of the pressure drop $\Delta p_{ic1} = f(Q)$ and $\Delta p_{ic2} = f(Q)$ in internal channel of the motor (no cavitation)—motor supplied with oil [23].

The results of the research on the motor supplied with oil shows (Figure 15) that in a low low-pressure working chamber *LPC* (Figure 10), during cavitation, the minimum value of the pressure p_{LPC} was –0.75 bar.

7.4. Compression of Liquid in Working Chamber

Experimental data (Figures 12 and 17) show that, in a motor with zero overlap in the commutation units, for one direction of the flow rate (from port *A* to port *B*), $\Delta p_{ic1} \neq \Delta p_{ic2}$. Therefore, there is a pressure increase Δp_{HPC} caused by the compression of liquid in a closed working chamber. Characteristics of $\Delta p_{HPC} = f(n)$ in the motor supplied with water and supplied with oil are shown in Figure 18. These characteristics are proposed to describe by the empirical formula in the form:

- For water:

$$\Delta p_{HPC} = 0.635{\cdot}10^{-4}{\cdot}\mathrm{n}^{1.3} \tag{16}$$

- For oil:

$$\Delta p_{HPC} = 1.1{\cdot}10^{-4}{\cdot}\mathrm{n}^{1.14} \tag{17}$$

Therefore, the assumption described by Formula (7) is not true.

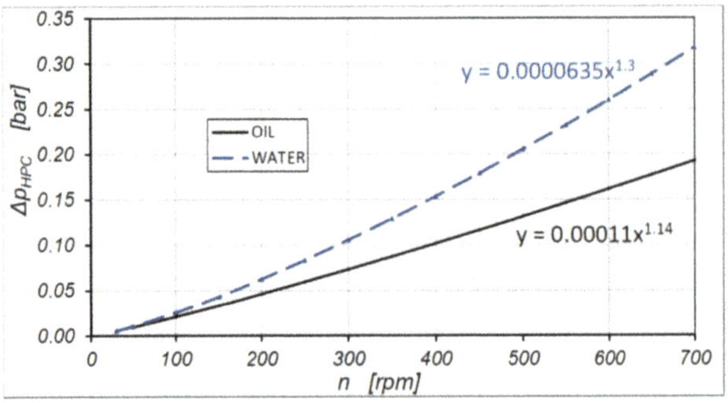

Figure 18. Characteristics of the pressure increase $\Delta p_{HPC} = f(n)$ calculated according to Formula (9)—motor supplied with oil and water.

7.5. Pressure Drop in Internal Channel

Characteristic of $\Delta p_{ich} = f(Q)$ in the motor supplied with water is shown in Figure 19, and in the motor supplied with oil is shown in Figure 20.

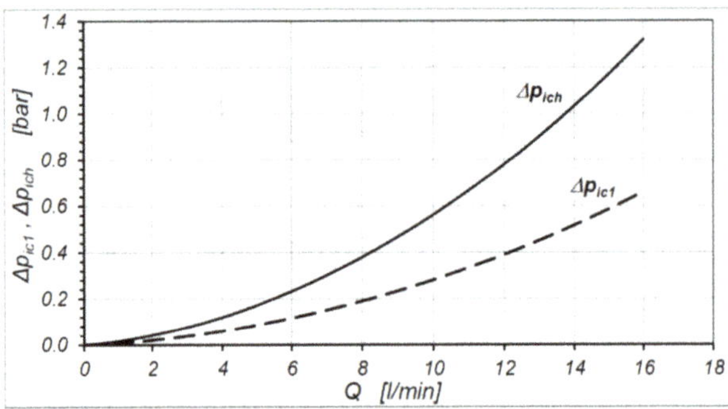

Figure 19. Characteristics of the total pressure drop $\Delta p_{ich} = f(Q)$ in the motor and the pressure drop $\Delta p_{ic1} = f(Q)$ in one internal channel of the motor—motor supplied with water [23].

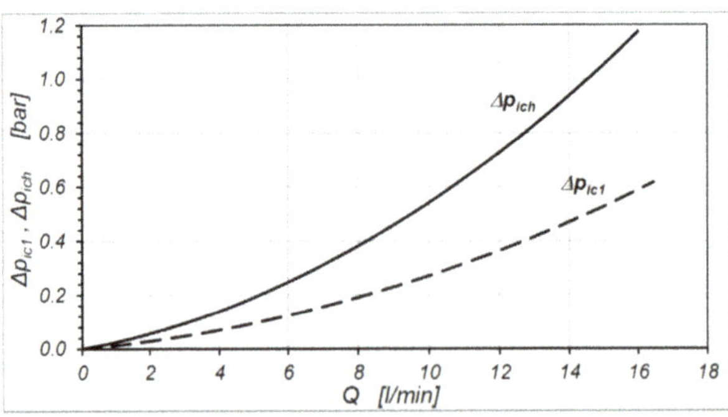

Figure 20. Characteristics of the total pressure drop $\Delta p_{ich} = f(Q)$ in the motor and the pressure drop $\Delta p_{ic1} = f(Q)$ in one internal channel of the motor—motor supplied with oil [23].

Characteristics of $\Delta p_{ic1} = f(Q)$, shown in Figures 19 and 20, can be described by Equation (13). The value of constants C_t and C_l are:

(a) For water: $C_t = 7.88 \times 10^8$ and $C_l = 409.8 \times 10^8$;

(b) For oil: $C_t = 6.73 \times 10^8$ and $C_l = 19.2 \times 10^8$.

Whereas the total pressure drop in the motor, according to Formulas (13) and (3), is:

(a) For water:

$$\Delta p_{ich} = 1576{\cdot}\rho{\cdot}Q^2 + 81960{\cdot}\nu{\cdot}\rho{\cdot}Q \text{ MPa} \tag{18}$$

(b) For oil:

$$\Delta p_{ich} = 1346{\cdot}\rho{\cdot}Q^2 + 3840{\cdot}\nu{\cdot}\rho{\cdot}Q \text{ MPa} \tag{19}$$

The ratio $\Delta p_{ich,W}/\Delta p_{ich,O}$ is shown in Figure 21 ($\Delta p_{ich,W}$ and $\Delta p_{ich,O}$—pressure drop in the internal channels of the motor supplied with water an oil, respectively).

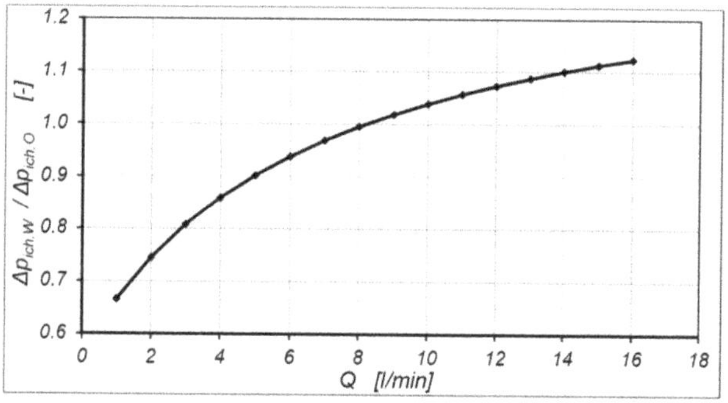

Figure 21. Characteristics of the ratio of pressure drops $\Delta p_{ich,W}/\Delta p_{ich,O} - f(Q)$—result of experiment.

In the range of a low flow rate Q (up to 8 L/min) a smaller pressure drop Δp_{ich} in the motor supplied with water was observed. However, in the range of a high flow rate, a smaller pressure drop was in the motor supplied with oil. This difference can be explained as follows. At a low flow rate, the share of the laminar flow component is larger than at a high flow rate. Because the viscosity of water is about 40 times lower than the viscosity of oil, a lower pressure drop in the motor supplied with water is at a low flow rate. However, at a high flow rate, the share of the laminar flow component is very small and the turbulent component is dominant. The biggest influence on the pressure drop during turbulent flow is the density of the liquid. Therefore, at a high flow rate, the pressure drop in the internal channels is smaller in the motor supplied with oil [29].

8. CFD Calculations

8.1. Method of Calculations

Numerical calculations of the liquid flow in the motor were carried out using the ANSYS Workbench program. The calculation module CFX and calculation model SST (Shear Stress Transport) were used. Some researchers, such as [33–38], applied the k-ε model to the calculation of the flow in hydraulic elements. However, the SST model contains features of the k-ε model and the k-ω model. Furthermore, the SST model ensures a faster convergence of results. For issues such as calculating the flow in the internal channels of a hydraulic motor, the SST model is sufficient.

8.2. The Geometry of Liquid in the Motor and Simplifying the Calculation Model

The geometry of liquid in a hydraulic motor is the geometry of the internal channels in the motor and the geometry of the working chambers. The following simplifications concerning models of the geometry of liquid in a satellite motor are used:

(a) The commutation plate (6) (Figures 9 and 11) is rotated relative to plate (7) by 30°. In this way, the *OC* holes (Figure 11) in the commutation plate (6) are opposite the *OC* holes in the commutation plate (7). Then, there is a free flow of liquid through the working mechanism and this mechanism does not rotate (is stationary);

(b) All of the calculations were carried out for one casual position of the working mechanism;

(c) Six working chambers were omitted—in these chambers the flow of liquid does not occur. Only four *OC* holes in the commutation plate allowed liquid flow by the working chambers (Figure 10);

(d) A leak in the gaps of the working mechanism was omitted;

(e) Tooth profiles in satellites, the curvature and the rotor were omitted. The satellites have been replaced by cylinders with diameters equal to the pith diameter of the satellite. In the rotor and curvature, the tooth profiles were replaced by a smooth profile according to the pitch line of tooth (Figure 22);

Figure 22. Simplified satellite working mechanism [28].

(f) An outline of the threads was omitted and replaced by cylinders with diameters equal to pith diameters of threads;

(g) Necks in the body and in the collectors were removed—there is no liquid flow in these elements and these elements have no significant influence on the pressure drop;

(h) The roughness of the walls was omitted.

The simplified calculation model of the satellite motor is shown in Figure 23.

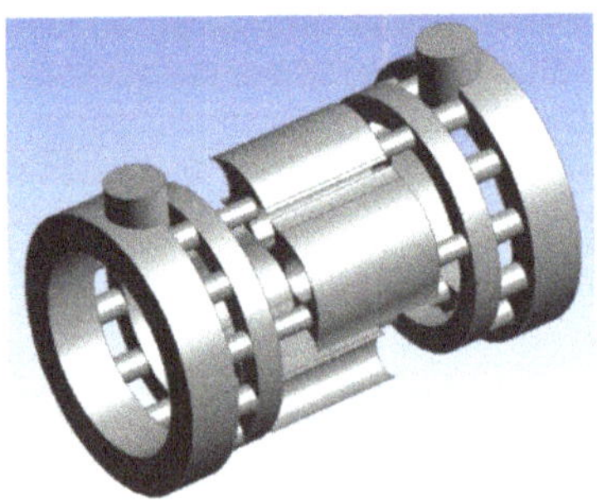

Figure 23. Simplified model of liquid volume in the motor [28].

8.3. The Mesh

In order to obtain reliable results from the simulation, it is necessary to create the best elements mesh. The uneven division into finite elements is a common issue. In areas where the wanted function changes rapidly, there is a denser mesh of elements. In contrast, in areas where the function changes slowly, there is a less dense mesh. The size of the mesh elements was selected by comparing the results of calculations with different sizes of finite elements. Finally, a mesh was chosen that gave an error of less than 5% in relation to the denser mesh.

The following discretization parameters were used to create the mesh:

- General size of finite elements is 0.6 mm;
- On the edges (like at changes in the diameter of the channel) more dense mesh is applied (the size of a single element is 0.25 mm).

Furthermore, a boundary layer was defined on all channel walls. The thickness of this layer depends on the dimensions of the channel. For channels with a smaller diameter, a layer with a smaller thickness was assigned. Therefore, in *OC* holes in the commutation plates (Figure 11) and in holes in low- and high-pressure manifolds (3) and (4) (Figure 9), ten boundary layers with a thickness of 0.2 mm were defined (increment coefficient 1.2). In the other channels, 18 boundary layers with a thickness of 0.3 mm were defined (increment coefficient also 1.2).

In this way, the mesh of elements was obtained (Figure 24). This mesh consisted of 3,992,282 finite elements and 1,367,996 nodes.

Figure 24. The mesh of elements of liquid in the internal channels of the motor [28].

8.4. Boundary Conditions

The boundary conditions are as follows:

- Defining the inflow (Figure 25a)—the wall that simulates inflow of the working liquid. As a kind of boundary condition, "Mass Inlet" was selected;
- Defining the outflow (Figure 25b)—the wall that simulates outflow of the working liquid. The value of the pressure on the outflow wall was assumed to be 0 bar (p = 0 bar);
- Defining the walls in contact with the liquid (Figure 25c)—the other walls of the model that define the shape and boundaries of the liquid. They were assigned a boundary condition in the form of "wall" with a slip limitation.

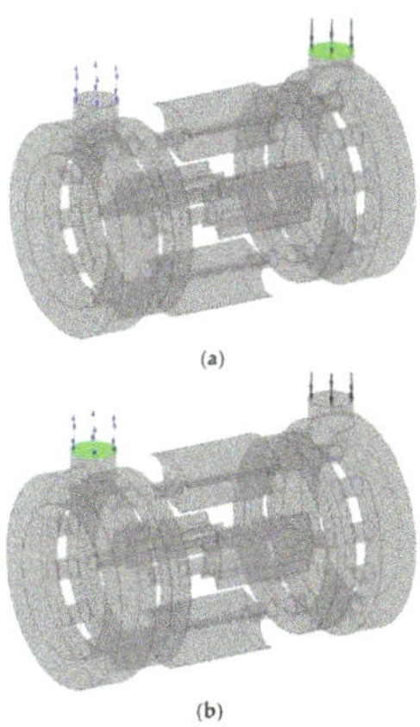

Figure 25. *Cont.*

(c)

Figure 25. Defined boundary conditions in a model of the liquid [28]: (**a**) inflow, (**b**) outflow, (**c**) wall.

In order to determine the characteristics of the pressure drop in the internal channels of the motor, the calculations were carried out in several steps for different values of the flow rate (Table 1). All calculations were conducted in one calculation series.

Table 1. Calculation steps and the corresponding flow rate.

No. of Step	1	2	3	4	5	6	7	8	9	10
[L/min]	2.0	4.0	6.0	8.0	10.0	12.0	14.0	16.0	18.0	20.0
[kg/s]	0.033	0.066	0.100	0.133	0.166	0.200	0.233	0.266	0.300	0.333

8.5. Parameters of the Liquids

Defining the parameters of water is limited to loading these parameters from the database offered by Ansys and determining the water temperature. Therefore, for water:

- The density ρ = 999.8 kg/m^3;
- The kinematic viscosity ν = 0.89 cSt;
- The temperature T = 25 °C.

Ansys does not have oil parameters in its database. Therefore, it is necessary to define the parameters for oil. In the calculations, it was assumed:

- The density ρ = 862 kg/m^3;
- The kinematic viscosity ν = 40 cSt;
- The temperature T = 43 °C.

It was assumed that water and oil are incompressible and their properties are constant.

8.6. The Conditions of the Simulations

The target value for the convergence of results has been defined—the residual value was set to 0.0001.

It was assumed that the simulation is stationery.

Before starting the calculations, it was necessary to define the number of processors that were used to perform computer calculations. In the described case, eight processors were used.

8.7. Results of the Calculations

In a later part of this article, an example of the velocity distribution and the pressure drop in the internal channels of the motor for a water and oil flow rate equal to 10 L/min are presented.

The distribution of water and oil velocity in the internal channels of the motor are shown in Figures 26–28, but the pressure distribution is shown in Figures 29 and 30. respectively.

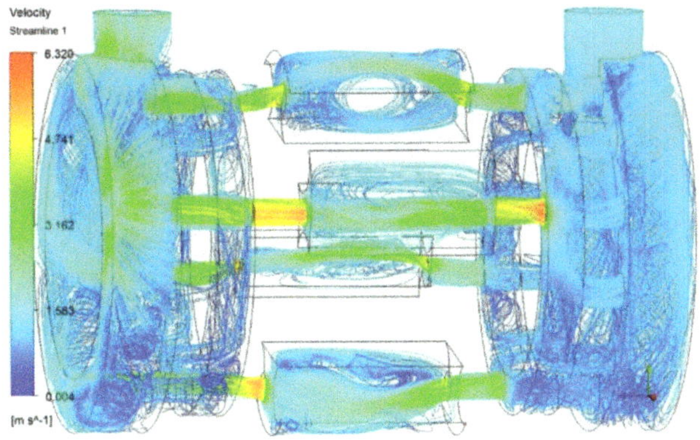

Figure 26. Distribution of water velocity in the internal channels of the motor (flow rate $Q = 10$ L/min) [28].

Figure 27. Distribution of oil velocity in the internal channels of the motor (flow rate $Q = 10$ L/min) [28].

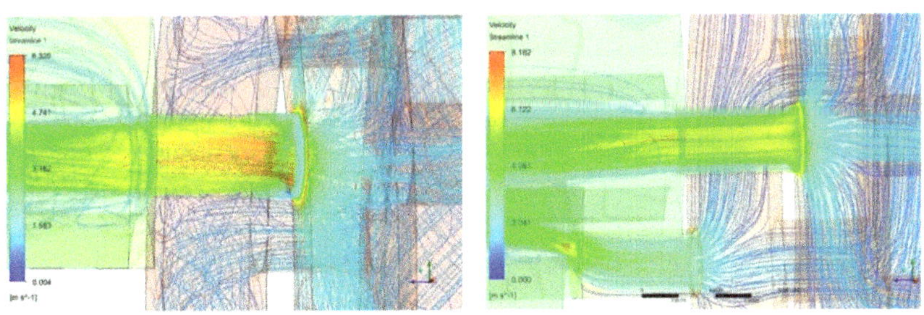

Figure 28. Distribution of water velocity (**left figure**) and oil velocity (**right figure**) in hole *OC* in commutation plate [28].

Figure 29. Distribution of water pressure in the internal channels of the motor (flow rate $Q = 10$ L/min) [28].

Figure 30. Distribution of oil pressure in the internal channels of the motor (flow rate $Q = 10$ L/min) [28].

The liquid flow area is the smallest in the holes of the commutation unit plates. Therefore, the highest pressure drop and the highest liquid flow velocities can be found in these holes (Figures 31 and 32).

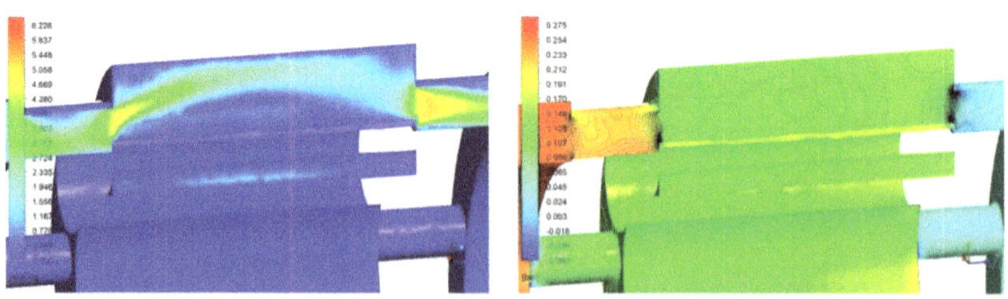

Figure 31. Distribution of water velocity (**left figure**) and distribution of water pressure (**right figure**) in the commutation holes *OC* and in the working chamber of the motor (flow rate $Q = 10$ L/min). The holes *OC* in the commutation plates are partially obscured by the satellite [28].

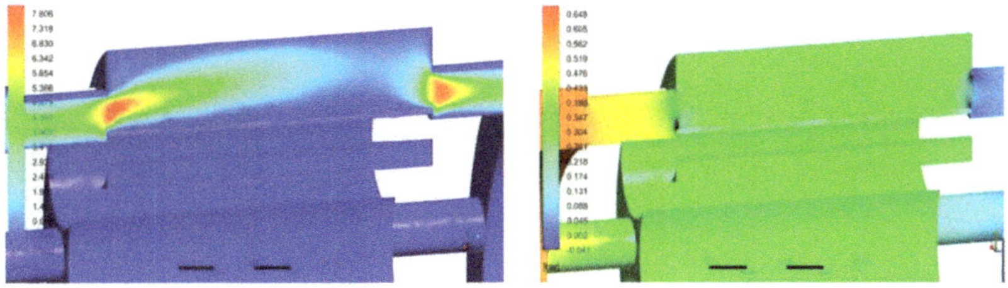

Figure 32. Distribution of oil velocity (**left figure**) and distribution of oil pressure (**right figure**) in the commutation holes *OC* and in the working chamber of the motor (flow rate $Q = 10$ L/min). The holes *OC* in the commutation plates are partially obscured by the satellite [28].

In order to determine the characteristics of the pressure drop as a function of the flow path in the internal channels of the motor, a line was defined as in Figure 33 (by the inflow and outflow ports, the inflow and outflow manifolds, the holes in the commutation unit plate and the working chamber). It should be noted that the holes in the commutation unit plate are completely open.

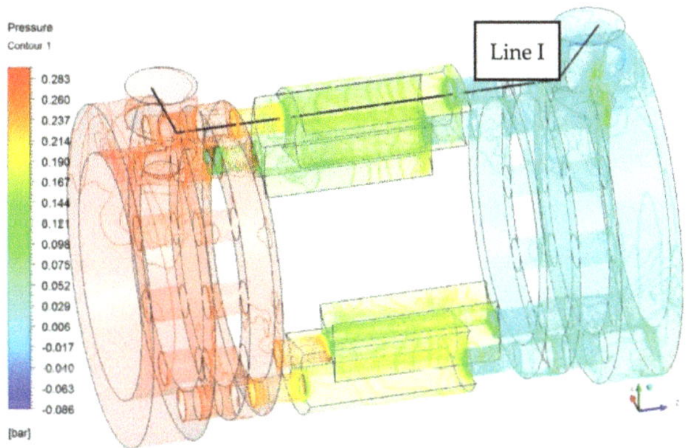

Figure 33. The course of "Line I" helping to determine the characteristic of the pressure drop in the internal channels of the motor [28].

Characteristics of the average liquid flow velocity and the average pressure in the internal channel of the motor are shown in Figures 34 and 35. These characteristics show that the largest pressure differences were between the hole of the commutation unit plate and the working chamber.

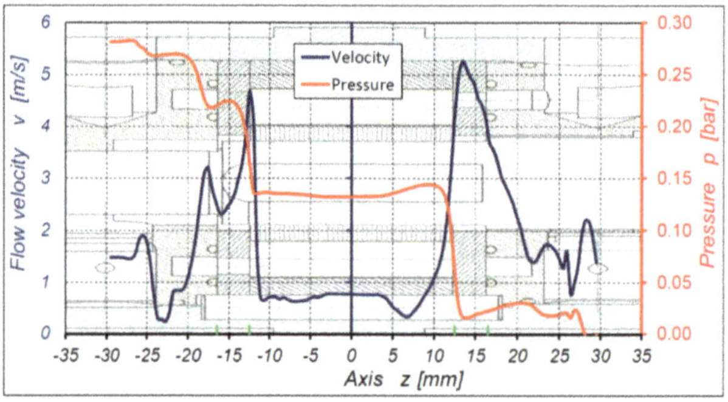

Figure 34. Characteristics of the average water flow velocity v and water pressure p in the internal channels of the motor as a function of the path "z" (the "z" path coincides with the axis of the motor shaft). Results of CFD calculation for the flow rate Q = 10 L/min. The *OC* holes in the commutation unit plates are completely open.

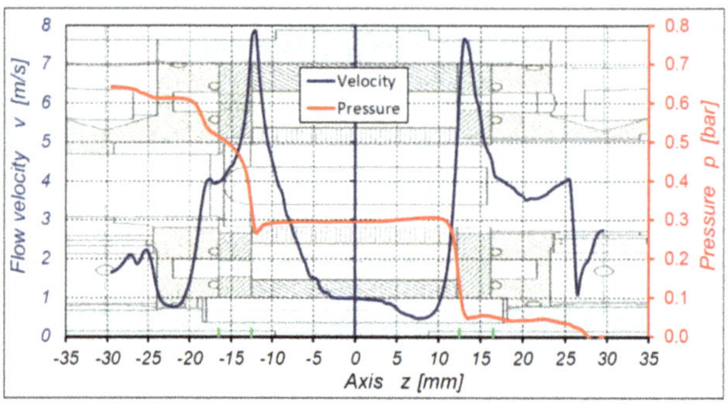

Figure 35. Characteristics of the average oil flow velocity v and oil pressure p in the internal channels of the motor as a function of the path "z" (the "z" path coincides with the axis of the motor shaft). Results of CFD calculation for the flow rate Q = 10 L/min. The *OC* holes in the commutation unit plates are completely open.

8.8. Pressure Drop in the Internal Channel of the Motor

It was assumed that the *OC* holes in commutation plate 6 (according to Figure 9) are opposite the *OC* holes in commutation plate 7. This caused some of the holes in commutation plates to be obscured by the satellite (each hole to a different degree). Thus, the fluid flow fields in holes of the commutation plates were different. As a result, the flow velocities and pressure drop in these holes were also different. Therefore, the pressure in the working chambers were also different. Thus, the statement that the pressure drop in the internal channel should be understood as the pressure difference in the motor ports and in any given working chamber gives an incorrect result. For confirmation, the pressure drop $\Delta p_{ic1\text{-}II}$ in internal low-pressure channels, the pressure drop $\Delta p_{ic2\text{-}II}$ in internal high-pressure channels and the total pressure drop $\Delta p_{ich\text{-}II}$ along line 1 (from Figure 33) were calculated (Figures 36 and 37).

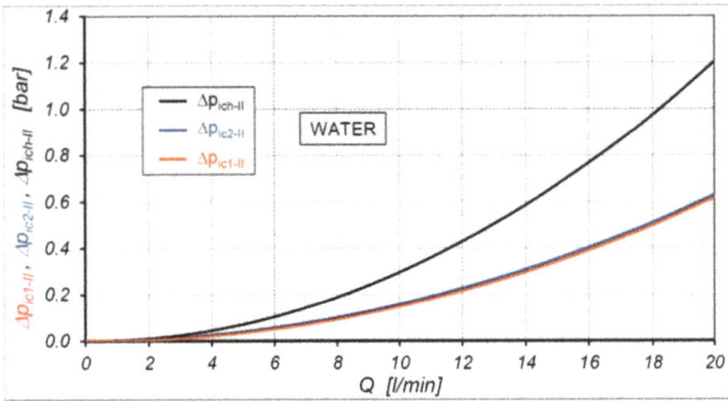

Figure 36. Characteristics of the pressure drop $\Delta p_{ic1\text{-}!!} = f(Q)$. $\Delta p_{ic2\text{-}!!} = f(Q)$ and $\Delta p_{ich\text{-}!!} = f(Q)$ in the internal channels of the motor supplied with water—results of CFD calculations.

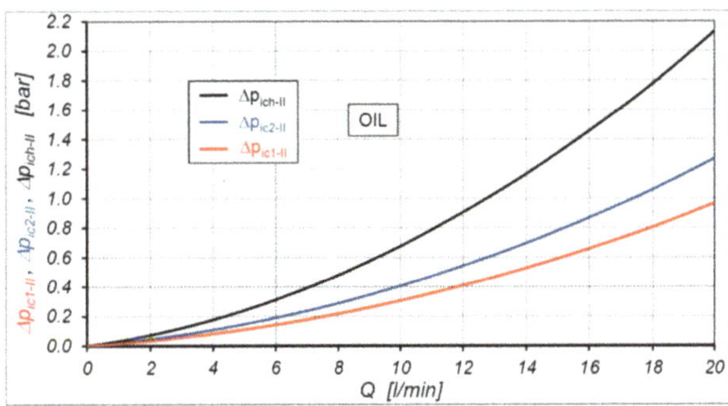

Figure 37. Characteristics of the pressure drop $\Delta p_{ic1\text{-}!!} = f(Q)$, $\Delta p_{ic2\text{-}!!} = f(Q)$ and $\Delta p_{ich\text{-}!!} = f(Q)$ in the internal channels of the motor supplied with oil—results of CFD calculations.

The ratio of the pressure drops $\Delta p_{ich,W}/\Delta p_{ich,O}$ in the internal channels of the motor supplied with water and supplied with oil is shown in Figure 38.

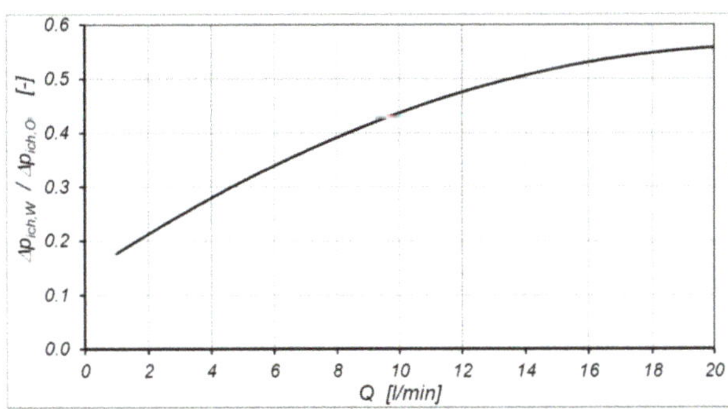

Figure 38. Characteristic of $\Delta p_{ich,W}/\Delta p_{ich,O} = f(Q)$. $\Delta p_{ich,W}$ and $\Delta p_{ich,O}$—pressure drop in the internal channels of the motor supplied with water and supplied with oil. Result of simulation.

In contrast to the results of experiment, the results of the CFD calculations showed that in all ranges of flow rate the pressure drop in the internal channels of the motor supplied with water is smaller than that supplied with oil. This is a significant difference in results of simulation compared to the results of the experiment. Most likely, this difference is the result of the adopted simplifications in the model—mainly by stopping the satellite mechanism and setting the holes in commutation plates opposite each other.

9. Conclusions

The results of the experimental research showed that:

(a) The type of liquid and the parameters of the liquid (kinematic viscosity and density) have a big influence on the pressure drop in the internal channels of the hydraulic motor;
(b) The flow in the internal channels of the motor is not a fully developed turbulent flow. This applies to both water and mineral oil;
(c) Low flow rate in the motor (up to 8 L/min) for water is a lower pressure drop than for oil, but regarding high flow rate oil is a lower pressure drop;
(d) The value of the pressure p_{LPC} in the suction chamber (low-pressure chamber *(LPC)*) (Figure 10) of the satellite motor depends on the type of liquid (in the motor supplied with oil the pressure p_{LPC} is -0.75 bar, but in the motor supplied with water the p_{LPC} is -0.85 bar);
(e) It is possible to test the pressure drop in internal channels of a hydraulic motor according to the proposed method without measuring the pressure in the working chambers. Then, the presented above values of pressure p_{LPC} should be taken instead of value $p_{LPC} = -1$ bar;
(f) The pressure drop in both the internal low-pressure channels (inflow channels) and in the internal high-pressure channels (outflow channels) in the hydraulic motor can be described with Formula (13);
(g) If the volume of the working chamber tends to the minimum, then in this chamber additional increase of pressure Δp_{HPC} occurs. This pressure depends on the overlap in the commutation unit and also depends on the rotational speed of the motor. Therefore, the Δp_{HPC} is not the component of pressure losses in the internal channels of the motor. The Δp_{HPC} has influence only on mechanical losses in the working mechanism of the motor.

The results of the CFD calculation have shown that it is possible to carry out the numerical calculation of flow in the internal channel of the motor but only with substantial assumptions and simplifications. This simplification does not fully reflect the phenomena occurring in the motor.

The CFD calculation showed that the biggest pressure drop is in the holes of the commutation unit plates, regardless of the type of liquid.

The values of the pressure drop in the internal channels of the motor obtained from the experiment (Figures 19 and 20) differ from the results obtained by CFD calculations (Figures 36 and 37). The differences are shown in Figure 39.

The main reasons for the differences are:

(a) The simplifying of the calculation model by stopping the satellite mechanism and setting the holes in commutation plates opposite each other;
(b) The impossibility of simulating the pressure difference in the working chambers for a rotating satellite mechanism;
(c) Difficultly of researching dynamic phenomena that occur in the working chambers during the operation of the satellite mechanism.

As the final conclusions, it should be stated that:

(a) The CFD calculation can only estimate the pressure drop in the internal channels of the motor. The underestimation can be up to 50%;
(b) The CFD method has some advantages over the experimental method. Namely, it allows the velocity and pressure distribution in the internal channels with an irregular, complicated shape to be determined.

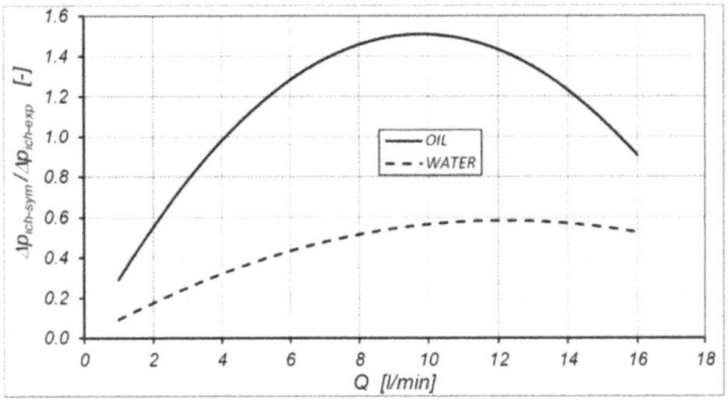

Figure 39. Characteristics of $\Delta p_{ich\text{-}sym}/\Delta p_{ich\text{-}exp} = f(Q)$. $\Delta p_{ich\text{-}sym}$—pressure drop in the internal channels calculated numerically (CFD); $\Delta p_{ich\text{-}exp}$—pressure drop in the internal channels obtained experimentally. Motor supplied with water and with oil.

In the future, it should be attempted to simulate flow in a satellite motor without major simplification, i.e., with a working (rotating) satellite mechanism.

Author Contributions: Conceptualization, P.S.; methodology, P.S.; software, P.P.; validation, P.P. and P.S.; formal analysis, P.S.; investigation, P.S.; resources, P.S. and P.S.; data curation, P.S.; writing—original draft preparation, P.S.; writing—review and editing, P.S.; visualization, P.P.; supervision, P.S.; project administration, P.S.; funding acquisition, P.S. All authors have read and agreed to the published version of the manuscript.

Funding: This research was funded by the National Centre for Research and Development in Poland, grant number LIDER/35/102/L-2/10/NCBiR/2011.

Conflicts of Interest: The authors declare no conflict of interest.

References

1. Guzowski, A.; Sobczyk, A. Reconstruction of hydrostatic drive and control system dedicated for small mobile platform. In *Fluid Power Systems Technology*; American Society of Mechanical Engineers: New York, NY, USA, 2014. [CrossRef]
2. Pobedza, J.; Sobczyk, A. Properties of high-pressure water hydraulic components with modern coatings. In *Advanced Materials Research*; Trans Tech Publications Ltd.: Stafa-Zurich, Switzerland, 2014; Volume 849. [CrossRef]
3. Litwin, W.; Olszewski, A. Water-lubricated sintered bronze journal bearings—Theoretical and experimental research. *Tribol. Trans.* **2014**, *57*, 114–122. [CrossRef]
4. Walczak, P.; Sobczyk, A. Simulation of water hydraulic control system of Francis turbine. In *Fluid Power Systems Technology*; American Society of Mechanical Engineers: New York, NY, USA, 2014. [CrossRef]
5. Dymarski, C.; Dymarski, P. Developing methodology for model tests of floating platforms in low-depth towing tank. *Arch. Civ. Mech. Eng.* **2016**, *16*, 159–167. [CrossRef]
6. Lubinski, J.; Sliwinski, P. Multi parameter sliding test result evaluation for the selection of material pair for wear resistant components of a hydraulic motor dedicated for use with environmentally friendly working fluids. In *Solid State Phenomena*; Trans Tech Publications Ltd.: Stafa-Zurich, Switzerland, 2015; Volume 225. [CrossRef]
7. Urbanczyk, J. Research of Hydraulic Motors for Needs of Small Mechanization in the Mining Industry. Ph.D. Thesis, AGH University of Science and Technology, Cracow, Poland, 1999.
8. Sliwinski, P. The methodology of design of axial clearances compensation unit in hydraulic satellite displacement machine and their experimental verification. *Arch. Civ. Mech. Eng.* **2019**, *19*, 1163–1182. [CrossRef]
9. Maczyszyn, A. Energy Analysis of Rotary Positive Displacement Machines Used in Hydrostatic Transmissions. Ph.D. Thesis, Gdansk University of Technology, Gdansk, Poland, 2014.
10. Osinski, P.; Deptula, A.; Partyka, M. Discrete optimization of a gear pump after tooth root undercutting by means of multi-valued logic trees. *Arch. Civ. Mech. Eng.* **2013**, *13*, 422–431. [CrossRef]
11. Stosiak, M.; Zawislak, M.; Nishta, B. Studies of resistances of natural liquid flow in helical and curved pipes. *Pol. Marit. Res.* **2018**, *3*, 123–130. [CrossRef]
12. Osinski, P.; Warzynska, U.; Kollek, W. The influence of gear micropump body asymmetry on stress distribution. *Pol. Marit. Res.* **2017**, *24*, 60–65. [CrossRef]
13. Jasinski, R. Problems of the starting and operating of hydraulic components and systems in low ambient temperature (Part III). *Pol. Marit. Res.* **2009**, *4*, 22–31. [CrossRef]

14. Jasinski, R. Problems of the starting and operating of hydraulic components and systems in low ambient temperature (Part IV). *Pol. Marit. Res.* **2017**, *3*, 45–57. [CrossRef]
15. Paszota, Z. *Energy Losses in Hydrostatic Drive*; LABERT Academic Publishing: Saarbrücken, Germany, 2016.
16. Saheban Alahadi, M.J.; Shirneshan, A.; Kolahdoozan, M. Experimental investigation of the effect of grooves cut over the piston surface on the volumetric efficiency of a radial hydraulic piston pump. *Int. J. Fluid Power* **2017**, *18*, 181–187. [CrossRef]
17. Balawender, A. Energy analysis and methodics of testing of low-speed hydraulic motors. In *Scientific Book of the Gdansk University of Technology*; Mechanika, No. 54; Gdansk University of Technology Publishing House: Gdansk, Poland, 1988.
18. Zaluski, P. Influence of the Position of the Swash Plate Rotation Axis on the Volumetric Efficiency of Axial Piston Pumps. Ph.D. Thesis, Gdansk University of Technology, Gdansk, Poland, 2017.
19. Zaluski, P. *Experimental Research of an Axial Piston Pump with Displaced Swash Plate Axis of Rotation*; Lecture Notes in Mechanical Engineereing. Advances in Hydraulic and Pneumatic Drives and Control 2020; Springer Nature Switzerland AG: Cham, Switzerland, 2021. [CrossRef]
20. Sliwinski, P. The influence of water and mineral oil on volumetric losses in hydraulic motor. *Pol. Marit. Res.* **2017**, *S1*, 213–223. [CrossRef]
21. Sliwinski, P. The flow of liquid in flat gaps of satellite motors working mechanism. *Pol. Marit. Res.* **2014**, *2*, 125–135. [CrossRef]
22. Sliwinski, P. The influence of water and mineral oil on the leaks satellite motor commutation unit clearances. *Pol. Marit. Res.* **2017**, *3*, 58–67. [CrossRef]
23. Sliwinski, P. *Satellite Displacement Machines. Basic of Design and Analysis of Power Loss*; Gdansk University of Technology Publishers: Gdansk, Poland, 2016.
24. Patrosz, P. Influence of properties of hydraulic fluid on pressure peaks in axial piston pumps' chambers. *Energies* **2021**, *14*, 3764. [CrossRef]
25. Patrosz, P. *Influence of Gaps' Geometry Change on Leakage Flow in Axial Piston Pumps*; Lecture Notes in Mechanical Engineereing. Advances in Hydraulic and Pneumatic Drives and Control 2020; Springer Nature Switzerland AG: Cham, Switzerland, 2021. [CrossRef]
26. Skorek, G. Laboratory tests of pressure losses in a displacement pump. *Hydraulika i Pneumatyka* **2005**, *2*, 12–15.
27. Balawender, A. Physical and mathematical model of losses in hydraulic motors. In *Developments in Mechanical Engineering*; Gdansk University of Technology Publishers: Gdansk, Poland, 2005.
28. Maciejewski, M. Komputerowe Obliczenia Spadków Ciśnienia w Kanałach Wewnętrznych Silnika Satelitowego. Master's Thesis, Faculty of Mechanical Engineering, Gdansk University of Technology, Gdansk, Poland, 2013.
29. Sliwinski, P.; Patrosz, P. The influence of water and mineral oil on pressure losses in hydraulic motor. In *Lecture Notes in Mechanical Engineering*; Cavas-Martínez, F., Chaari, F., Gherardini, F., Haddar, M., Ivanov, V., Kwon, Y.W., Trojanowska, J., di Mare, F., Eds.; Springer International Publishing: Geneva, Switzerland, 2020. [CrossRef]
30. Sliwinski, P. Determination of the theoretical and actual working volume of a hydraulic motor. *Energies* **2020**, *13*, 5933. [CrossRef]
31. Sliwinski, P. Determination of the theoretical and actual working volume of a hydraulic motor—Part II (The method based on the characteristics of effective absorbency of the motor). *Energies* **2021**, *14*, 1648. [CrossRef]
32. Sliwinski, P. Pressure Losses and Power Balance in the Unloaded Satellite Pump. *Hydraul. Pneum* **2013**, *1–2*, 1–5.
33. Amirante, R.; Moscatelli, P.; Catalano, L. Evaluation of the flow forces on a direct (single stage) proportional valve by means of a computational fluid dynamic analysis. *Energy Convers. Manag.* **2007**, *48*, 942–953. [CrossRef]
34. Chattopadhyay, H.; Kundu, A.; Saha, B.; Gangopadhyay, T. Analysis of flow structure inside a spool type pressure regulating valve. *Energy Convers. Manag.* **2012**, *53*, 196–204. [CrossRef]
35. Domagala, M. CFD analysis of a flow control valve. In Proceedings of the 5th FPNI PhD Symposium, Cracow, Poland, 1–5 July 2008.
36. Lisowski, E.; Czyzycki, W.; Rajda, J. Three dimensional CFD analysis and experimental test of flow force acting on the spool of solenoid operated directional control valve. *Energy Convers. Manag.* **2013**, *70*, 220–229. [CrossRef]
37. Lisowski, E.; Filo, G.; Rajda, J. Pressure compensation using flow forces in a multi-section proportional directional control valve. *Energy Convers. Manag.* **2015**, *103*, 1052–1064. [CrossRef]
38. Lisowski, E.; Rajda, J. CFD analysis of pressure loss during flow by hydraulic directional control valve constructed from logic valves. *Energy Convers. Manag.* **2013**, *65*, 285–291. [CrossRef]

energies

MDPI

Article

The Effect of Aggressive Driving on Vehicle Parameters

Emilia M. Szumska * and Rafał Jurecki

Department of Automotive Engineering and Transport, Faculty of Mechatronics and Mechanical Engineering, Kielce University of Technology, 25-314 Kielce, Poland; r.jurecki@tu.kielce.pl

* Correspondence: eszumska@tu.kielce.pl; Tel.: +48-41-342-4718

Received: 2 November 2020; Accepted: 14 December 2020; Published: 17 December 2020

Abstract: Driver behavior is one of the most relevant factors affecting road safety. Many traffic situations require a driver to be able to recognize possible danger. In numerous works, aggressive driving is understood as unsafe and as a hazard entailing the risk of potential crashes. However, traffic safety is not the only thing affected by a vehicle operator's driving style. A driver's behavior also impacts the operating costs of a vehicle and the emission of environmental air pollutants. This is confirmed by numerous works devoted to the examination of the effect of driving style on fuel economy and air pollution. The objective of this study was to investigate the influence of aggressive driving on fuel consumption and emission of air pollutants. The simulation was carried out based on real velocity profiles collected in real-world tests under urban and motorway driving conditions. The results of simulations confirm that an aggressive driving style causes a significant increase in both fuel consumption and emission of air pollutants. This is particularly apparent in urban test cycles, where an aggressive driving style results in higher average fuel consumption and in pollutant emissions as much as 30% to 40% above the average compared to calm driving.

Keywords: aggressive driving; hybrid electric vehicle; driver behavior

1. Introduction

A driver's experience, skills, powers of observation, and ability to recognize potential hazards allow a driver to avoid possible traffic incidents. Driver behavior has been a focus of investigations since the late 1940s; the earliest research described the relationship between a driver's personality and road safety [1]. Driving style, according to [2], can be characterized by three aspects:

1. The individual manner of driving, differing among individuals;
2. A regular way of driving, reflecting regular behavior while driving;
3. A reflection of conscious choices made by a driver.

Many authors differentiate drivers based on driving style. Drivers are classified based on vehicle kinematic parameters, including acceleration, speed, lane crossing, steering angle, and positions of the accelerator and brake pedals [3–6]. Methods based on visual features are also used to differentiate driving styles. Some researchers monitored and analyzed the driver's head position, facial expression, ocular state (e.g., how long the driver's eyes are open or closed, and their eye blink rate), and mouth using cameras or smartphones [7–9]. Previous studies have also been devoted to the physical or mental factors that impact a driver's style and cause unusual driving behavior. Examples of these studies can be found in [10–12], among others.

Most classifications of driver behavior distinguish the aggressive driving style among other types. Aggressive driving refers to speeding, rapid acceleration, late braking, frequent lane changes, tailgating, and driving through red lights [13,14]. Such factors influence road safety. Many works have shown the relationship between aggressive driving and the risk of traffic hazards [15,16]. Aggressive driving also influences vehicle parameters such as fuel and energy consumption, and air-pollutant emissions. The results presented in studies [17–19] confirmed the increase in fuel consumption and air-pollutant emissions during aggressive driving.

In one study [17], the authors distinguished four types of drivers: aggressive, conservative, professional, and beginner. The driver classification was based on 23 personal factors, including gender, age, experience, and acceleration and deceleration values at low and high speeds during real-world tests in urban conditions. The impact of driver behavior on fuel consumption and air-pollutant emissions was then estimated. A similar method was presented [18], in which city bus drivers were tested in urban, suburban, and motorway driving conditions. The driver type was classified using two methods, the first based on speed profile characteristics such as vehicle speed, acceleration, standard deviation of speed, and longitudinal jerk; and the second based on characteristics of vehicle operation such as number, average, and standard deviations of accelerator pedal depression, steering wheel angle, and time of gear shift. The second part of the study showed the impact of driving style on fuel consumption, using a model-based analysis method and a statistical analysis method, based on real driving data.

Research [19] reported the extended car-following model with respect to the driver's bounded rationality. The test was conducted for two different traffic situations. The results of numerical tests were used to explore the impact of the driver's bounded rationality on air-pollutant emissions and fuel consumption. The driver's bounded rationality has positive effects on driving behavior, and reduces total fuel consumption and CO, HC, and NO_x emissions.

According to another study [20], fuel consumption can vary from 78.5% to 137.3% for gasoline vehicles, and from 116.3% to 128.3% for diesel vehicles, comparing aggressive driving to calm driving. Results [21] demonstrated that aggressive driving influences fuel consumption regardless of the road grade. During aggressive driving in urban areas, the air-pollutant emissions were as much as 40% higher than during calm driving [22,23].

Hybrid electric vehicles (HEVs) are perceived as an alternative to conventional-drive vehicles. A hybrid powertrain system combines conventional-drive components with electric-drive elements [24,25]. The electric drive alone may be engaged when idling or driving at low speed. This solution can significantly lower air-pollutant emissions and fuel consumption, especially in urban areas, where traffic congestion is common. Hybrid powertrains can employ regenerative braking, in which kinetic energy generated during braking is converted into electrical energy that is stored in the battery pack until needed. In this process, the electric traction motor works as a generator to recover energy that would otherwise be lost to the brake discs in the form of heat. Hybrid-drive vehicles provide a driving range comparable to that of conventional-drive vehicles, but without the additional infrastructure typical of conventional-drive vehicles. Numerous studies have shown that the use of hybrid electric vehicles reduces fuel consumption and emissions. Supporting examples can be found in [26–28], among others.

The literature features many definitions and methods to describe an aggressive driving style. In the observation method, driver characteristics are developed based on the observation of a driver's behavior by researchers, who determine what behavior they consider is aggressive. For example, in one study [29], the behaviors used to distinguish an aggressive driver were short honking, cutting in front of another vehicle while passing, and passing one or more vehicles by driving on the shoulder and then cutting in front of the vehicles. The results demonstrated a relationship among aggressive driving, congestion, and different time periods of driving. Increases in congestion contribute to more aggressive driving, while aggressive driving diminishes during the weekend and non-rush hours.

Research on aggressive driving behavior was also described [30], presenting an observational study of aggressive driving during certain traffic events. The tests were conducted using a driving simulator. Another study [31] presented the impact of aggressive and nonaggressive driving behavior on crash injuries in traffic accidents. The model indicated that many key indicators characteristic of aggressive drivers correlated with a wide variety of factors related to the crash, including severity of driver injury, vehicle type, driver profile, spatial and temporal characteristics, roadway attributes, and traffic volume.

Another method to identify an aggressive driver is through a questionnaire designed to identify the respondent's aggressive road behavior, including verbal and physical aggression, and violations of highway codes [32,33]. According to [34], violations of traffic regulations (speeding, not stopping at red lights, driving while under the influence of alcohol) are perceived as more aggressive than racing, voicing insults, cutting off other drivers, flashing headlights, shouting, tailgating, and making rude gestures.

Another method of identifying a driving style is based on vehicle-motion parameters. Motion sensors, such as accelerometers and gyroscopes, are used to record a vehicle's motion parameters, allowing for the collection of data, including the vehicle's speed; its lateral, longitudinal, and vertical accelerations; and its instantaneous positions [35]. In many studies, a driving style is determined by the speed profile and the lateral and longitudinal acceleration values. For example, one study [36] applied an online approach to monitoring a vehicle's running state and identifying aggressive driving under normal driving conditions using kinematic parameters collected by a recorder mounted in the vehicle. A value of longitudinal acceleration above or below 3.5 m/s^2 was adopted as the indicator of aggressive driving. Another study [37] presented a method for discovering unsafe driving behavior using acceleration analysis. The definition of the proper range of kinematic parameters was based on recorded readings of speeds and acceleration rates. Speed and acceleration values not within the range of the aforesaid parameters were considered as unsafe driving behavior.

According to [38], aggressive drivers can be identified based on their vehicle's longitudinal jerk (the change rate of acceleration with respect to time). The authors reported that jerk analysis seems to perform well in identifying aggressive drivers. Issues related to defining aggressive driver behavior based on vehicle motion parameters have also been described [39,40].

Compared to the previously presented studies, in which the research was based on multiple measurement runs for many drivers, in our study, one driver was involved. The driver navigated the same test route twice, the first time driving calmly and the second time driving aggressively. The method of driving was not imposed—the driver chose what they considered to be either calm or aggressive driving. To assess how the driver would subjectively change their driving behavior, the driver was not limited by any quantitative parameters. The method of driving, and its possible change, resulted only from the willingness of the driver and the traffic situations.

In this study, we aimed to examine the impact of aggressive driving on fuel consumption and air-pollutant emissions. The analysis was conducted based on velocity profiles collected in real-world tests in urban and motorway driving conditions. Collected speed profiles were used in simulation tests. The presented tests are preliminary tests to determine the types of driver behavior based on vehicle motion parameters collected in real-world tests.

2. Materials and Methods

2.1. Data Collection

The test vehicle was an Audi A6 passenger vehicle (A6, Audi, Ingolstadt, Germany) (Figure 1a). During the experiment, the mass of the car was 1920 kg; according to the manufacturer, the gross vehicle weight is 2475 kg. The vehicle was powered by a 92 kW diesel engine with a capacity of 2967 cm^3.

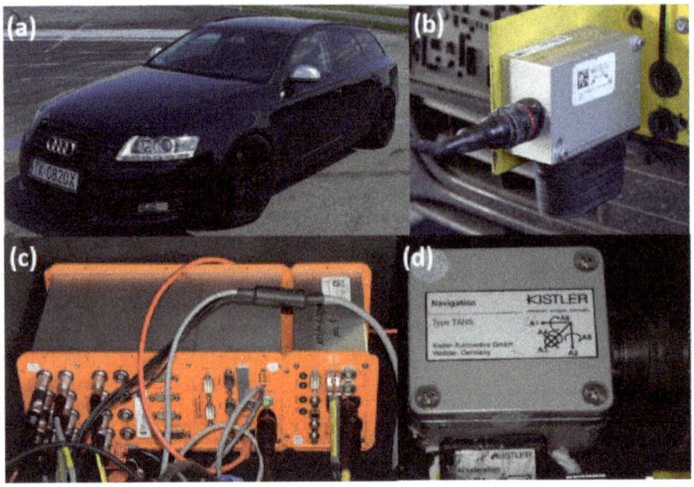

Figure 1. Measurement equipment used in real-world conditions test: (**a**) test vehicle, (**b**) optoelectronic sensor, (**c**) data acquisition system, and (**d**) three-axis linear acceleration sensor.

The vehicle movement parameters were collected using measurement equipment consisting of:

- An S-350 Aqua Datron® optoelectronic sensor for measuring longitudinal speed (Figure 1b);
- A uEEP-12 Datron® Data Acquisition Station (Figure 1c) with ARMS® data acquisition and analysis software; and
- A three-axis linear acceleration sensor (TAA Datron® and Navigation Sensor Modules), combining a solid-state, three-axis gyro with a three-axis linear accelerometer TANS Datron® for measuring longitudinal and lateral accelerations (Figure 1d).

The equipment allowed for the collection of the following comprehensive vehicle-movement data: drive time, distance traveled, instantaneous speed, instantaneous acceleration, and instantaneous localization. Movement parameters were recorded during test drives on a motorway and in urban traffic in Kraków, Poland.

As mentioned above, the driver was given no requirements or instructions on how to drive. What might be either calm or aggressive driving was left to the discretion of the driver. The driving tactic they adopted resulted from their individual experience and subjective assessment. The recorded velocity and acceleration profiles in both runs were then used in computer simulations. The impact of driving style on fuel consumption and air-pollutant emissions were assessed based on the simulation results.

2.2. Test Cycles

The test cycles, based on the real-world vehicle movement parameters, were conducted during test drives on a working day under two driving conditions: motorway and urban area. Driver behavior was also considered. The tests were conducted for aggressive and calm driving styles. The speed profiles recorded during real driving tests on a motorway and in urban driving conditions are shown in Figure 2.

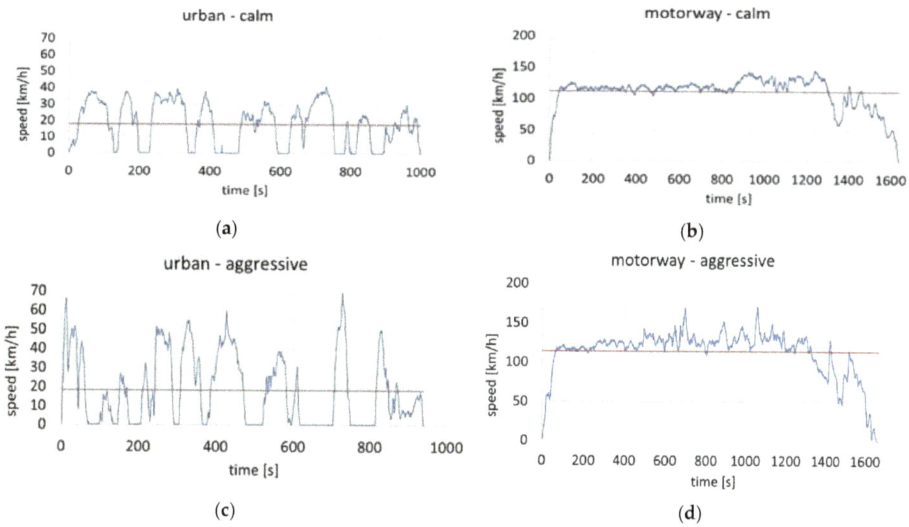

Figure 2. Velocity profiles of test cycles on a motorway and in urban driving condition (red line marks the average speed). (**a**) urban calm driving; (**b**) motorway calm driving; (**c**) urban aggressive driving (**d**) motorway aggressive driving.

A calm driving style is understood as unhurried and patient driving behavior, moderate acceleration, anticipatory braking, and obedience of speed limits and traffic signs. An aggressive driving style is characterized by abrupt movements of the pedals and steering wheel, quick gear changes and acceleration, braking at the last possible moment, and frequent lane changes. Figure 3 shows the percentage distribution of velocity in examined test cycles.

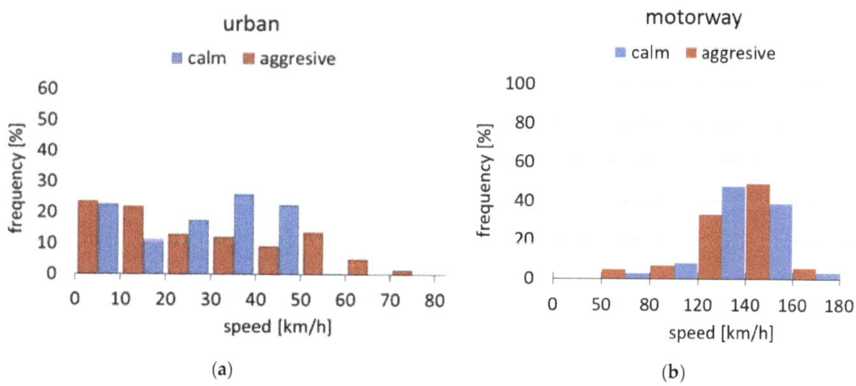

Figure 3. Percentage distribution of velocity during the test cycles. (**a**) urban driving; (**b**) motorway driving.

As shown in Figure 3, during the motorway test, cycles reflecting aggressive driving at speeds in excess of the speed limit constituted more than 40% of the total driving. In the urban test cycle, 23% of the total driving was in excess of the speed limit. The longitudinal acceleration profiles of the test cycles are presented in Figure 4.

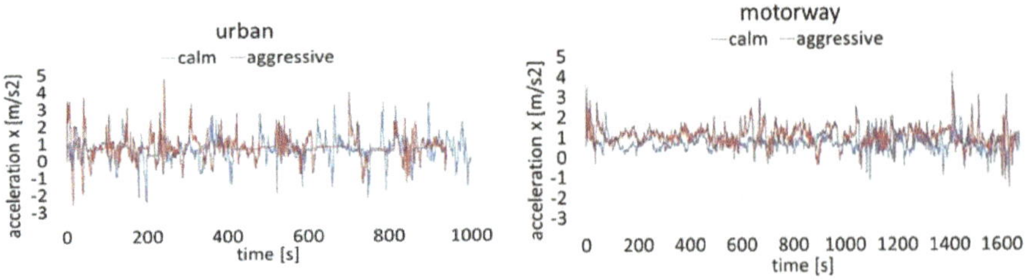

(a) (b)

Figure 4. Longitudinal acceleration profiles of the test cycles. (**a**) urban driving; (**b**) motorway driving.

The average speed and acceleration values obtained for motorway driving conditions were similar, regardless of whether the driving behavior was calm or aggressive. The aggressive driving style on the motorway included speeding, frequent lane changes, abrupt braking, and rapid acceleration. Figure 3 shows the percentage distribution of acceleration during the test routes. Calm driving consisted of slight fluctuations in longitudinal acceleration values. In calm driving, the longitudinal acceleration values that ranged from −1 to 1 m/s^2-constituted 86% of the total acceleration values (Figure 5). Aggressive driving behavior on the test route was characterized by a maximum longitudinal acceleration of 4.21 m/s^2 and a maximum longitudinal deceleration of 1.51 m/s^2. The longitudinal acceleration values that ranged from −1 to 1 m/s^2 constituted 44% of the total acceleration values for aggressive driving.

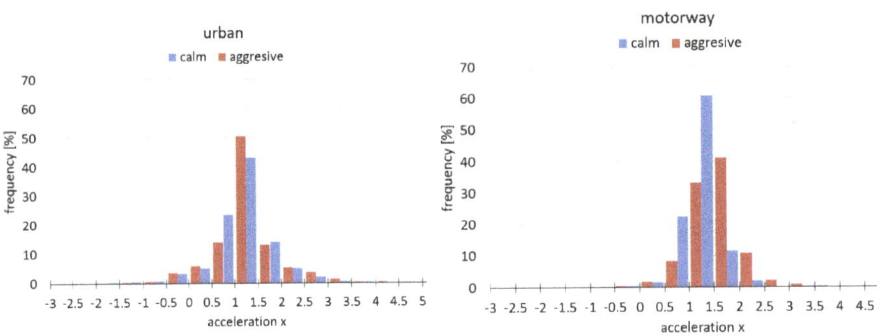

(a) (b)

Figure 5. Percentage distribution of longitudinal acceleration profiles over the test cycles. (**a**) urban driving; (**b**) motorway driving.

Driving in the urban area included obeying the corresponding speed limits, and frequent acceleration and braking. The short distance between traffic lights prevented the driver from maintaining a constant speed. Aggressive driving was identified by sudden movements when accelerating from a stop, abrupt braking when traffic lights changed, speeding, and crossing through an intersection against the red light. The test cycle for calm driving estimated the longitudinal acceleration values at −2.42–3.35 m/s^2. In aggressive driving, the longitudinal acceleration showed a wider range (−2.58 to 4.78m/s^2).

2.3. Simulation Test

The simulation tests were conducted using the ADVISOR (ADvanced Vehicle SImulatOR) (2003, National Renewable Energy Laboratory, Golden, CO, USA) program, which is widely employed to simulate vehicles of various drive configurations, including conventional, hybrid, electric, and hydrogen-cell drive. ADVISOR operates as part of MATLAB/Simulink (2015a, MathWorks, Natick, MA, USA) software. The vehicle model is built by selecting parameters from a complex database that contains various vehicle types and drive systems, as well as the particular elements of a given drive. In creating a vehicle model, the user can implement models of new vehicles and their components, then select a drive cycle. Using the assumed drive unit configuration and specified drive cycle, the program estimates the energy consumption and performance of the analyzed drive train. Figure 6 shows a parallel hybrid vehicle model developed in ADVISOR (2015a, MathWorks, Natick, MA, USA).

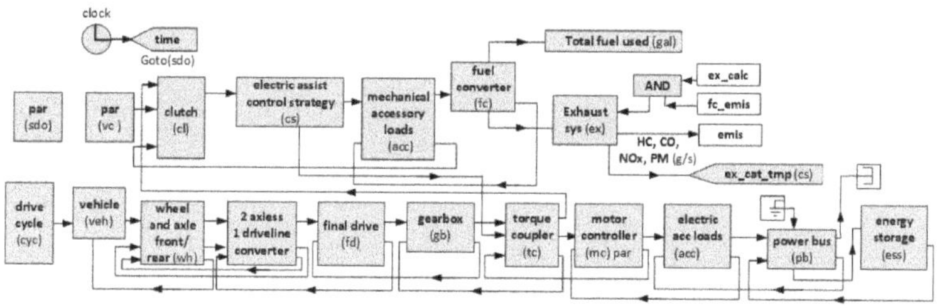

Figure 6. Model of parallel hybrid vehicle in ADVISOR.

The vehicle models available in ADVISOR were modified, and passenger car models with conventional and parallel hybrid electric (HEV) drive were developed. Two types of engine were considered: diesel and gasoline-powered internal combustion engines. The front area of the analyzed vehicles was 2.66 m^2, the rolling resistance coefficient was 0.009, and the aerodynamic resistance coefficient was 0.44. For all simulation cases, the curb weight was 1200 kg plus a load of 150 kg. For the hybrid vehicle, the weight was further increased by the battery weight. The selected parameters of the vehicles used in the simulation are presented in Table 1.

Table 1. Parameters of vehicles used in simulation tests.

	Conventional		Hybrid	
	Diesel	**Gasoline**	**HEV Diesel**	**HEV Gasoline**
Engine power (kW)	95	95	65	65
Electric machine power (kW)	–	–	75	75
Battery capacity (kWh)	–	–	4.6	4.6
Weight (kg)	–	–	64	64

In the case of hybrid electric vehicles, the simulation was conducted for various capacities of energy storage. The battery's initial state of charge before any trip was 70%.

3. Simulation Results

The results of the experiment showed that driving style has a major effect on fuel economy. The simulation results showed that aggressive driving has an incremental impact on average fuel consumption in urban driving conditions. Figure 7 shows the effect of driving style on average fuel consumption and percentage differences for various route types.

In urban conditions, aggressive driving of both conventional and hybrid-drive vehicles demonstrated an average fuel consumption approximately 30% higher than for calm driving. In motorway driving conditions, calm driving resulted in a 3% reduction in fuel consumption.

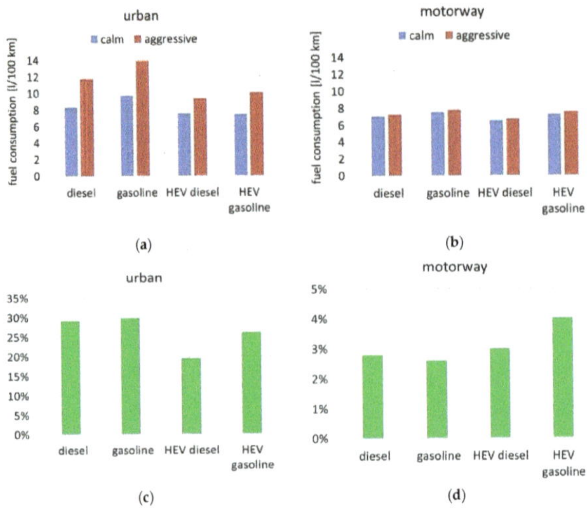

Figure 7. Average fuel consumption during test cycles (**a**) in urban conditions, (**b**) in motorway, and percentage difference (**c**) urban test cycles, (**d**) motorway test cycles.

The hybrid vehicle containing a gasoline engine experienced higher fuel consumption during aggressive driving compared to calm driving. Consumption was 26% higher in urban testing and 4% higher in motorway testing. The hybrid vehicle equipped with a diesel engine also showed higher fuel consumption during aggressive driving, with 19% higher average fuel consumption in urban testing and 3% higher average fuel consumption in motorway testing.

The hybrid vehicles showed an average fuel consumption in the urban tests of 10–28% less than diesel- and gasoline-powered vehicles. However, in the motorway tests, these differences were smaller, amounting to 3–7%. Figures 8 and 9 show the instantaneous fuel consumption and its percentage distribution for conventional-drive vehicles during the tests.

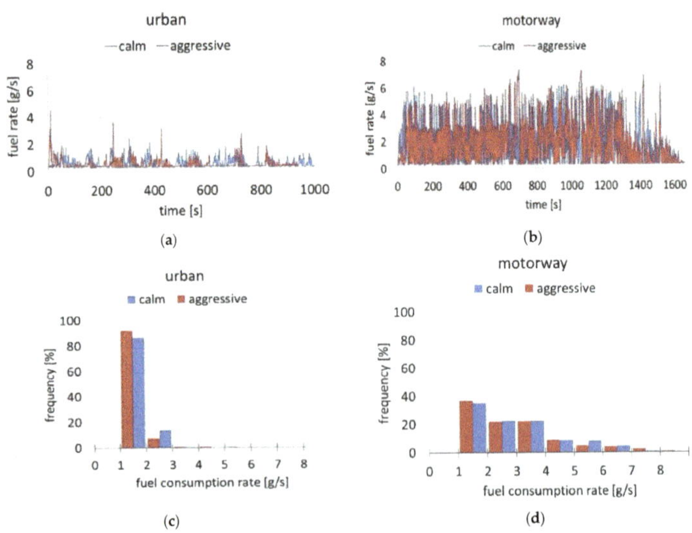

Figure 8. Instantaneous fuel consumption (**a**) in urban conditions, (**b**) in motorway, and its percentage distribution for the gasoline vehicle in (**c**) urban test cycles, (**d**) motorway test cycles.

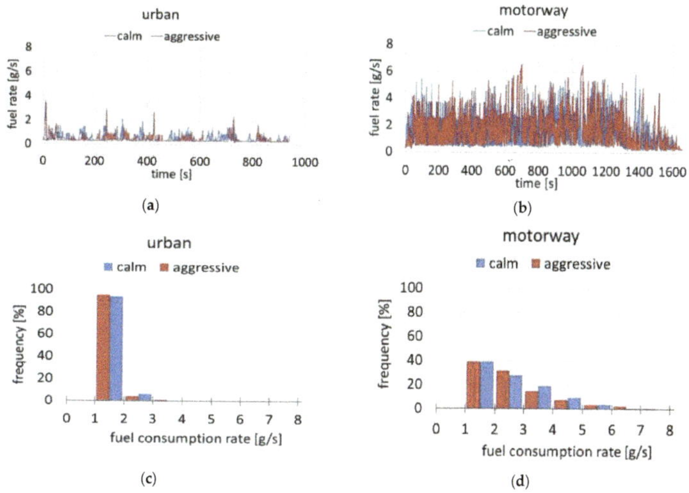

Figure 9. Instantaneous fuel consumption (**a**) in urban conditions, (**b**) in motorway, and its percentage distribution for the diesel vehicle in (**c**) in urban conditions, (**d**) in motorway.

In the instantaneous fuel-consumption profiles presented in Figure 6, fuel-consumption profiles differ for aggressive and calm driving styles. Rapid accelerations during aggressive driving caused a visible rise in fuel consumption. The rise in fuel consumption is particularly evident in the instantaneous fuel-consumption profiles of aggressive driving during urban testing. In some instances, the fuel-consumption rate showed peaks in excess of 4 g/s. Motorway tests resulted in a wider range of instantaneous fuel consumption compared to tests in the urban area, in an estimated range of 0.2–7.33 g/s for conventional vehicles and 0–4.86 g/s for hybrids. This is due to the high-speed operation of the engine. Figures 10 and 11 show the instantaneous fuel consumption and its percentage distribution for hybrid-drive vehicles during the tests.

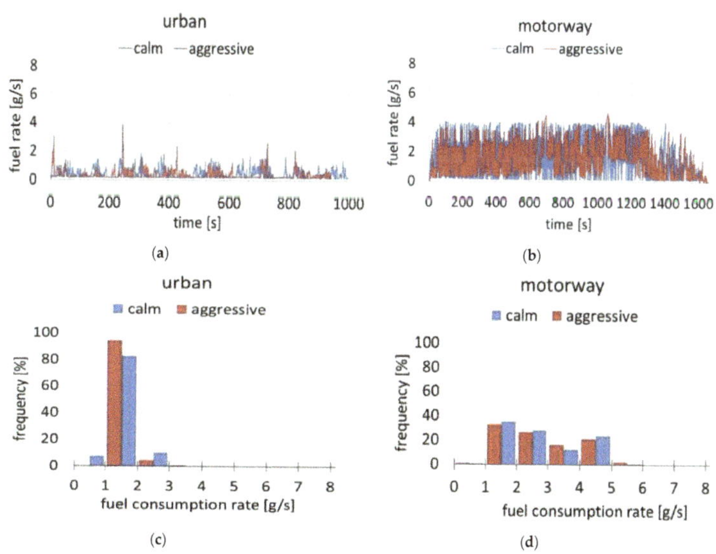

Figure 10. Fuel consumption rate estimated for HEV gasoline (**a**) in urban conditions, (**b**) in motorway; for HEV diesel (**c**) in urban conditions; (**d**) in motorway.

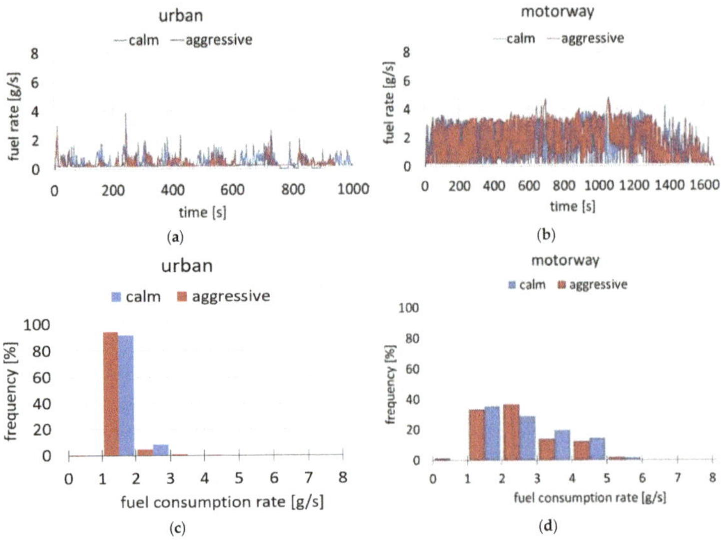

Figure 11. Fuel consumption rate estimated for HEV gasoline (**a**) in urban conditions, (**b**) in motorway; for HEV diesel (**c**) in urban conditions; (**d**) in motorway.

Hybrid-drive vehicles provide an option to switch off their combustion engine when it is not needed. This ability may reduce fuel consumption and pollutant emissions. However, during routes requiring higher energy demand, the internal combustion engine works to meet the traction requirements and directs part of the energy supply to recharge the batteries. Figure 12 presents the battery state of charge (SOC) of the hybrid-drive vehicles for both test cycles.

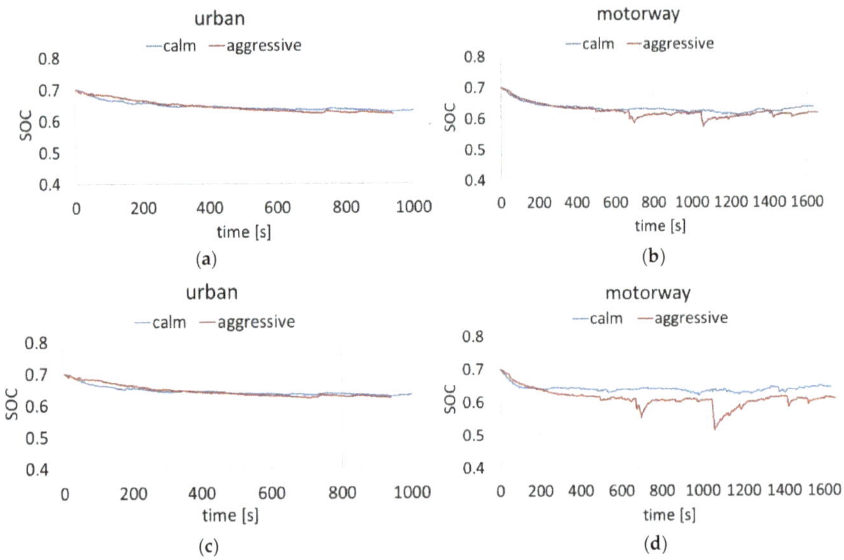

Figure 12. State of charge (SOC) estimated for HEV gasoline (**a**) in urban conditions, (**b**) in motorway; for HEV diesel (**c**) in urban conditions; (**d**) in motorway.

State of charge is the level of charge of an electrochemical battery in relation to its capacity. The units of SOC are expressed in percentage points: 0% = empty; 100% = fully charged. In the present study, the initial battery state of charge was 70%. Rapid acceleration during aggressive driving imposed a relatively high energy demand. As shown in Figure 12, the battery energy level decreased significantly in sudden acceleration movements.

The hybrid powered by the gasoline engine showed a minor change in SOC during the urban test cycles (1%). Larger temporary drops in the battery SOC were recorded during the motorway driving cycles (up to as much as 13%). The SOCs of HEV gasoline and HEV diesel batteries were almost identical during calm driving, whereas during aggressive driving, the energy level of the HEV diesel battery dropped by as much as 19%. Neither the gasoline nor the diesel hybrid showed rapid decreases in battery energy level during calm driving. However, more aggressive driving produced sudden temporary drops, related mainly to rapid acceleration.

Figure 13 shows the effect of driving style on average emissions of carbon oxides (CO_x), nitrogen oxides (NO_x), particulate matter (PM_x), and hydrocarbons (HC) for various route types. The results presented in Figure 13 show that aggressive driving in urban conditions causes a notable increase in the emission of pollutants.

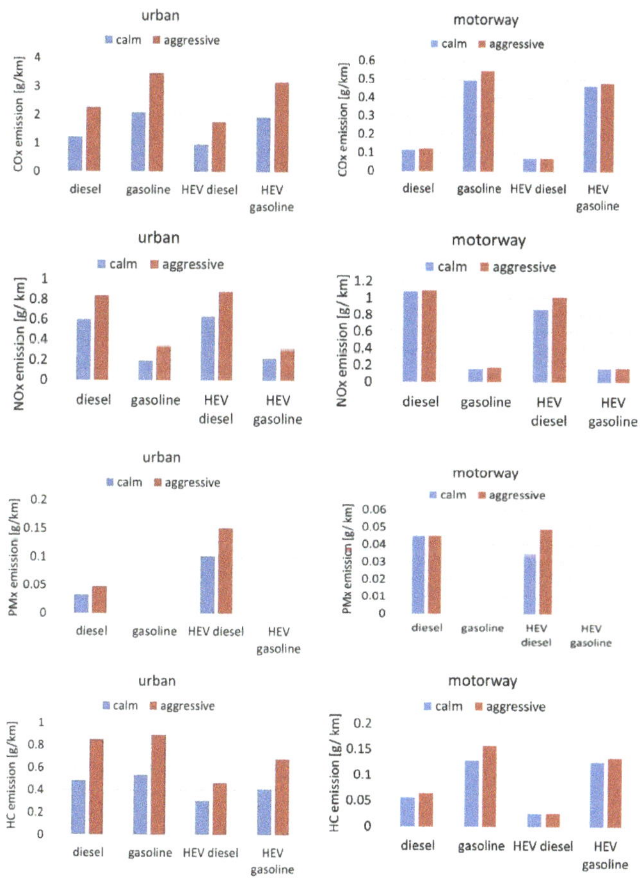

Figure 13. Emissions during the test cycles.

In urban conditions, aggressive driving contributed significantly to an increase in emissions. Compared to calm driving, during aggressive driving, the examined vehicles noted higher CO_x emissions (39–46%), NO_x emissions (29–41%), HC emissions (39–43%), and PM_x emissions (approximately 33%) (Table 2). The driving style had a lesser impact on air-pollutant emissions in motorway driving conditions. Compared to calm driving, aggressive driving produced 3–9% higher CO_x emissions, 2–14% higher NO_x emissions, 5–18% HC higher emissions, and 30% higher PM_x emissions. Figure 14 shows the effect of driving style on total pollutant emissions for various route types.

Table 2. Percentage increase in air-pollutant emissions caused by the aggressive driving style.

	Urban				Motorway			
	CO_x	NO_x	PM_x	HC	CO_x	NO_x	PM_x	HC
Diesel	43%	46%	29%	32%	13%	3%	0%	0%
Gasoline	40%	40%	46%	–	18%	9%	10%	–
HEV diesel	35%	45%	28%	33%	4%	0%	14%	29%
HEV gasoline	39%	39%	31%	–	5%	3%	2%	–

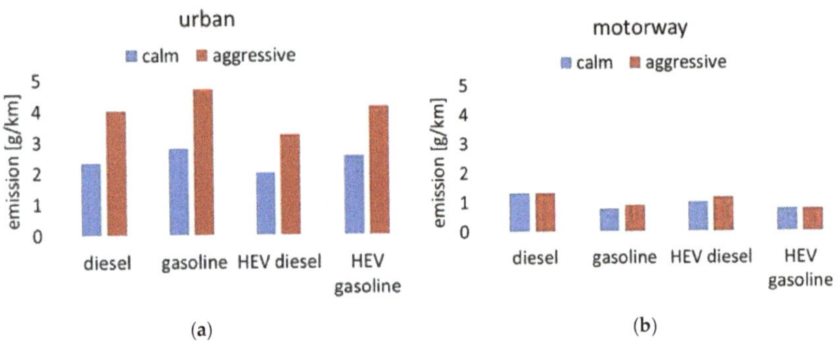

Figure 14. Total average emissions during the test cycles in (**a**) urban conditions; (**b**) motorway.

Calm driving in the urban area resulted in a reduction in average emissions of 40–42% for conventional-drive vehicles and 38% for hybrids. The pollutant-emission values estimated for tests conducted on the motorway did not differ widely from the emissions estimates for the urban tests. The results showed that for aggressive driving, average pollutant emissions were 2–14% higher than during calm driving. Hybrids were shown to have pollutant emissions lower than those of conventional-drive vehicles. In urban tests, the conventional-drive vehicles recorded emissions that were 9–14% higher than those of the hybrids. In motorway driving conditions, the conventional-drive vehicles showed average pollutant emissions that were 2–13% higher than those for hybrids.

4. Conclusions

The literature features many definitions of aggressive driving; however, there is no standard definition. Typically, aggressive driving is defined as behavior that includes verbal and physical aggression, abrupt maneuvers, and violations of highway codes. In the presented study, the aggressive driving style was understood to be a subjective feeling for the driver. The driver decided what maneuvers and behavior were to be considered as aggressive and dangerous. The calm and aggressive driving speed profiles recorded during the real-world tests served as input data for the simulation tests of fuel consumption and emissions. The aim of this study was to investigate the influence of aggressive driving style on fuel consumption and emissions of hybrid and conventional passenger vehicles with drive systems.

The presented results demonstrated that aggressive driving caused a significant increase in fuel consumption and pollutant emissions. This was particularly evident for urban driving conditions, during which the average fuel consumption for aggressive driving was up to 30% higher than during calm driving. Notably, the driver, thanks to the onboard computer, was aware of the increases in instantaneous fuel consumption during aggressive driving.

Similarly, compared to calm driving, aggressive driving caused an increase in average pollutant emissions of approximately 40%. Driving calmly in an urban area could reduce air-pollutant emissions by 40–42% for conventional-drive vehicles and 38% for hybrids. In motorway tests, the differences in fuel consumption and pollutant emission were not as significant as they are in the urban area. The results showed that aggressive driving resulted in higher average fuel consumption of up to 4%, with 14% higher average pollutant emissions.

During aggressive urban driving, the increase in CO_X emission was as high as 46%, and during motorway driving, the increase was up to 9%. Similarly, compared to calm driving, the NOx emissions for aggressive driving were 46% higher than in urban driving, and up to 14% higher than in motorway driving. During aggressive driving in urban driving conditions, the HC and PM_X emissions were 35–43% and 33% higher, respectively.

Replacing the conventional-drive vehicles with hybrids resulted in a considerable reduction in fuel consumption and pollutant emissions. The presented analysis indicated that hybrid-drive vehicles record lower emissions, and their use could improve fuel economy in both urban and motorway driving conditions.

This paper presented a preliminary analysis of driving-style research as part of a research project examining driver behavior. In subsequent studies, a larger group of drivers will be tested in different traffic conditions and in different vehicles. The classification of driver behavior allows for the determination of aggressive driving, which, if lessened, may result in a reduction in vehicle operating costs and harmful exhaust emissions.

Author Contributions: Conceptualization, E.M.S. and R.J.; methodology, E.M.S. and R.J.; software, E.M.S.; validation, E.M.S. and R.J.; formal analysis, E.M.S. and R.J.; investigation, E.M.S.; resources, E.M.S. and R.J.; data curation, E.M.S.; writing—original draft preparation, E.M.S. and R.J.; writing—review and editing, E.M.S. and R.J.; visualization, E.M.S. and R.J.; supervision, E.M.S. and R.J.; project administration, E.M.S. and R.J.; funding acquisition, E.M.S. and R.J. Both authors have read and agreed to the published version of the manuscript

Funding: The research was carried out as part of the Innovative system research project supporting the motor vehicle insurance risk assessment dedicated to UBI (Usage Based Insurance) No. POIR.04.01.04 00 0004/19 00 financed by the National Centre for Research and Development.

Conflicts of Interest: The authors declare no conflict of interest.

References

1. Tillmann, W.A.; Hobbs, G.E. The accident-prone automobile driver. *Am. J. Psychiatry* **2006**, *106*, 321–331. [CrossRef] [PubMed]
2. Sagberg, F.; Bianchi Piccinini, G.F.; Engström, J.A. Review of Research on Driving Styles and Road Safety. *Hum. Factors* **2015**, *57*, 1248–1275. [CrossRef] [PubMed]
3. Deng, Z.; Chu, D.; Wu, C.; He, Y.; Cui, J. Curve safe speed model considering driving style based on driver behaviour questionnaire. *Transp. Res. Part F Traffic Psychol. Behav.* **2019**, *65*, 536–547. [CrossRef]
4. Shahverdy, M.; Fathy, M.; Berangi, R.; Sabokrou, M. Driver behavior detection and classification using deep convolutional neural networks. *Expert Syst. Appl.* **2020**, *149*, 113240. [CrossRef]
5. Constantinescu, Z.; Marinoiu, C.; Vladoiu, M. Driving Style Analysis Using Data Mining Techniques. *Int. J. Comput. Commun. Control* **2010**, *5*, 654–663. [CrossRef]
6. Li, G.; Li, S.E.; Cheng, B.; Green, P. Estimation of driving style in naturalistic highway traffic using maneuver transition probabilities. *Transp. Res. Part C Emerg. Technol.* **2017**, *74*, 113–125. [CrossRef]
7. Cyganek, B.; Gruszczyński, S. Hybrid computer vision system for drivers' eye recognition and fatigue monitoring. *Neurocomputing* **2014**, *126*, 78–94. [CrossRef]

8. Li, G.; Wang, Y.; Zhu, F.; Sui, X.; Wang, N.; Qu, X.; Green, P. Drivers' visual scanning behavior at signalized and unsignalized intersections: A naturalistic driving study in China. *J. Saf. Res.* **2019**, *71*, 219–229. [CrossRef]
9. Lee, B.; Chung, W. Driver Alertness Monitoring Using Fusion of Facial Features and Bio-Signals. *IEEE Sens. J.* **2012**, *12*, 2416–2422. [CrossRef]
10. Przepiorka, A.M.; Hill, T.; Blachnio, A.P.; Sullman, M.J.M.; Taylor, J.E.; Mamcarz, P. Psychometric properties of the Driving Behaviour Scale (DBS) among Polish drivers. *Transp. Res. Part F Traffic Psychol. Behav.* **2020**, *73*, 29–37. [CrossRef]
11. Tement, S.; Plohl, N.; Horvat, M.; Musil, B.; Jakus, G.; Sodnik, J. Driving demands, stress reactivity and driving behavior: An interactional approach. *Transp. Res. Part F Traffic Psychol. Behav.* **2020**, *69*, 80–90. [CrossRef]
12. Hyodo, S.; Yoshii, T.; Satoshi, M.; Hirotoshi, S. An analysis of the impact of driving time on the driver's behavior using probe car data. *Transp. Res. Procedia* **2017**, *21*, 169–179. [CrossRef]
13. Kovaceva, J.; Isaksson-Hellman, I.; Murgovski, N. Identification of aggressive driving from naturalistic data in car-following situations. *J. Saf. Res.* **2020**, *73*, 225–234. [CrossRef] [PubMed]
14. Măirean, C.; Havârneanu, C.-E. The relationship between drivers' illusion of superiority, aggressive driving, and self-reported risky driving behaviors. *Transp. Res. Part F Traffic Psychol. Behav.* **2018**, *55*, 167–174. [CrossRef]
15. Havârneanu, C.E.; Jilavu, A.R.; Havârneanu, G.M. Digit Ratios as Correlates of Accident Involvement and Aggressive Driving—A Pilot Study. *Procedia Soc. Behav. Sci.* **2014**, *127*, 575–580. [CrossRef]
16. Eboli, L.; Mazzulla, G.; Pungillo, G. How to define the accident risk level of car drivers by combining objective and subjective measures of driving style. *Transp. Res. Part F Traffic Psychol. Behav.* **2017**, *49*, 29–38. [CrossRef]
17. Zheng, F.; Li, J.; van Zuylen, H.; Lu, C. Influence of driver characteristics on emissions and fuel consumption. *Transp. Res. Procedia* **2017**, *27*, 624–631. [CrossRef]
18. Ma, H.; Xie, H.; Huang, D.; Xiong, S. Effects of driving style on the fuel consumption of city buses under different road conditions and vehicle masses. *Transp. Res. Part D Transp. Environ.* **2015**, *41*, 205–216. [CrossRef]
19. Tang, T.-Q.; Huang, H.-J.; Shang, H.-Y. Influences of the driver's bounded rationality on micro driving behavior, fuel consumption and emissions. *Transp. Res. Part D Transp. Environ.* **2015**, *41*, 423–432. [CrossRef]
20. Tzirakis, E.; Zannikos, F.; Stournas, S. Impact of driving style on fuel consumption and exhaust emissions: Defensive and aggressive driving style. In Proceedings of the 10th International Conference on Environmental Science and Technology, Kos Island, Greece, 5–7 September 2007; pp. 1497–1504.
21. Faria, M.V.; Duarte, G.O.; Varella, R.A.; Farias, T.L.; Baptista, P.C. How do road grade, road type and driving aggressiveness impact vehicle fuel consumption? Assessing potential fuel savings in Lisbon, Portugal. *Transp. Res. Part D Transp. Environ.* **2019**, *72*, 148–161. [CrossRef]
22. González, N.; Kindelán, J.C.; Zapata, F. Influence of driving style on fuel consumption and Emissions in diesel-powered passenger car. In Proceedings of the 18th International Symposium Transport and Air Pollution, TAP 2010, Dübendorf, Switzerland, 18–19 May 2010; pp. 1–6.
23. Szumska, E.; Jurecki, R.; Pawełczyk, M. Evaluation of the use of hybrid electric powertrain system in urban traffic conditions. *Eksploat. I Niezawodn. Maint. Reliab.* **2020**, *22*, 154–160. [CrossRef]
24. Bayindir, Ç.K.; Gözüküçük, M.A.; Teke, A. A comprehensive overview of hybrid electric vehicle: Powertrain configurations, powertrain control techniques and electronic control units. *Energy Convers. Manag.* **2011**, *52*, 1305–1313. [CrossRef]
25. Ehsani, M.; Gao, Y.; Longo, S.; Ebrahimi, K. *Modern Electric, Hybrid Electric, and Fuel Cell Vehicles*; CRC Press: Boca Raton, FL, USA, 2018. [CrossRef]
26. Wang, Y.; Hao, C.; Ge, Y.; Hao, L.; Tan, J.; Wang, X.; Zhang, P.; Wang, Y.; Tian, W.; Lin, Z.; et al. Fuel consumption and emission performance from light-duty conventional/hybrid-electric vehicles over different cycles and real driving tests. *Fuel* **2020**, *278*, 118340. [CrossRef]
27. Hiermann, G.; Hartl, R.F.; Puchinger, J.; Vidal, T. Routing a mix of conventional, plug-in hybrid, and electric vehicles. *Eur. J. Oper. Res.* **2019**, *272*, 235–248. [CrossRef]
28. Zahabi, S.A.H.; Miranda-Moreno, L.; Barla, P.; Vincent, B. Fuel economy of hybrid-electric versus conventional gasoline vehicles in real-world conditions: A case study of cold cities in Quebec, Canada. *Transp. Res. Part D Transp. Environ.* **2014**, *32*, 184–192. [CrossRef]

29. Shinar, D.; Compton, R. Aggressive driving: An observational study of driver, vehicle, and situational variables. *Accid. Anal. Prev.* **2004**, *36*, 429–437. [CrossRef]
30. Abou-Zeid, M.; Kaysi, I.; Al-Naghi, H. Measuring aggressive driving behaviour using a driving simulator: An exploratory study. In Proceedings of the 3rd International Conference on Road Safety and Simulation, Indianapolis, IN, USA, 14–16 September 2011; pp. 1–19.
31. Islam, M.; Mannering, F. A temporal analysis of driver-injury severities in crashes involving aggressive and non-aggressive driving. *Anal. Methods Accid. Res.* **2020**, *27*, 100128. [CrossRef]
32. Webster, G.D.; DeWall, C.N.; Pond, R.S., Jr.; Deckman, T.; Jonason, P.K.; Le, B.M.; Smith, C.V. The brief aggression questionnaire: Psychometric and behavioral evidence for an efficient measure of trait aggression. *Aggress. Behav.* **2014**, *40*, 120–139. [CrossRef]
33. Konopka, K.; Rajchert, J.; Dominiak-Kochanek, M. The Influence of Aggression-Evoking Cues on Aggressive Cognitions in Males and Females: Different Procedures—Similar Effects. *Curr. Psychol.* **2020**, *39*, 128–141. [CrossRef]
34. Alonso, F.; Esteban, C.; Montoro, L.; Serge, A. Conceptualization of aggressive driving behaviors through a Perception of aggressive driving scale (PAD). *Transp. Res. Part F Traffic Psychol. Behav.* **2019**, *60*, 415–426. [CrossRef]
35. Lin, N.; Zong, C.; Tomizuka, M.; Song, P.; Zhang, X.; Li, G. An Overview on Study of Identification of Driver Behavior Characteristics for Automotive Control. *Math. Probl. Eng.* **2014**, 569109. [CrossRef]
36. Ma, Y.; Tang, K.; Chen, S.; Khattak, A.J.; Pan, P. On-line aggressive driving identification based on in-vehicle kinematic parameters under naturalistic driving conditions. *Transp. Res. Part. C Emerg. Technol.* **2020**, *114*, 554–571. [CrossRef]
37. Eboli, L.; Mazzulla, G.; Pungillo, G. Combining speed and acceleration to define car users' safe or unsafe driving behaviour. *Transp. Res. Part. C Emerg. Technol.* **2016**, *68*, 113–125. [CrossRef]
38. Feng, F.; Bao, S.; Sayer, J.R.; Flannagan, C.; Manser, M.; Wunderlich, R. Can vehicle longitudinal jerk be used to identify aggressive drivers? An examination using naturalistic driving data. *Accid. Anal. Prev.* **2017**, *104*, 25–136. [CrossRef] [PubMed]
39. Miles, D.E.; Johnson, G.L. Aggressive driving behaviors: Are there psychological and attitudinal predictors? *Transp. Res. Part. F Traffic Psychol. Behav.* **2003**, *6*, 147–161. [CrossRef]
40. Krahé, B. Predictors of women's aggressive driving behavior. *Aggress. Behav. Off. J. Int. Soc. Res. Aggress.* **2005**, *31*, 537–546. [CrossRef]

MDPI
St. Alban-Anlage 66
4052 Basel
Switzerland
Tel. +41 61 683 77 34
Fax +41 61 302 89 18
www.mdpi.com

Energies Editorial Office
E-mail: energies@mdpi.com
www.mdpi.com/journal/energies